Nonlinear Equations in
the Applied Sciences

This is volume 185 in
MATHEMATICS IN SCIENCE AND ENGINEERING
Edited by William F. Ames, *Georgia Institute of Technology*

A list of recent titles in this series appears at the end of this volume.

NONLINEAR EQUATIONS IN THE APPLIED SCIENCES

Edited by

W.F. Ames
SCHOOL OF MATHEMATICS
GEORGIA INSTITUTE OF TECHNOLOGY
ATLANTA, GEORGIA

C. Rogers
DEPARTMENT OF MATHEMATICAL SCIENCES
LOUGHBOROUGH UNIVERSITY OF TECHNOLOGY
LOUGHBOROUGH, LEICESTERSHIRE
ENGLAND

ACADEMIC PRESS, INC.
Harcourt Brace Jovanovich, Publishers

Boston   San Diego   New York
London   Sydney   Tokyo   Toronto

ACADEMIC PRESS, INC.
1250 Sixth Avenue, San Diego, CA 92101

United Kingdom Edition published by
ACADEMIC PRESS LIMITED
24–28 Oval Road, London NW1 7DX

Library of Congress Cataloging-in-Publication Data

Nonlinear equations in the applied sciences / edited by W.F. Ames, C.
  Rogers.
            p.   cm. — (Mathematics in sciences and engineering; v. 185)
      Includes bibliographical references and index.
      ISBN 0-12-056752-0 (alk. paper)
      1. Nonlinear theories.   I. Ames, William F.   II. Rogers, C.
  III. Series.
  QA427.N65   1992
  515—dc20   91-18319
                                                  CIP

PRINTED IN THE UNITED STATES OF AMERICA

92 93 94 95   9 8 7 6 5 4 3 2 1

Contents

R. W. Ogden

Contributors

Numbers in parentheses indicate the pages on which the authors' contributions begin.

Karen A. Ames (1), Department of Mathematical Sciences, University of Alabama in Huntsville, Huntsville, Alabama 35899

W. F. Ames (31), School of Mathematics, Georgia Institute of Technology, Atlanta, Georgia 30332

Shui-Nee Chow (79), Center for Dynamical Systems and Nonlinear Studies, School of Mathematics, Georgia Institute of Technology, Atlanta, Georgia 30332

Andrea Donato (149), Department of Mathematics, University of Messina, 98010 Sant'Agata, Messina, Italy

A. S. Fokas (175), Department of Mathematics and Computer Science and the Institute for Nonlinear Studies, Clarkson University, Potsdam, New York 13699-5815

Benno Fuchssteiner (211), Faculty of Mathematics, University of Paderborn, D 4790 Paderborn, Germany

Greg King (257), Nonlinear Systems Laboratory, Mathematics Institute, University of Warwick, Coventry CV4 7AL, England

B. G. Konopelchenko (317), Institute of Nuclear Physics, Siberian Division of the Academy of Sciences USSR, 630090 Novosibirsk, USSR

R. H. Martin, Jr. (363), Department of Mathematics, North Carolina State University, Raleigh, North Carolina 27695-8205

M. C. Nucci (399), School of Mathematics, Georgia Institute of Technology, Atlanta, Georgia 30332

R. W. Ogden (437), Department of Mathematics, University of Glasgow, Glasgow G12 8QW, Scotland

M. Pierre (363), University of Nancy, Nancy 54000, France

C. Rogers (317), Department of Mathematical Sciences, Loughborough University of Technology, Loughborough, Leicestershire LE11 3TU, England

Ian Stewart (257), Nonlinear Systems Laboratory, Mathematics Institute, University of Warwick, Coventry CV4 7AL, England

Masahiro Yamashita (79), Department of Mathematics and Statistics, Wright State University, Dayton, Ohio 45435

Preface

Nature pays no attention to the linearity requirement of much of classical mathematics. Consequently, the equations describing various phenomena are usually nonlinear—an ambiguous negative definition. There are a considerable number of classes of nonlinear equations that have received attention. The present collection of eleven articles addresses several physically motivated systems for which much theory is available. These include reaction-diffusion systems, elasticity and other areas of mechanics, and nonlinear waves. Theories available for wider classes of equations include discussions of Lie symmetries, improperly posed problems, integrable nonlinear equations, Bäcklund and reciprocal transformations, Hamilton structures and integrability, and Riccati type pseudo-potentials. A detailed study of the geometry of the Melnikov vector of dynamical systems and an article on popular symmetric chaos round out the volume.

While some theory is necessary, in general the authors remember that the main purpose of the book is to address real situations. Thus a plethora of applications is available for the reader to see how far some of the studies of the nonlinear world have progressed. We hope that the developing nonlinear world will be somewhat more understandable as a result of this volume. The stimulation of additional research is also one of our objectives.

W. F. Ames, Atlanta, Georgia, USA
C. Rogers, Loughborough, England

Improperly Posed Problems for Nonlinear Partial Differential Equations

Karen A. Ames
Department of Mathematical Sciences
University of Alabama in Huntsville
Huntsville, AL 35899

Dedicated to Lawrence E. Payne.

1 Introduction

The study of improperly posed problems has received considerable recent attention in response to the realization that a number of physical situations lead to these kinds of mathematical models. Many classes of improperly posed problems include nonlinear partial differential equations. In this Chapter, we review some of the questions confronted in the investigation of a particular class of such problems, namely ill–posed Cauchy problems.

A boundary or initial value problem for a partial differential equation is understood to be well posed in the sense of Hadamard if it possesses a unique solution that depends continuously on the prescribed data. This definition is made precise by indicating the space to which the solution must belong and the measure in which continuous dependence is desired. A problem is called improperly posed (or ill–posed) if it fails to have a global solution, a unique solution, or a solution that depends continuously on the data. We shall discuss the questions of uniqueness, existence, and stabilization of solutions to ill–posed problems for nonlinear partial differential equations in the subsequent sections.

Although we intend to focus on improperly posed Cauchy problems for nonlinear equations, a few comments about these kinds of ill–posed problems for linear equations should be included here. A number of interesting challenges arise in the study of such problems despite the "cloak" of linearity. Mathematicians now know that some linear partial differential equations without singular points have no nontrivial solutions, e.g. the classical construction of Lewy (1957). In addition, the literature contains examples that belie the hypothesis that ill–posed Cauchy problems for linear equations have at most one solution. While it is typical that solutions (if they exist) of such problems are unique, examples of nonunique continuation for linear elliptic and parabolic equations have been constructed by Plis (1960, 1963) and Miller (1973, 1974). Even in those cases where unique solutions exist, they may not depend continuously on all of the data. In attempting to derive stability inequalities for these linear problems, we are faced with the difficulty that the spaces in which we are guaranteed continuous data dependence are nonlinear. Consequently, we again encounter the obstacles inherent in the study of nonlinear problems.

Recently, considerable progress has been made in the investigation of improperly posed problems for partial differential equations. We refer the reader to Payne's monograph (1975) as a primary source on various topics in this field as well as substantial bibliography of the work done prior to 1975. In the last twenty years, research on ill–posed problems has intensified and moved in so many diverse directions that it would be virtually impossible to give a comprehensive, up–to–date account of all the work that has appeared on even the particular class of problems treated here. It is no surprise then that the few existing books on improperly posed problems deal with specific classes of these problems. In addition to Payne's volume, there are contributions from the Russian researchers Lavrentiev (1967), Tikhonov and Arsenin (1977), Morozov (1984), and Lavrentiev, Romanov, and Sisatskii (1986). We also mention the book by Lattès and Lions (1969) that describes a regularization method for approximating solutions to improperly posed problems. Collections of papers or research notes in this area include *Symposium on Non–Well-Posed Problems and Logarithmic Convexity* (edited by R. J. Knops) in 1973, *Improperly Posed Boundary Value Problems* (Carasso and

Stone, 1975), *Instability, Nonexistence and Weighted Energy Methods in Fluid Dynamics and Related Theories* (Straughan, 1982), and *Inverse and Ill–Posed Problems* (edited by Engl and Groetsch, 1987).

Because many ill–posed problems for partial differential equations have not usually yielded to standard methods of analysis, a variety of techniques have been developed in order to study these problems. We will focus on three of these methods since they have proven to be useful in treating nonlinear problems, namely logarithmic convexity, weighted energy, and concavity. In Section 3 we shall make a few remarks concerning the quasireversibility method when we address the question of existence of solutions.

The first applications of logarithmic convexity arguments to improperly posed problems for partial differential equations have been attributed to Pucci (1955), John (1955, 1960), and Lavrentiev (1956). A detailed treatment of this method can be found in the comprehensive work of Agmon (1966). Basically, this procedure employs second order differential inequalities to investigate the properties of solutions. Solution properties can also be obtained via the weighted energy method. The development of this technique, which was utilized by M. H. Protter early in the 1950s, as well as examples of its applications in fluid dynamics are covered in the monograph of Straughan (1982). Weighted energy arguments have more recently been employed to study ill–posed problems for nonlinear equations by Bell (1981a,b), Bennett (1986), Lavrentiev et. al. (1986), Payne and Straughan (1990a) and Straughan (1983) among others. Modifications of the original ideas of Protter and his co–workers (see Lees and Protter, 1961; Murray, 1972; Murray and Protter, 1973) has generated improved results in varous problems. We cite the work of Payne (1985) and Ames, Levine and Payne (1987). Finally, the concavity method has proved to be useful in establishing nonexistence theorems for problems in such areas as nonlinear elasticity and nonlinear continuum mechanics. Extensive use of this technique has been made by Hills and Knops (1974), Knops, Levine, and Payne (1974), Knops and Straughan (1976), Levine (1973; 1974a,b,c), Levine and Payne (1974a,b; 1976) and Straughan (1975b, 1976). We refer the reader to Payne (1975) for a detailed description of these three methods. In the following three sections, we shall illustrate how such techniques can be exploited to study the topics of uniqueness, existence, and continuous data dependence of solutions to improperly posed Cauchy problems for nonlinear partial differential equations. We emphasize that this is not intended to be an exhaustive treatment of the subject but is meant to give some indication of the difficulties encountered in our attempts to understand what constitutes a well–posed problem for a differential equation.

2   Uniqueness

Mathematicians who investigated ill–posed Cauchy problems for partial differential equations before 1950 dealt primarily with questions of uniqueness for linear equations since, as mentioned earlier, such problems typically have at most one solution. The literature of the 1950's and early 1960's is rich with uniqueness studies, some of which involve nonlinear equations. Serrin (1963), for example, established uniqueness for classical solutions of the Navier–Stokes equations on a bounded spatial domain backward in time using weighted energy arguments based on work by Lees and Protter (1961). Knops and Payne (1968) gave an alternative proof of backward uniqueness for the same problem when they addressed the question of continuous dependence on the final data. A number of uniqueness studies for various problems associated with the Navier–Stokes equations backward in time have appeared since then (cf. Straughan (1982), Payne and Straughan (1990c and references therein)). Uniqueness questions for ill–posed problems that arise in nonlinear elasticity have also received recent attention (Knops and Payne, 1979, 1983).

In this section we present a nonlinear example of nonuniqueness. The initial–boundary value problem we consider is the following:

$$u_t = \Delta u - u|u|^{-\alpha} \quad \text{in } \Omega \times [0, \infty) \tag{2.1}$$

$$u = 0 \quad \text{on } \partial\Omega \times [0, \infty) \tag{2.2}$$

$$u(x, 0) = f(x) \quad x \in \Omega. \tag{2.3}$$

Here Ω is a bounded open set in $\mathbb{R}^n$ with smooth boundary $\partial\Omega$, Δ is the Lapace operator, $f(x)$ is a prescribed function and the fixed exponent α satisfies $0 < \alpha < 1$. Equation (2.1) might be regarded as a model for heat conduction in a body with a sublinear heat sink. We shall now show that problem (2.1)–(2.3) exhibits nonuniqueness of solutions backward in time.

Define the functional $\phi_p(t)$ by

$$\phi_p(t) = \int_\Omega u^{2p}\,dx \tag{2.4}$$

where p is a positive integer. Differentiating (2.4), we obtain

$$\frac{d\phi_p}{dt} = 2p\int_\Omega u^{2p-1}u_t\,dx.$$

Substitution of the differential equation and integration by parts leads to the expression

$$\frac{d\phi_p}{dt} = -2p(2p-1)\int_\Omega u^{2p-2}|\mathrm{grad}\ u|^2 dx - 2p\int_\Omega u^{2p}|u|^{-\alpha}dx.$$

Thus,

$$\frac{d\phi_p}{dt} \le -2p\int_\Omega u^{2p}|u|^{-\alpha}dx.$$

If we now let

$$\tilde{u}(t) = \max_\Omega |u(x,t)|, \tag{2.5}$$

it follows that

$$\frac{d\phi_p}{dt} \le -2p[\tilde{u}(t)]^{-\alpha}\phi_p.$$

Integrating this inequality from 0 to t, we obtain

$$\phi_p(t) \le \phi_p(0)e^{-2p\int_0^t [\tilde{u}(s)]^{-\alpha}ds}. \tag{2.6}$$

Let use now raise both sides of (2.6) to the $\frac{1}{2p}$ power and then let $p \to \infty$. We find that

$$\tilde{u}(t) \le \tilde{u}(0)e^{-\int_0^t [\tilde{u}(s)]^{-\alpha}ds}. \tag{2.7}$$

The assumption that $\tilde{u}(t) > 0$ allows use to write (2.7) in the form

$$\frac{1}{[\tilde{u}(0)]^\alpha} \leq \frac{1}{[\tilde{u}(t)]^\alpha} e^{-\alpha \int_0^t [\tilde{u}(s)]^{-\alpha} ds}$$

and an integration from 0 to t results in the inequality

$$\frac{t}{[\tilde{u}(0)]^\alpha} \leq \frac{1}{\alpha} \left[1 - e^{-\alpha \int_0^t [\tilde{u}(s)]^{-\alpha} ds} \right] \leq \frac{1}{\alpha}.$$

Consequently, we conclude that

$$t \leq \frac{1}{\alpha} [\tilde{u}(0)]^\alpha. \qquad (2.8)$$

Observing that (2.8) cannot hold for all time, we see that if a solution to (2.1)–(2.3) exists, it must decay to zero in a finite time T satisfying the upper bound

$$T \leq \frac{1}{\alpha} [\tilde{u}(0)]^\alpha.$$

We thus have an example for which we fail to have a unique solution backward in time. Similar results (Payne, 1975) can be obtained for the Cauchy problem for nonlinear equations of the form

$$u_t = \Delta u - f(u)$$

where the point function f satisfies the condition

$$u f(u) \geq c|u|^{2-\alpha}$$

for $0 < \alpha < 1$ and a constant $c > 0$.

We see that a loss of uniqueness may be one price we have to pay for introducing nonlinear terms into a mathematical model. We note here that the uniqueness studies appearing in the literature on ill–posed problems are frequently concerned with the determination of criteria necessary to ensure unique solutions to the problems under consideration. These results are often consequences of unique continuation theorems (cf. Payne, 1975).

3 Existence

We begin our discussion of existence by remarking that throughout this chapter, the term "existence" refers to global existence. For some of the problems considered here, local existence can be deduced from the Cauchy-Kovalevski theorem.

The question of global existence of solutions to ill–posed problems is difficult to treat with much generality. It is frequently the case that very strong regularity and compatibility conditions must be imposed on the data in order for a solution to exist. Such conditions cannot always be guaranteed in practice because of the possibility of error in data measurement. From a practical viewpoint, these problems probably require a theorem asserting the existence of a solution that satisfies an appropriate stabilizing condition and exhibits data that are close in some measure to the given values.

Attempts to deal with the existence question have often followed the strategy of modifying either the concept of a solution or the mathematical model. For many physical situations, it is sufficient to define the "solution" as a domain function that belongs to an appropriate class and best approximates the prescribed data. Alternatively, changing the underlying model has led to the development of techniques to generate approximate solutions, including Tikhonov regularization (Tikhonov, 1963) and the quasi reversibility method (Lattès and Lions, 1969). The primary aim of these methods is the numerical solution of classes of ill–posed initial–boundary value problems. Since existence theorems for such problems are generally not available, approximate solutions may be as close as we can come to establishing the existence of a "solution."

We remark that Lattès and Lions (1969) formulate and briefly discuss quasi reverisiblity methods for several nonlinear problems of evolution, including the Navier–Stokes equations. They give a few simple numerical examples but one surmises from their sketchy development that nonlinearity further complicates the task of constructing approximate solutions to ill–posed problems. One issue we encounter in attempting to compute solutions of these problems (both linear and nonlinear) is the question of stability. Not only do we need to derive stability estimates but we also need to develop numerical methods that effectively incorporate the prescribed global bounds usually required to stabilize ill–posed problems. We shall not pursue these topics here but refer the reader to the work of Miller (1973), Showalter (1974, 1975), and Ewing (1975).

We leave the problem of generating approximate solutions and turn to the related topic of identifying those nonlinear equations for which certain types of data are incompatible with global existence of solutions. Nonexistence results for initial–boundary value problems for classes of nonlinear parabolic equations have been obtained by Kaplan (1963), Fujita (1966, 1970), Levine and Payne (1974), Straughan (1975b), and more recently by Ames, Payne and Straughan (1989). Operator equation analogues of such problems have been treated by Levine (1973) and Tsutsumi (1971, 1972), among others. Similar studies for nonlinear hyperbolic equations have been done by Sattinger (1968), Glassey (1973, 1981), Levine (1974a, b), Knops et. al. (1974) and John (1979). The recent review article by Levine (1990) collects a number of the available nonexistence results and provides an existence list of references on the subject. These investigations have relied on a variety of methods to establish global nonexistence theorems. We focus here on one particular method, the concavity method, in order to illustrate the type of results that can be obtained.

Let us use as our model the problem

$$\frac{\partial^2 u}{\partial t^2} = \Delta u + u^\beta \quad \text{in } \Omega \times [0, T) \tag{3.1}$$

$$u = 0 \quad \text{on } \partial\Omega \times [0, T) \tag{3.2}$$

$$u(x, 0) = f(x) \quad u_t(x, 0) = g(x) \quad \text{for } x \in \Omega \tag{3.3}$$

where Ω is a bounded domain in $\mathbb{R}^n$, Δ is the Laplace operator, $\beta \geq 2$ is a positive integer, and T is understood to be either infinity or the limit of the existence interval. For any classical solution of (3.1)–(3.3) we define a functional

$$F(t) = \int_\Omega u^2 dx + k(t + t_0)^2 \equiv \|u\|^2 + k(t + t_0)^2 \tag{3.4}$$

for nonnegative constants k and t_0 to be chosen. The idea of the concavity method is to show that $F^{-\alpha}$ for some $\alpha > 0$ is a concave function of t. If we choose k and t_0 so that $F(0) > 0$, it will follow that $F(t) > 0$ for all t in the existence interval and thus $F^{-\alpha}$ will be concave if it satisfies the inequality

$$FF'' - (1 + \alpha)(F')^2 \geq 0 \tag{3.5}$$

for all such t. We shall now show that (3.4) satisfies the inequality (3.5).

Differentiation of (3.4) leads to

$$F'(t) = 2 \int_\Omega uu_t dx + 2k(t + t_0) \tag{3.6}$$

and then

$$F''(t) = 2 \int_\Omega u_t^2 dx + 2 \int_\Omega uu_{tt} dx + 2k. \tag{3.7}$$

Substituting the differential equation, we find that

$$F''(t) = 4(1+\alpha)\|u_t\|^2 - 2(1+2\alpha)\|u_t\|^2 - 2\|\text{grad } u\|^2 + 2 \int_\Omega u^{\beta+1} dx + 2k. \tag{3.8}$$

If we multiply (3.1) by u_t and integrate with respect to t, we obtain

$$\frac{1}{2}\|u_\tau\|^2 \Big|_0^t = -\frac{1}{2}\|\text{grad } u\|^2 \Big|_0^t + \int_0^t \int_\Omega u_\tau u^\beta dx d\tau$$

or

$$E(t) \equiv \frac{1}{2}\|u_t\|^2 + \frac{1}{2}\|\text{grad } u\|^2 - \int_0^t \int_\Omega u_\tau u^\beta dx d\tau = E(0) \tag{3.9}$$

where

$$E(0) = \frac{1}{2}\|g\|^2 + \frac{1}{2}\|\text{grad } f\|^2. \tag{3.10}$$

Let us next use (3.9) to eliminate the $-2(1 + 2\alpha)\|u_t\|^2$ term in expression (3.8). We thus have

$$\begin{aligned}
F''(t) = {} & 4(1 + \alpha)\|u_t\|^2 + 4\alpha\|\text{grad } u\|^2 + 2 \int_\Omega u^{\beta+1} dx \\
& -4(1 + 2\alpha) \int_0^t \int_\Omega u_\tau u^\beta dx d\tau \\
& -4(1 + 2\alpha)E(0) + 2k
\end{aligned} \tag{3.11}$$

or, observing that $\int_0^t \int_\Omega u_\tau u^\beta dx d\tau = \frac{1}{\beta+1} \int_\Omega u^{\beta+1} |_0^t dx$,

$$
\begin{aligned}
F''(t) = {} & 4(1+\alpha)\|u_t\|^2 + 4\alpha\|\text{grad } u\|^2 + \frac{2(\beta-1-4\alpha)}{\beta+1} \int_\Omega u^{\beta+1} dx \\
& -4(1+2\alpha)\left\{ E(0) + \frac{1}{\beta+1} \int_\Omega f^{\beta+1} dx \right\} + 2k. \qquad (3.12)
\end{aligned}
$$

We now form the expression $FF'' - (1+\alpha)(F')^2$ using (3.4), (3.6) and (3.12). After observing that the term $4\alpha F\|\text{grad } u\|^2$ is clearly nonnegative, we obtain the inequality

$$
\begin{aligned}
FF'' - (1+\alpha)(F')^2 \geq {} & 4(1+\alpha)s^2 + \frac{2}{\beta+1}F(\beta-1-4\alpha) \int_\Omega u^{\beta+1} dx \\
& -2(1+2\alpha)F\{k + 2\tilde{E}(0)\} \qquad (3.13)
\end{aligned}
$$

where

$$
s^2 = \left[\|u\|^2 + k(t+t_0)^2 \right] \left[\|u_t\|^2 + k \right] - \left[\int_\Omega u u_t dx + k(t+t_0) \right]^2 \tag{3.14}
$$

and

$$
\tilde{E}(0) = E(0) + \frac{1}{\beta+1} \int_\Omega f^{\beta+1} dx.
$$

Since $s^2 \geq 0$ by the Cauchy-Schwarz inequality, we need to ensure that the second and third terms on the right side of the inequality in (3.13) are nonnegative in order to obtain (3.5). The second term will be nonnegative if we choose $\alpha = \frac{\beta-1}{4} > 0$. We then have the inequality

$$
FF'' - (1+\alpha)(F')^2 \geq -(\beta+1)F\{k + 2\tilde{E}(0)\}. \tag{3.15}
$$

Using (3.15) we can now obtain the following result:

Theorem 1 *A global solution of (3.1)–(3.3) cannot exist if (a) $\tilde{E}(0) < 0$ or (b) $\tilde{E}(0) = 0$ and $\int_\Omega fg dx > 0$.*

To prove part (a) of the theorem we choose $k = -2\tilde{E}(0) > 0$. Then (3.15) becomes $FF'' - (1+\alpha)(F')^2 \geq 0$ and therefore $(F^{-\alpha})'' \leq 0$. If we integrate this inequality twice, we find that

$$
F^{-\alpha}(t) \leq F^{-\alpha}(0) - \alpha F^{-\alpha-1}(0)F'(0)t
$$

from which it follows that

$$F^\alpha(t) \geq \frac{F^\alpha(0)}{1 - \alpha t \left\{ \frac{F'(0)}{F(0)} \right\}}. \qquad (3.16)$$

Recalling that $F'(0) = 2 \int_\Omega fg dx + 2kt_0$, we choose t_0 so that $F'(0) > 0$ and observe that the right side of (3.16) becomes unbounded at $t^* = \frac{F(0)}{\alpha F'(0)}$. Consequently, $F(t)$ must cease to exist at some time less than or equal to t^*. It follows that the solution u cannot exist for all time.

To establish (b) we take $k = 0$. In this case, we have $F'(0) = 2 \int_\Omega fg dx > 0$ by assumption. Similar arguments as those used in part (a) lead to the global nonexistence result. In both situations we also obtain an upper bound t^* for the time at which the solution ceases to exist. Such upper bounds can usually be read directly from the inequality that establishes nonexistence of solutions in most of the methods used to determine these type of results. Lower bounds for this time are more difficult to obtain but can be found for some problems, e.g., the Cauchy problem for a nonlinear heat equation (see Payne, 1975).

We remark that there exist nonexistence results in the literature for (3.1)–(3.3) if $E(0) > 0$ but the arguments needed to derive them are somewhat more detailed (Knops et. al. 1974; Levine 1974b,c; Straughan, 1975b). Similar results can be demonstrated if we replace the nonlinearity u^β in (3.1) by any nonlinear function $h(u)$ that satisfies $h(0) = 0$ and the condition

$$\int_\Omega uh(u)dx \geq 2(1 + 2\alpha) \int_\Omega \int_0^u h(\xi)d\xi dx. \qquad (3.17)$$

In fact, such problems can be analyzed in a more abstract setting (see Payne, 1975 for a model problem). We observe that global nonexistence theorems for nonlinear hyperbolic equations are typically desired for weak solutions since a number of these equations may not possess classical solutions. Concavity arguments can be easily modified to deal with appropriately defined weak solutions.

The concavity technique extends to problems where the nonlinearity appears in the boundary condition. Nonexistence results, for example, can be obtained for the problem

$$\frac{\partial^2 u}{\partial t^2} = \Delta u \quad \text{in } \Omega \times [0,T) \tag{3.18}$$

$$\frac{\partial u}{\partial n} = h(u) \quad \text{on } \partial\Omega \times [0,T) \tag{3.19}$$

$$u(x,0) = f(x) \quad u_t(x,0) = g(x) \quad x \in \Omega \tag{3.20}$$

where $\frac{\partial}{\partial n}$ represents the outward normal derivative to Ω and h is some nonlinear function. Concavity arguments applied to the functional

$$F(t) = \|u\|^2 + k(t + t_0)^2$$

lead to a similar nonexistence theorem provided we replace (3.17) by the condition

$$\int_{\partial\Omega} uh(u)dS \geq 2(1 + 2\alpha) \int_{\partial\Omega} \int_0^u h(\xi)d\xi dS. \tag{3.21}$$

Further details can be found in Levine and Payne (1974a,b).

4   Continuous Dependence

In this section we address the third requirement for a problem in partial differential equations to be well posed, namely continuous dependence of the solution on the data. This subject has received considerable attention since 1960 with the result that there now exist many methods for deriving continuous dependence inequalities in various measures (see Payne, 1975 for a historical summary and bibliography). It has been shown that for many improperly posed problems, if the class of admissible solutions is suitably restricted, then solutions in this class will depend continuously on the data in the appropriate norms. In many physically interesting problems, stability, in the modified sense of Hölder (John, 1955, 1960), is guaranteed when a uniform bound for the solution or some functional of the solution is prescribed. Since such a bound need not be sharp, it is reasonable to anticipate that it can be obtained from the physics of a problem via observation and measurement. We emphasize here that this type of continuous dependence should not be confused with asymptotic stability or with the usual notion of continuous dependence defined for well posed problems.

The derivation of continuous dependence inequalities has remained an important pursuit because of the central role such estimates play in obtaining numerical solutions to ill–posed problems for both linear and nonlinear equations. We point out that such inequalities also imply uniqueness of solutions.

Our discussion of stability in the context of improperly posed problems will focus on three important types of errors that can be introduced in the derivation and analysis of a mathematical model for a physical process:

1. errors in measuring data (initial or boundary values, coefficients in the equations or boundary operators, parameters),

2. errors in characterizing the spatial geometry of the underlying domain and the initial–time geometry,

3. errors in formulating the model equation.

In order to recover continuous dependence in ill posed problems, we must constrain the solution in some mathematically and physically realizable manner. It would be optimal if this constraint could simultaneously stabilize the problem against all possible errors that might arise as we set up the model. However, it is usually the case that the appropriate constraint is difficult to determine. Moreover, such a restriction will turn an otherwise linear problem into a nonlinear one and thus caution must be exercised in any attempt to study the effects of various errors separately and to then superpose these effects.

4.1   Continuous Dependence on Cauchy Data

Questions of continuous dependence of solutions on Cauchy data have received the most attention in the literature. A number of such studies have appeared for various nonlinear ill posed problems: the final value problem for the Navier–Stokes equations (Knops and

Payne, 1968; Cannon and Knightly, 1970; Galdi and Straughan, 1988); the conduction–diffusion Boussinesq equations backward in time (Straughan, 1975a, 1977); first and second order nonlinear operator equations (Levine, 1970a, b); and non–standard Cauchy problems for nonlinear parabolic equations (Bell, 1981a,b; Payne, 1985; Payne and Straughan, 1990a).

The logarithmic convexity and the weighted energy methods have been particularly successful in deriving continuous data dependence inequalities for nonlinear ill posed problems. To illustrate the type of results that can be expected, we shall briefly outline here how the logarithmic convexity technique leads to Hölder continuous dependence of solutions on the Cauchy data for the problem

$$v_t + \Delta v = G(t, v) \quad \text{in } \Omega \times [0, T) \tag{4.1}$$

$$v = 0 \quad \text{on } \partial\Omega \times [0, T) \tag{4.2}$$

$$v(x, 0) = f(x) \quad x \in \Omega. \tag{4.3}$$

The nonlinear function $G(t, v)$ is assumed to satisfy a uniform Lipschitz condition. The stability question may be studied by assuming that v_1 and v_2 are two solutions of (4.1)–(4.3) corresponding to the initial data f_1 and f_2, respectively. Let us define the functional

$$\phi(t) = \int_0^t \int_\Omega (v_1 - v_2)^2 dx d\eta + (T + \mu - t) \int_\Omega (f_1 - f_2)^2 dx \tag{4.4}$$

for $t \in [0, T)$ and a positive constant μ. A judicious use of inequalities yields the second order differential inequality

$$\phi\phi'' - (\phi')^2 \geq -\kappa_1 \phi^2 - \kappa_2 \phi\phi' \tag{4.5}$$

for explicit, nonnegative constants κ_1 and κ_2. The change of variable $\sigma = e^{-\kappa_2 t}$ transforms (4.5) into

$$\frac{d^2}{d\sigma^2} \left[\ln(\phi\sigma^{-\kappa_1/\kappa_2^2}) \right] \geq 0. \tag{4.6}$$

We thus see that $\ln \phi(t) + \frac{\kappa_1}{\kappa_2} t$ is a convex function of $e^{-\kappa_2 t}$. Integration of (4.6) by means of Jensen's inequality results in

$$\phi(t) \leq C[\phi(0)]^{1-\delta(t)}[\phi(T)]^{\delta(t)}, \quad 0 \leq t \leq T \tag{4.7}$$

where

$$C = \exp\left\{\frac{\kappa_1}{\kappa_2}(\delta T - t)\right\}$$

$$\delta(t) = \frac{1 - e^{-\kappa_2 t}}{1 - e^{-\kappa_2 T}}.$$

If we assume that $w = v_1 - v_2$ belongs to that class of functions satisfing the a priori bound $\phi(T) \leq M$ for some positive constant M, then we obtain from inequality (4.7) the desired Hölder continuous dependence result, namely,

$$\phi(t) \leq k M^\delta [\phi(0)]^{1-\delta} \tag{4.8}$$

on compact sub–intervals of $[0, T)$. Here k is a computable constant and $0 \leq \delta < 1$.

We point out that the concavity inequality $(F^{-\alpha})'' \leq 0$ together with the condition $F'(0) > 0$ that we discussed in section 3 also leads to a Hölder stability inequality for $t < t^*$, the time at which the solution ceases to exist. More specifically, we obtain

$$F^{-\alpha}(t) \geq \left(1 - \frac{t}{t_0}\right) F^{-\alpha}(0) + \frac{t}{t_0} F^{-\alpha}(t_0) \quad 0 \leq t \leq t_0 < t^*$$

or

$$F^\alpha(t) \leq \frac{[F(0)F(t_0)]^\alpha}{\left[\left(1 - \frac{t}{t_0}\right) F^\alpha(t_0) + \frac{t}{t_0} F^\alpha(0)\right]}. \tag{4.9}$$

Inequality (4.9) implies Hölder continuous dependence on the Cauchy data in the class of solutions for which $F(t_0)$ is bounded.

Logarithmic convexity arguments do not appear adaptable to a number of nonlinear improperly posed Cauchy problems. Some of these have been analyzed by the weighted energy method from which, until recently, the usual stability result obtained has been a weak logarithmic continuous dependence (e.g. Bell, 1981a,b). Payne (1985) was able to modify the method in some specific problems and obtain Hölder continuous dependence on the data for solutions in the appropriate constraint sets. One of the problems he considered was the non–characteristic Cauchy problem for a nonlinear heat equation:

$$c(x,t)u_{xx} - u_t = H(x,t,u,u_x) \quad \text{in } \Omega \subset \mathbb{R}^2 \tag{4.10}$$

$$u(0,t) = f(t) \quad \text{on } \Gamma \qquad (4.11)$$

$$u_x(0,t) = g(t) \quad \text{on } \Gamma. \qquad (4.12)$$

Here Ω is an open region in the first quadrant of the (x,t) plane with a part Γ of its boundary lying on a segment $0 \le t \le T$ of the t-axis. The nonlinear function H is assumed to be Lipschitz in its last two arguments and the coefficient $c(x,t)$ is continuous and positive for all $(x,t) \in \Omega$. Payne (1985) derived $\mathcal{L}^2$ Hölder continuous dependence estimates for classical solutions that are suitably constrained.

More recently, Payne and Straughan (1990a) adapted a weighted energy argument to study a one–space dimension problem for a piece of cold ice, given data on only part of its boundary. They considered the non–standard problem

$$\frac{\partial u}{\partial t} = \frac{\partial}{\partial x}\left\{(1 + \epsilon_1 u + \epsilon_2 u^2)\frac{\partial u}{\partial x}\right\} \quad x > 0, t \in (-T,T)$$

$$(4.13)$$

$$u(0,t) = h(t), \quad \frac{\partial u}{\partial x}(0,t) = g(t) \quad t \in (-T,T)$$

where $\epsilon_2 > \frac{1}{4}\epsilon_1^2$ and obtained pointwise and $\mathcal{L}^2$ Hölder continuous dependence inequalities on the Cauchy data. As they point out, such estimates play an important role in the numerical solution of these problems.

4.2   Continuous Dependence on Geometry

One source of error that has not received much attention even with regard to well posed problems is that attributable to inaccuracies in describing the geometry of the problem. Such errors include both errors in characterizing the geometry of the physical domain and in the initial time geometry. These latter errors arise since initial data is typically not measured at precisely the same time.

The first results on the question of continuous dependence on spatial geometry for ill posed problems were obtained by Crooke and Payne (1984). These investigators were successful in stabilizing solutions of the initial–boundary value problem for the backward heat equation with Dirichlet data under perturbations of the spatial domain. Persens (1986) extended the analysis to the case in which the domains vary with time and to the analogous problem with Neumann boundary data. He also studied the Dirichlet initial–boundary value problem of linear elastodynamics with indefinite strain energy under perturbations of the spatial geometry and the Cauchy problem for the Poisson equation under variations in the Cauchy surface. Extension of the results of Crooke and Payne (1984) to the exterior problem has recently been accomplished by Payne and Straughan (to appear). All of the aforementioned studies deal with ill posed problems for linear equations and essentially comprise the aggregate of known results. Despite the fact that the equations involved in these investigations are linear, we again emphasize that the stabilization of solutions to ill posed problems against errors in spatial geometry can occur provided we restrict the class of admissible solutions. Such restrictions effectively transform these linear problems into nonlinear problems.

The first treatment of errors in the initial time geometry for an ill posed problem appeared in the work of Knops and Payne (1969) who investigated this question in the context of linear elastodynamics and then later improved their original continuous dependence results (see Knops and Payne, 1988). Additional studies of continuous dependence on initial time geometry can be found in Song (1988) and more recently in Payne and Straughan (1990b) whose analysis for the heat equation on an exterior region is adaptable to several other parabolic systems.

We refer the reader to the surveys of Payne (1987a,b; 1989) for a more complete discussion of the current state of research on continuous dependence on both spatial and initial time geometry.

4.3   Continuous Dependence on Modeling

While the task of stabilizing ill posed problems under perturbations of the geometry is difficult, it is not as formidable as that of regularizing against errors made in formulating the mathematical model, e.g. errors made in treating a fluid as a continuum, in assuming inexact physical laws, or in approximating the model equation. These latter errors are more serious primarily because it is impossible to characterize them precisely. As Bennett (1986) illustrated, constraints that stabilize a problem against other sources of error may fail to stabilize it against modeling errors. Consequently, predictions for the physical process that are based on the improperly posed mathematical problem may not be reliable unless they can be verified by actual experiments.

Perhaps the first attempt at some kind of continuous dependence on modeling investigation for ill posed problems was made by Payne and Sather (1967) who compared the solution of a backward heat conduction problem with that of a perturbed Cauchy problem for an elliptic equation. Adelson (1973, 1974) later considered a class of problems in which both the perturbed and unperturbed problems are Cauchy problems for quasilinear elliptic equations. He was interested in the following system:

$$\begin{aligned} \epsilon b L v + v &= u \\ L u &= F(x, \epsilon, u, v) \end{aligned} \quad \text{in } \Omega \subset \mathbb{R}^n \qquad (4.14)$$

$$u, v, \text{grad } u, \text{grad } v \quad \text{prescribed on } \Gamma. \qquad (4.15)$$

Here ϵ is a small positive parameter, b is a constant, L is a strongly elliptic operator, F is assumed to be Lipschitz in its last three arguments, and Γ is a portion of the boundary of Ω. Adelson (1973, 1974) asked several questions about (4.14)–(4.15) including

  i) can we find a reasonable approximation solution of this sytem in some neighborhood Ω_a of Γ by appropriately constraining u and/or v?

  ii) if ϵ is "very small", can we obtain a good enough approximation by setting $\epsilon = 0$ and approximating the solution w of the simpler Cauchy problem

$$L w = F(x, 0, w, w) \quad \text{in } \Omega \qquad (4.16)$$

$$w, \text{grad } w \quad \text{prescribed on } \Gamma,$$

assuming the data of w and v are close?

Using $\mathcal{L}^2$ constraints, Adelson was able to provide affirmative answers to these questions for the case $b < 0$. For $b > 0$ the necessary constraints on v were more severe and the resulting estimates less sharp. The details of this analysis are quite complicated but a little ingenuity led Adelson to a continuous dependence inequality of the form

$$\|v - w\|_{\Omega_a} \le c\epsilon^{\nu(a)} \qquad (4.17)$$

where $\nu(a)$ depends on the imposed constraint, the sign of b, and the geometry of Ω_a.

Additional results in the category of continuous dependence on modeling were obtained by Bennett (1986). He considered a number of different ill posed problems, the analysis of which required a range of methods. Two of the less complicated examples he investigated follow.

A. Cauchy problem for the minimal surface equation

Bennett compared the solution of

$$\left[(1 + |\nabla u|^2)^{-1/2} u_{,j}\right]_{,j} = 0 \quad \text{in } \Omega \subset \mathbb{R}^n \qquad (4.18)$$

$$\begin{aligned} u &= \epsilon f(x) \\ \text{grad } u &= \epsilon \underline{h}(x) \end{aligned} \quad \text{on } \Gamma$$

where Γ is a smooth portion of the boundary of Ω, with that of the problem

$$\Delta w = 0 \quad \text{in } \Omega$$

$$\begin{aligned} w &= f^*(x) \\ \text{grad } w &= \underline{h}^*(x) \end{aligned} \quad \text{on } \Gamma \qquad (4.19)$$

He showed that if

$$\int_{\Gamma} \left[|f - f^*|^2 + |\underline{h} - \underline{h}^*|^2\right] dS \le c_1 \epsilon^a \qquad (4.20)$$

for some $a > 0$ and if u and w are suitably constrained, then in a region Ω_β adjoining Γ, $v = u - \epsilon w$ satisfies a continuous dependence inequality for $a = 2$ of the form

$$\int_{\Omega_b} v^2 dx \le k_1 \epsilon^{4\nu(\beta)} \quad 0 < \beta < \beta_1 < 1$$

where ν is a smooth function such that $\nu(0) = 1$, $\nu(\beta_1) = 0$, and $\nu'(\beta) < 0$. We note that the constraints required to stabilize problem (4.18), regarded as a modeling perturbation of (4.19), are much stronger than the usual $\mathcal{L}^2$ constraint that stabilizes other ill posed problems against various sources of error. We do not go into details about these constraints but refer the reader to Bennett's comprehensive account of such nonstandard problems.

B. An end problem for a one dimensional nonlinear heat equation

In this example, we wish to relate the solution of

$$\frac{\partial u}{\partial t} = \frac{\partial}{\partial x}\left[\rho\left(\left|\frac{\partial u}{\partial x}\right|^2\right)\frac{\partial u}{\partial x} \right] \quad 0 < x < x_0, \quad t > t_1$$

$$\begin{aligned} u &= \epsilon f(t) \\ \tfrac{\partial u}{\partial x} &= \epsilon g(t) \end{aligned} \quad \text{on } x = 0, \quad t_1 < t < t_2 \qquad (4.21)$$

to that of the ill posed problem

$$\frac{\partial w}{\partial t} - \frac{\partial^2 w}{\partial x^2} = 0 \quad 0 < x < x_0, \quad t > t_1$$

$$\begin{aligned} w &= f^*(t) \\ \tfrac{\partial w}{\partial x} &= g^*(t) \end{aligned} \quad \text{on } x = 0, \quad t_1 < t < t_2 \qquad (4.22)$$

Here again the data were assumed to be "close" in the appropriate measure and $\rho(|u_x|^2)$ satisfied certain order conditions. Bennett obtained a Hölder continuous dependence inequality on the perturbation parameter ϵ for solutions belonging to the appropriate constraint classes.

The papers of Ames (1982a,b) might also be interpreted as continuous dependence on modeling investigations. Ames compared solutions of Cauchy problems for certain classes of first and second order operator equations with solutions of associated perturbed problems. Neither the original problem nor the perturbed problem was required to be well posed. Hölder stability inequalities relating solutions of the perturbed and unperturbed problems were obtained by Ames using logarithmic convexity arguments. The class of problems and type of derivable results can be exemplified with a comparison of the solution of

$$u_t + Mu = F(t, u) \quad t \in [0, T] \qquad (4.23)$$
$$u(0) = f_1$$

with the solution of the perturbed problem

$$w_t + Mw + \epsilon M^2 w = F(t, w) \quad t \in [0, T] \qquad (4.24)$$
$$w(0) = f_2.$$

Here ϵ is a small positive parameter, M is a symmetric, time independent linear operator (bounded or unbounded) defined on a Hilbert space, and the nonlinear term F satisfies a uniform Lipschitz condition in its second argument. Ames showed that if u and w belong to the appropriate constraint sets and if the Cauchy data are close (in a suitably defined sense), then the difference $u - w$ in a certain measure is of order $\epsilon^{\mu(t)}$ for $0 \le t < T$ where $\mu(t) > 0$.

We emphasize that the studies attempting to address the question of continuous dependence on modeling must be regarded only as mathematical illustrations. Unfortunately, we cannot interpret the results of these investigations as answers to the modeling questions that arise in actual physical problems. In fact, the question of stabilizing improperly posed problems for partial differential equations against errors in modeling might well be the major unresolved issue in the field of ill posed problems.

Bibliography

Adelson, L. (1973) *Singular perturbations of improperly posed problems.* SIAM J. Math. Anal., 4, 344–366.

Adelson, L. (1974) *Singular perturbation of an improperly posed Cauchy problem*, SIAM J. Math. Anal., 5, 417–424.

Agmon, S. (1966) *Unicité et Convexité dans les Problèmes Différentiels*, University of Montreal Press, Montreal.

Ames, K. A. (1982a) *On the comparison of solutions of related properly and improperly posed Cauchy problems for first order operator equations*, SIAM J. Math. Anal., 13, 594–606.

Ames, K. A. (1982b) *Comparison results for related properly and improperly posed Cauchy problems for second order operator equations*, J. Differential Equations, 44, 383–399.

Ames, K. A., Levine, H. A. and Payne, L. E. (1987) *Improved continuous dependence results for a class of evolutionary equations*, Inverse and Ill–Posed Problems (ed. H. W. Engl and C. W. Groetsch) Academic Press, 443–450.

Ames, K. A., Payne, L. E. and Straughan, B. (1989) *On the possibility of global solutions for variant models of viscous flow on unbounded domains*, Int. J. Engng. Sci., 27, 755–766.

Bell, J. B. (1981a) *The noncharacteristic Cauchy problem for a class of equations wtih time dependence I, Problems in one space dimension*, SIAM J. Math. Anal., 12, 759–777.

Bell, J. B. (1981b) *The noncharacteristic Cauchy problem for a class of equations with time dependence II, Multidimensional problems*, SIAM J. Math. Anal., 12, 778–797.

Bennett, A. D. (1986) *Continuous dependence on modeling in the Cauchy problem for second order nonlinear partial differential equations*, Ph.D. Thesis, Cornell University.

Cannon, J. R. and Knightly, G. H. (1970) *Some continuous dependence theorems for viscous fluid motions*, SIAM J. Appl. Math., 18, 627–640.

Carasso, A. and Stone, A. P. (1975) *Improperly Posed Boundary Value Problems*, Pitman Research Notes in Mathematics, Vol. 1, Pitman Press, London.

Crooke, P. S. and Payne, L. E. (1984) *Continuous dependence on geometry for the backward heat equation*, Math. Methods in the Applied Sciences, 6, 433–448.

Engl, H. W. and Groetsch, C. W., ed. (1987) *Inverse and Ill–Posed Problems*, Notes and Reports in Mathematics in Science and Engineering, Vol. 4, Academic Press, Inc.

Ewing, R. E. (1975) *The approximation of certain parabolic equations backward in time by Sobolev equations*, SIAM J. Math. Anal., 6, 283–294.

Fujita, H. (1966) *On the blowing up of solutions to the Cauchy problem for $u_t = \Delta u + u^{1+\alpha}$*, J. Fac. Sci. Univ. Tokyo, 13, 109–124.

Fujita, H. (1970) *On some nonexistence and nonuniqueness theorems for nonlinear parabolic equations*, Proc. Symp. Pure Math., 18, 105–113.

Galdi, G. P. and Straughan, B. (1988) *Stability of solutions to the Navier–Stokes equations backward in time*, Arch. Rational Mech. Anal., 101, 107–114.

Glassey, R. T. (1973) *Blow up theorems for nonlinear wave equations*, Math. Z., 132, 183–203.

Glassey, R. T. (1981) *Finite time blow up for solutions of nonlinear wave equations*, Math. Z., 177, 323–340.

Hills, R. N. and Knops, R. J. (1974) *Concavity and the evolutionary properties of a class of general materials*, Proc. Roy. Soc. Edinburgh, 72, 239–243.

John, F. (1955) *A note on "improper" problems in partial differential equations*, Comm. Pure Appl. Math., 8, 494–495.

John, F. (1960) *Continuous dependence on data for solutions of partial differential equations with a prescribed bound*, Comm. Pure. Appl. Math., 13, 551–585.

John, F. (1979) *Blow up of solutions of nonlinear wave equations in three dimensions*, Man. Math. 28, 235–268.

Kaplan, S. (1963) *On the growth of solutions of quasilinear parabolic equations*, Comm. Pure Appl. Math., 16, 305–330.

Knops, R. J. ed (1973) *Symposium on Non–Well–Posed Problems and Logarithmic Convexity*, Springer Lecture Notes, #316.

Knops, R. J., Levine, H. A. and Payne, L. E. (1974) *Nonexistence, instability and growth theorems for solutions of a class of abstract nonlinear equations with applications to nonlinear elastodynamics*, Arch. Rational Mech. Anal., 55, 52–72.

Knops, R. J. and Payne, L. E. (1968) *On the stability of solutions of the Navier–Stokes equations backward in time*, Arch. Rational Mech. Anal., 29, 331–335.

Knops, R. J. and Payne, L. E. (1969) *Continuous data dependence for the equations of classical elastodynamics*, Proc. Camb. Phil. Soc., 66, 481–491.

Knops, R. J. and Payne, L. E. (1979) *Uniqueness and continuous dependence of the null solution in the Cauchy problem for a nonlinear elliptic system.*, Mathematical Research—Inverse and Improperly Posed Problems in Differential Equations (ed. G. Anger), Vol. 1, 151–160.

Knops. R. J. and Payne, L. E. (1983) *Some uniqueness and continuous dependence theorems for non–linear elastodynamics in exterior domains*, Appl. Anal., 15, 33–51.

Knops, R. J. and Payne, L. E. (1988) *Improved estimates for continuous data dependence in linear elastodynamics*, Proc. Camb. Phil. Soc., 103, 535–559.

Knops, R. J. and Straughan, B. (1976) *Non–existence of global solutions to non–linear Cauchy problems arising in mechanics*, Trends in Applications of Pure Mathematics to Mechanics, Pitman, London.

Lattès, R. and Lions, J. L. (1969) *The Method of Quasireversibility, Applications to Partial Differential Equations*, American Elsevier, New York.

Lavrentiev, M. M. (1956) *On the Cauchy problem for the Laplace equation*, Izvest. Akad. Nauk SSR, Ser. Math., 120, 819–842.

Lavrentiev, M. M. (1967) *Some Improperly Posed Problems in Mathematical Physics*, Springer-Verlag, New York.

Lavrentiev, M. M., Romanov, V. G. and Sisatskii, S. P. (1986) *Ill-Posed Problems of Mathematical Physics and Analysis, Translations of Mathematical Monographs*, 64, American Mathematical Society, Providence.

Lees, M. and Protter, M. H. (1961) *Unique continuation for parabolic differential equations and differential inequalities*, Duke Math. J., 28, 369–382.

Levine, H. A. (1970a) *Logarithmic convexity and the Cauchy problem for some abstract second order differential inequalities*, J. Differential Equations, 8, 34–55.

Levine, H. A. (1970b) *Logarithmic convexity, first order differential inequalities and some applications*, Trans. Amer. Math. Soc., 152, 299–319.

Levine, H. A. (1973) *Some nonexistence and instability theorems for solutions of formally parabolic equations of the form $Pu_t = -Au + F(u)$*, Arch. Rational Mech. Anal., 51, 371–386.

Levine, H. A. (1974a) *Instability and nonexistence of global solutions to nonlinear wave equations of the form $Pu_{tt} = -Au + F(u)$*, Trans. Amer. Math. Soc., 192, 1–21.

Levine, H. A. (1974a) *Some additional remarks on the nonexistence of global solutions to nonlinear wave equations*, SIAM J. Math. Anal. 5, 138–146.

Levine, H. A. (1974c) *A note on a nonexistence theorem for nonlinear wave equations*, SIAM J. Math. Anal., 5, 644–648.

Levine, H. A. (1990) *The role of critical exponents in blow up theorems*, SIAM Review, 32, 262–288.

Levine, H. A. and Payne, L. E. (1974a) *Nonexistence theorems for the heat equation with nonlinear boundary conditions and for the porous medium equation backward in time*, J. Differential Equations, 16, 319–334.

Levine, H. A. and Payne, L. E. (1974b) *Some nonexistence theorems for initial boundary value problems with nonlinear boundary constraints*, Proc. Amer. Math. Soc., 46, 277–284.

Levine, H. A. and Payne, L. E. (1976) *Nonexistence for global weak solutions for classes of nonlinear wave and parabolic equations*, J. Math. Anal. Appl., 55, 329–334.

Lewy, H. (1957) *An example of a smooth linear partial differential equation without solution*, Ann. of Math., 66, 155–158.

Miller, K. (1973a) *Non–unique continuation for certain ODE's in Hilbert space and for uniformly parabolic elliptic equations in self–adjoint divergence form*, Symp. on Non–well–posed Problems and Logarithmic Convexity, Springer Lecture Notes # 316, 85–101.

Miller, K. (1973b) *Stabilized quasi–reversibilite and other nearly–best–possible methods for non–well–posed problems*, Symp. on Non–well–posed Problems and Logarithmic Convexity, Springer Lecture Notes #316, 161–176.

Miller, K. (1974) *Nonunique continuation for uniformly parabolic and elliptic equations in self–adjoint divergence form with Hölder continuous coefficients*, Arch. Rational Mech. Anal., 54, 105–117.

Morozov, V. A. (1984) *Methods for Solving Incorrectly Posed Problems*, Springer-Verlag, Berlin–Heidelberg–New York.

Murray, A. (1972) *Uniqueness and continuous dependence for the equations of elastodynamics without strain energy function*, Arch. Rat. Mech. Anal., 47, 195–204.

Murray, A. C. and Protter, M. H. (1973) *The asymptotic behavior of solutions of second order systems of partial differential equation*, J. Differential Equations, 13, 57–80.

Payne, L. E. (1975) *Improperly Posed Problems in Partial Differential Equations*, Regional Conference Series in Applied Mathematics, Vol. 22, SIAM.

Payne, L. E. (1985) *Improved stability estimates for classes of ill-posed Cauchy problems*, Applicable Analysis, 19, 63–74.

Payne, L. E. (1987a) *On stabilizing ill posed problems against errors in geometry and modeling*, Inverse and Ill Posed Problems (ed. H. W. Engl and C. W. Groetsch), Academic Press, 443–450.

Payne, L. E. (1987b) *On geometric and modeling perturbations in partial differential equations*, L.M.S. Symposium on Non-Classical Continuum Mech. Proc. ed. by R. J. Knops and A. A. Lacey, Cambridge Univ. Press, 108–128.

Payne, L. E. (1989) *Continuous dependence on geometry with applications in continuum mechanics*, Continuum Mechanics and its Applications, ed. G. A.C. Graham and S. K. Malitz, 877–890.

Payne, L. E. and Sather, D. (1967) *On singular perturbations of non–well–posed problems*, Annali di Mat. Pura ed. Appl., 75, 219–230.

Payne, L. E. and Straughan, B. (1990a) *Error estimates for the temperature of a piece of cold ice, given data on only part of the boundary*, Nonlinear Anal., Theory, Meths. Applicns., 14, 443–452.

Payne, L. E. and Straughan, B. (1990b) *Effects of errors in the initial–time geometry on the solution of the heat equation in an exterior domain*, Q. J. Mech. Appl. Math., 43, 75–86.

Payne, L. E. and Straughan, B. (1990c) *Improperly posed and non-standard problems for parabolic partial differential equations*,

Ian Sneddon 70th Anniversary Volume, (G. Eason & R. W. Ogden, eds.), Ellis-Horwood Pub. Co., 273–299.

Payne, L. E.and Straughan, B. *Continuous dependence on geometry for the backward heat equation in an exterior domain,* Proc. Lond. Math. Soc. (to appear).

Persens, J. (1986) *On stabilizing ill posed problems in partial differential equations under perturbations of the geometry of the domain,* Ph.D. Thesis, Cornell University.

Plis, A. (1960) *Nonuniqueness in Cauchy's problem for differential equations of elliptic type,* J. Math. Mech., 9, 557–562.

Plis, A. (1963) *On nonuniqueness in the Cauchy problem for an elliptic second order differential equation,* Bull. Acad. Polon. Sci., 11, 95–100.

Pucci, C. (1955) *Sui problemi di Cauchy non "ben posti,"* Rend. Accad. Naz. Lincei, 18, 473–477.

Sattinger, D. (1968) *On global solutions of nonlinear hyperbolic equations,* Arch. Rational Mech. Anal., 30, 148–172.

Serrin, J. (1963) *The initial value problem for the Navier–Stokes equations* in Nonlinear Problems, Univ. Wisconsin Press, Madison.

Showalter, R. E. (1974) *The final value problem for evolution equations,* J. Math. Anal. Appl., 47, 563–572.

Showalter, R. E. (1975) *Quasi-reversibility of first and second order parabolic evolution equations,* Pitman Research Notes in Mathematics, Vol. 1, 76–84.

Song, J. C. (1988) *Some stability crtieria in fluid and solid mechanics,* Ph.D. Thesis, Cornell University.

Straughan, B. (1975a) *Uniqueness and stability for the conduction–diffusion solution to the Bousssinesq equations backward in time,* Proc. Roy. Soc. A, 347, 435–446.

Straughan, B. (1975b) *Further global nonexistence theorems for abstract nonlinear wave equations*, Proc. Amer. Math. Soc., 48, 381–390.

Straughan, B. (1976) *Global nonexistence of solutions to Ladyzhenskaya's variants of the Navier–Stokes equations backward in time*, Proc. R. Soc. Edinburgh, 75A, 165–170.

Straughan, B. (1977) *Uniqueness and continuous dependence theorems for the conduction–diffusion solution to the Boussinesq equations on an exterior domain*, J. Math. Anal. Appl., 57, 203-233.

Straughan, B. (1982) *Instability, Nonexistence and Weighted Energy Methods in Fluid Dynamics and Related Theories*, Pitman Research Notes in Mathematics Vol. 74, Pitman Press, London.

Straughan, B. (1983) *Backward uniqueness and unique continuation for solutions to the Navier–Stokes equations on an exterior domain*, J. Math. pures et appl., 62, 49–62.

Tikhonov, A. N. (1963) *On the solution of ill–posed problems and the method of regularisation*, Dokl. Akad. Nauk SSSR, 151, 501–504.

Tikhonov, A. N. and Arsenin, V. Y. (1977) *Solution of Ill–Posed Problems*, Wiley Press.

Tsutsumi, M. (1971) *Some nonlinear evolution equations for second order*, Proc. Japan. Acad., 47, 450–455.

Tsutsumi, M. (1972) *Existence and non-existence of global solutions for nonlinear parabolic equations*, Publ. Research Inst. Math. Sci., Kyoto Univ., 8, 211–229.

Symmetry in Nonlinear Mechanics

W. F. Ames
School of Mathematics
Georgia Institute of Technology
Atlanta, GA 30332

1   Introduction

The mathematical models of many problems in mechanics consist of
nonlinear partial differential equations. This chapter is concerned
with invariance of those equations under groups of Lie transformations and the resulting qualitative properties and symmetry reductions.

A systematic investigation of continuous transformation groups
was carried out by Lie (1882-1899). His original goal was the creation
of a theory of integration for ordinary differential equations analogous
to the Abelian theory for the solution of algebraic equations. He investigated the fundamental concept of the invariance group admitted
by a given system of differential equations. Today, the mathematical approach whose object is the construction and analysis of the
full invariance group admitted by a system of differential equations
is called *group analysis* of differential equations. These groups, now
usually called Lie groups, and the associated Lie algebras have important real world applications. In particular, they bear upon the
concepts of homogeneity and isotropy of space and time, the dynamic
similarity of physical phenomena, as well as Galilean and Lorentzian
invariance.

The task of determining the largest group of Lie transformations
that leaves invariant a system of differential equations involves purely
mechanical, albeit lengthy procedures, well-suited to symbolic com-

putation. Once the group is known, one can study the action of the group on the set of solutions to the equations. Thus, the group invariance may be utilized to generate new solutions from known ones. Moreover, the admitted group introduces an algebraic structure that allows important classifications of solutions. Subalgebras may be analyzed and group invariants used in reductions of the original system. Exact solutions of these reduced systems may, on occasion, be derived.

The differential equations of engineering and the physical sciences often involve parameters or constitutive laws that are determined experimentally (here these are called *arbitrary elements*). The equations of the mathematical model should be simple enough to be amenable to analysis. The group approach suggests the acceptance of, as a simplicity criterion, the requirement that the arbitrary element be such that, with it, the model differential equation admits appropriate group invariance.

There is a considerable literature on group analysis. We must begin with the basic works of Lie (1888a, 1890, 1891, 1893). Elementary introductory books include those by Page (1897), Cohen (1911) and Ince (1956). They use Lie's theory of one-parameter groups with special application to ordinary differential equations. Birkhoff's monograph discusses the symmetry concepts in fluid mechanics in a group-theoretic manner (1960). Since that fine study, the application and generalization of group analysis has blossomed. By 1962, Ovsiannikov's first work had appeared and was subsequently translated by Bluman (Ovsiannikov, 1962) in 1967. In 1964, Hansen (Hansen, 1964) published a volume devoted to a number of engineering studies in diffusion and fluid mechanics. In 1965, the first book on nonlinear problems by Ames (1965) treated similarity by specific groups (dilation and translation) and gave applications in engineering. In 1968, Ames (1968, Ch. 2 and 3), provided an introduction to the group concept wtih a number of applications to ordinary differential equations in fluid mechanics.

More detailed group analysis, aimed primarily at group construction and analysis for partial differential equations, is to be found in a number of recent books (arranged chronologically). These include

Ames (1972), Bluman and Cole (1974), a second definitive volume by Ovsiannikov (1978) and a related work by Anderson and Ibragimov (1979). Barenblatt (1979), Na (1979) and Seshadri and Na (1985) include many applications in engineering, while Hill (1982b) has one-parameter group applications. Ibragimov (1983) is concerned with applications of mathematical physics and Winternitz (1983) with nonlinear equations. Applications in physics, geometry and mechanics are given by Sattinger and Weaver in (1986). An excellent general theory of the subject is presented by Olver (1986). Rogers and Ames (1989) have collected a large table of available groups. Bluman and Kumei (1989) give a comprehensive treatment of Lie groups in a book which can be used as a textbook.

Collections of papers, reporting group calculations, include *Group Theoretical Methods in Mechanics* (Ibragimov and Ovsiannikov, eds.) in 1978, *Group Theoretical Methods in Physics* (Beiglbock, Böhm and Takasugi, eds.) in 1979, *Symposium on Nonlinear Integrable Systems* (Jumbo and Miwa, eds., 1983), *Nonlinear Phenomena* (Wolf, ed.) in 1983, and *Symmetry Methods in Differential Equations* (Anderson and Olver, eds.) in 1987, and *Symmetries of Partial Differential Equations — Conservation Laws — Applications — Algorithms* in 1989 (A. M. Vinogradov, ed.). Lastly, in a more general context, the book of Konopelchenko (1987) is an outstanding compendium, with over 600 references on the literature of nonlinear integrable equations.

1.1   Wave Equations and Group Invariance

Some authors, in their studies of nonlinear wave equations, carry out a direct application of the scaling or other simple groups such as translation. This leads, when applicable to similar or travelling wave solutions, respectively. This time saving approach is carried out because in many physical models, that are based upon Newtonian mechanics, invariance under scaling (dilation), translation and rotation is to be expected.

1.2   Wave Propagation with Arbitrary Constitutive Laws

Many applications of the direct approach are found. For example
Suliciu et al (1973), Ames and Suliciu (1982), Ames and Donato
(1988) employ the method. Ames and Suliciu studied the system

$$\rho\frac{\partial v}{\partial t} - \frac{\partial \sigma}{\partial x} = 0 \tag{1.1}$$

$$\frac{\partial e}{\partial t} - \frac{\partial v}{\partial x} = 0 \tag{1.2}$$

$$\frac{\partial \sigma}{\partial t} - f\left(e, \sigma, \frac{\partial e}{\partial t}\right) = 0 \tag{1.3}$$

under the scaling, translation and spiral subgroups. Here ρ is (constant) density, v is velocity, σ is stress and e is strain. The function
f (an arbitrary element) characterizes the properties of the material.
This approach may, of course, omit some important operators. The
full Lie group was calculated by K. A. Ames (1989) in primitive form
(equations (1.1)–(1.3)) and in third order form by Hölzmuller (1991).
The generators for the three equations (1.1)–(1.3) are

Case 1: $f_{e_t} \neq 0$

$$\Gamma_1 = x\frac{\partial}{\partial x} + t\frac{\partial}{\partial t} + v\frac{\partial}{\partial v} + \sigma\frac{\partial}{\partial \sigma} + e\frac{\partial}{\partial e} \qquad \text{(Scaling)}$$

$$\Gamma_2 = \frac{\partial}{\partial x}, \qquad \Gamma_3 = \frac{\partial}{\partial t} \qquad \text{(Translation)}$$

$$\Gamma_4(\beta) = \beta(x,t)\frac{\partial}{\partial v}$$

$$\Gamma_5(\gamma) = \gamma(x,t)\frac{\partial}{\partial \sigma}$$

$$\Gamma_6(\alpha) = \alpha(x,t)\frac{\partial}{\partial e}$$

where α, β, γ are sufficiently smooth functions of x and t that satisfy
the system

$$\rho\frac{\partial \beta}{\partial t} - \frac{\partial \alpha}{\partial x} = 0,$$

$$\frac{\partial \alpha}{\partial t} - \frac{\partial \beta}{\partial x} = 0,$$

$$Ef_e + \Sigma f_\sigma + \frac{\partial \alpha}{\partial t} f_{e_t} = \frac{\partial \gamma}{\partial t}$$

and $E = c_1 e + \alpha(x,t)$, $\Sigma = c_1 \sigma + \gamma(x,t)$. This algebra is infinite dimensional. The first three generators are the independent generators of a three dimensional subalgebra.

Case 2: $f_{e_t} = 0$

Here the algebra has the seven generators

$$\Gamma_1 = 3x\frac{\partial}{\partial x} + 2t\frac{\partial}{\partial t} + v\frac{\partial}{\partial v} + 2\sigma\frac{\partial}{\partial \sigma} \tag{1.4}$$

$$\Gamma_2 = -2x\frac{\partial}{\partial x} - t\frac{\partial}{\partial t} - \sigma\frac{\partial}{\partial \sigma} + e\frac{\partial}{\partial e} \tag{1.5}$$

$$\Gamma_3 = \frac{\partial}{\partial x}, \quad \Gamma_4 = \frac{\partial}{\partial t}, \quad \Gamma_5 = \beta(x,t)\frac{\partial}{\partial v},$$

$$\Gamma_6(\gamma) = \gamma(x,t)\frac{\partial}{\partial \sigma}, \quad \Gamma_7(\alpha) = \alpha(x,t)\frac{\partial}{\partial e}$$

The first four operators are the independent generators of a four dimensional subalgebra. Generators (1.4) and (1.5) are scaling transformations which give rise to solutions of the form

$$v = t^{1/2}f(\eta), \quad \sigma = tg(\eta), \quad e = \text{constant}, \quad \eta = xt^{-3/2}$$

and

$$v = \text{constant}, \quad \sigma = th(\eta), \quad e = \frac{1}{t}k(\eta), \quad \eta = xt^{-2},$$

respectively.

1.3 Motions in Strings and Cables

A number of studies of nonlinear equations governing the motions of strings and cables have been published. For a perfectly flexible nonlinear elastic string or cable undergoing large deflection or deformation (Peters and Ames (1990)) the equations

$$u_t - v\psi_t - \dot\Sigma(\lambda)\lambda_x = 0 \tag{1.6}$$

$$u_x - v\psi_x - \lambda_t = 0 \tag{1.7}$$

$$v_t + u\psi_t - \Sigma(\lambda)\psi_x = 0 \tag{1.8}$$

$$v_x + u\psi_x - \lambda\psi_t = 0 \tag{1.9}$$

have been used to characterize the motions. In this system x is the
Lagrangian space coordinate embedded in the string, t is time, u
and v are the tangential and normal components of velocity, ψ is the
angle of inclination between the cable and the x axis, λ is the stretch
and $\Sigma = \Sigma(\lambda)$ is the general constitutive law. It is usually assumed
that $\Sigma(\lambda) > 0$, $\lambda > 1$ (string is in tension), $\dot{\Sigma}(\lambda) > 0$, and $\Sigma \neq k\lambda$ so
the system is hyperbolic.

For general $\Sigma(\lambda)$ the symmetry group has the six generators

$$\Gamma_1 = \frac{\partial}{\partial t}, \quad \Gamma_2 = \frac{\partial}{\partial x}, \quad \Gamma_3 = x\frac{\partial}{\partial x} + t\frac{\partial}{\partial t}$$

$$\Gamma_4 = \sin\psi\frac{\partial}{\partial u} + \cos\psi\frac{\partial}{\partial v}, \quad \Gamma_5 = -\cos\psi\frac{\partial}{\partial u} + \sin\psi\frac{\partial}{\partial v}$$

$$\Gamma_6 = \frac{\partial}{\partial\psi} \tag{1.10}$$

These generators indicate that the system admits translations in x,
t, and ψ as well as scaling (from Γ_3).

Let α_i be the parameter associated with each of the ith generators.
With $\alpha_3 = \alpha_6 = 1$, $\alpha_i = 0$ for $i \neq 3$ or 6 the independent variable
invariant is $\eta = x/t$ and the symmetry reduction of equations (1.6)–
(1.9) is

$$-f_2 + (\eta^2 - \dot{\Sigma}(f_4))f_4' = 0$$
$$f_1' - f_2 f_3' + \eta f_4' = 0$$
$$f_1 - \eta f_4 + (\eta^2 f_4 - \Sigma(f_4))f_3' = 0$$
$$f_2' - f_4 + (f_1 + \eta f_4)f_3' = 0$$

with the original variables given by

$$u(x,t) = f_1(\eta)$$
$$v(x,t) = f_2(\eta)$$
$$\psi(x,t) = \ln t + f_3(\eta)$$
$$\lambda(x,t) = f_4(\eta). \tag{1.11}$$

Among the interesting properties of equations (1.11) are that they
again arise in a host of other ways: with $\alpha_3 = \alpha_4 = \alpha_6 = 1$, the

other $\alpha_i = 0$; with $\alpha_3 = \alpha_5 = \alpha_6 = 1$, the other $\alpha_i = 0$; and with $\alpha_3 = \alpha_4 = \alpha_5 = \alpha_6 = 1$, the other $\alpha_i = 0$. Also, equations (1.11) may be rewritten in the form

$$f_1' = \frac{f_2(\eta f_4 - f_1)}{\eta^2 f_4 - \Sigma(f_4)} - \frac{\eta f_2}{\eta^2 - \dot{\Sigma}(f_4)}$$

$$f_2' = \frac{f_1^2 - f_4 \Sigma(f_4)}{\eta^2 f_4 - \Sigma(f_4)}$$

$$f_3' = \frac{\eta f_4 - f_1}{\eta^2 f_4 - \Sigma(f_4)}$$

$$f_4' = \frac{f_2}{\eta^2 - \dot{\Sigma}(f_4)}. \qquad (1.12)$$

In equations (1.12) a discontinuity can evolve if either of the denominators vanish. The dependence of the discontinuities on the constitutive laws is nicely displayed.

For specific arbitrary elements (constitutive laws), $\Sigma(\lambda)$, the symmetry group is often larger. Thus with $\Sigma(\lambda) = k\lambda^n$, $n \neq 1$ a different scaling generator

$$\Gamma_3' = wx\frac{\partial}{\partial x} + \mu t\frac{\partial}{\partial t} + \alpha u\frac{\partial}{\partial u} + \alpha v\frac{\partial}{\partial v} + \delta\lambda\frac{\partial}{\partial \lambda} \qquad (1.13)$$

is found, where $\delta = 2(w - \mu)/(n - 1)$, w, μ and α are arbitrary. With $w = \mu$ and $\alpha = 0$, Γ_3 is recovered. With (1.13) an infinite family of invariant solutions with the form

$$u(x,t) = t^{\alpha/\mu} f_1(\eta), \qquad v(x,t) = t^{\alpha/\mu} f_2(\eta)$$
$$\psi(x,t) = f_3(\eta), \qquad \lambda(x,t) = t^{\delta/\mu} f_4(\eta)$$
$$\eta = xt^{-w/\mu}$$

are possible.

The equations for the large amplitude vibrations of a travelling string have been formulated by Ames et al. (1968),

$$V^2 u_{xx} + 2V u_{xt} + u_{tt} = \frac{T}{m} u_{xx}(1 + u_x^2 + v_x^2)^{-1}, \qquad (1.14)$$

$$V^2 v_{xx} + 2V v_{xt} + v_{tt} = \frac{T}{m} v_{xx}(1 + u_x^2 + v_x^2)^{-1}, \qquad (1.15)$$

$$V_t + VV_x = \frac{1}{m[1 + u_x^2 + v_x^2]^{1/2}} \left[\frac{T}{(1 + u_x^2 + v_x^2)^{1/2}} \right]_x, \qquad (1.16)$$

$$(m(1 + u_x^2 + v_x^2)^{1/2})_t + (mV(1 + u_x^2 + v_x^2)^{1/2})_x = 0 \qquad (1.17)$$

$$T = \Phi(m, m_t) \qquad (1.18)$$

In this system V is the velocity in the axial direction (x), u and v are the plane and vertical components of displacement, T is tension and m is mass per unit length.

In the case $T = \Phi(m)$, only, the symmetry group (Ames et al. (1989)) consists of the nine generators

$$X_1 = \frac{\partial}{\partial t}, \qquad X_2 = \frac{\partial}{\partial x},$$

$$X_3 = \frac{\partial}{\partial u}, \qquad X_4 = \frac{\partial}{\partial v} \qquad \text{(Translations)}$$

$$X_5 = t\frac{\partial}{\partial t} + x\frac{\partial}{\partial x} + u\frac{\partial}{\partial u} + v\frac{\partial}{\partial v} \qquad \text{(Scaling)}$$

$$X_6 = t\frac{\partial}{\partial x} + \frac{\partial}{\partial v} \qquad \text{(Galilean boost)}$$

$$X_7 = v\frac{\partial}{\partial u} - u\frac{\partial}{\partial v} \qquad \text{(Rotation of Dependent Variables)}$$

$$X_8 = t\frac{\partial}{\partial u}, \qquad X_9 = t\frac{\partial}{\partial v}.$$

In the case $T = \hat{\Phi}(m)m_t^{2(A+c_5)/A}$, $\hat{\Phi}$ arbitrary, the symmetry group has eight generators

$$X_1 = \frac{\partial}{\partial x}, \qquad X_2 = \frac{\partial}{\partial t},$$

$$X_3 = \frac{\partial}{\partial u}, \qquad X_4 = \frac{\partial}{\partial v} \qquad \text{(Translations)}$$

$$X_5 = x\frac{\partial}{\partial x} + u\frac{\partial}{\partial u} + v\frac{\partial}{\partial v} + V\frac{\partial}{\partial V} \qquad \text{(Scaling)}$$

$$X_{6A} = -t\frac{\partial}{\partial t} + V\frac{\partial}{\partial V} \qquad \text{(Scaling)}$$

$$X_7 = t\frac{\partial}{\partial v}, \qquad X_8 = t\frac{\partial}{\partial u}$$

where A is the arbitrary real number appearing in the constitutive law and c_5 is the scalar associated with the scaling generator X_5.

The scaling generator in the $T = \Phi(m)$ case has been used by Ames and Donato (1988) to study the evolution of weak discontinuities of this system in the plane, in a state characterized by invariant solutions. Donato and Oliveri (1988) give a more convenient formulation of equations (1.14)–(1.18) in order to study the occurrence of shock waves. They determine, in terms of the initial conditions, the critical time when the weak discontinuity becomes unbounded. Also studied is the interaction of a weak discontinuity with a shock wave.

1.4   Quasilinear Wave Equations

In Ames et al. (1981) it was demonstrated how a number of physical problems from gas dynamics, shallow water waves, dynamics of a nonlinear string, elastic-plastic materials and electromagnetic transmission line satisfied the common quasilinear hyperbolic equation

$$u_{tt} = [f(u)u_x]_x, \tag{1.19}$$

where we assume $f \in C^2(\mathbb{R})$, $f > 0$ and $f' \neq 0$. The symmetry group was found by Ames et al. (1981). The group is presented in several special cases in dependence on $f(u)$.

Case A: $f(u)$ arbitray

$$\Gamma_1 = \frac{\partial}{\partial x}, \quad \Gamma_2 = \frac{\partial}{\partial t}, \quad \Gamma_3 = x\frac{\partial}{\partial x} + t\frac{\partial}{\partial t} \tag{1.20}$$

Case B: $f = c\exp(mu)$, c, m real and arbitrary.

$$\Gamma_1, \Gamma_2, \Gamma_3 \quad \text{of (1.20) and} \quad \Gamma_4 = 2\frac{\partial}{\partial u} + mx\frac{\partial}{\partial x}$$

Case C: $f = c(u + d)^m$, c, m real and arbitrary.

$$\Gamma_1, \Gamma_2, \Gamma_3 \quad \text{of (1.20) and} \quad \Gamma_4' = 2(u + d)\frac{\partial}{\partial u} + mx\frac{\partial}{\partial x} \tag{1.21}$$

Case C_1: $m = -4/3$.

$$\Gamma_1, \Gamma_2, \Gamma_3 \Gamma_4' \quad \text{of (1.21) and} \quad \Gamma_5' = 3x(u+d)\frac{\partial}{\partial u} - x^2\frac{\partial}{\partial x}$$

Case C_2: $m = -4$

$$\Gamma_1, \Gamma_2, \Gamma_3 \Gamma_4' \quad \text{of (1.21) and} \quad \Gamma_5' = t(u+d)\frac{\partial}{\partial u} + t^2\frac{\partial}{\partial t}$$

The group analysis has been extended to the equation

$$u_{tt} = [f(u,x)u_x]_x \tag{1.22}$$

by Torrisi and Valenti (1985). Şuhubi and Bakkaloğlu (1991) consider the equation

$$u_{tt} = [f(u_x, u_t, u, x, t)]_x \tag{1.23}$$

for the hyperbolic case, $(\partial f/\partial u_t)^2 + 4(\partial f/\partial u_x) > 0$, by means of the methods of the exterior calculus.

Lastly, the semilinear equation

$$u_{xt} = f(u) \tag{1.24}$$

has been analyzed by Pucci and Salvatori (1986). If

$$\Gamma = X\frac{\partial}{\partial x} + T\frac{\partial}{\partial t} + U\frac{\partial}{\partial u}$$

the authors find

$$X = X(x), \quad T = T(t), \quad U = au + b(x,t)$$

where

$$auf'(u) + bf'(u) - (a - X' - T')f(u) - b_{xt} = 0.$$

A number of special cases are detailed but only that of $f(u) = (hu + k)^{1-s}$, $s \neq 0, h \neq 0$ is given here. The symmetry generators are found to be

$$\Gamma_1 = -x\frac{\partial}{\partial x} + t\frac{\partial}{\partial t}, \quad \Gamma_2 = \frac{\partial}{\partial t}, \quad \Gamma_3 = \frac{\partial}{\partial x}$$

$$\Gamma_4 = \frac{s}{2}x\frac{\partial}{\partial x} + \frac{s}{2}t\frac{\partial}{\partial t} + \left(u + \frac{k}{h}\right)\frac{\partial}{\partial u}.$$

1.5   Acoustics

The Helmholtz equation in cylindrical coordinates is often employed
to study acoustic waves in fluids. Richards (1987) computed the
symmetry group for this linear equation

$$r^{-1}(ru_r)_r + r^{-2}u_{\theta\theta} + u_{zz} + H(r,\theta,z)u = 0 \qquad (1.25)$$

and found it to be

$$\begin{aligned}
\Gamma_1 &= \frac{\partial}{\partial z}, \quad \Gamma_2 = r\frac{\partial}{\partial r} + z\frac{\partial}{\partial z} \\
\Gamma_3 &= rz\frac{\partial}{\partial r} + \frac{1}{2}(z^2 - r^2)\frac{\partial}{\partial z} \\
\Gamma_4 &= \frac{\partial}{\partial \theta}, \quad \Gamma_5 = \bar{u}(r,\theta,z)\frac{\partial}{\partial u}
\end{aligned}$$

where $\bar{u}$ is any solution of equation (1.25). Admissible $H(r,\theta,z)$ are
determined by solutions of

$$(c_1 rz + c_2 r)H_r + c_4 H_\theta + \left[\frac{c_1}{2}(z^2 - r^2) + c_2 z + c_3\right]H_z + 2(c_1 z + c_2)H = 0.$$

Two dimensional subgroups and invariant solutions are also found
by Richards (1987).

The Khoklov-Zabolotskaya equation

$$\frac{1}{2}\frac{\partial^2 u^2}{\partial q_1^2} - \frac{\partial^2 u}{\partial q_1 \partial q_2} + \frac{\partial^2 u}{\partial q_3^2} + \frac{\partial^2 u}{\partial q_4^2} = 0 \qquad (1.26)$$

describes the propagation of bounded sound beams in a nonlinear
medium without dispersion and absorption. Equation (1.21) has
been studied by Sharomet (1989). He calculated the symmetries,
invariant solutions and conservation laws for this nonlinear acoustics
equation. The variables of equation (1.26) are as follows: The density
$\rho = \rho_0 + \mu u(q_1, q_2, q_3, q_4) + O(\mu u^2)$, with ρ_0 constant, defines u. The
parameter μ is small and

$$q_1 = \frac{t - x/c_0}{\sqrt{\gamma+1}}\sqrt{\rho_0 c_0}, \quad q_2 = \mu x, \quad q_3 = \mu^{1/2}\sqrt{\frac{2}{c_0}}y, \quad q_4 = \mu^{1/2}\sqrt{\frac{2}{c_0}}z,$$

where c_0 is a reference sound speed in the medium and γ is the adiabitic index.

For beams invariant under rotation of the (q_3, q_4) plane, equation (1.26) is replaced by

$$\frac{1}{2}\frac{\partial^2(u^2)}{\partial q_1^2} - \frac{\partial^2 u}{\partial q_1 \partial q_2} + \frac{\partial^2 u}{\partial q_3^2} + \frac{1}{q_3}\frac{\partial u}{\partial q_3} = 0. \tag{1.27}$$

Equations (1.26), (1.27) and the two dimensional analog of (1.26) are considered in the paper.

1.6   Some Comments on Linear Systems

The symmetry groups for the classical linear equations have been given in a variety of places (see e.g., Olver (1986) and Ibragimov (1983), Rogers and Ames (1989)). These Lie algebras are infinite dimensional.

A number of authors have noted that an nth order equation and the (equivalent) first order systems of n equations may not have the same symmetry group. The first detailed paper in the area was by Bluman and Kumei (1987). They studied the invariance properties of the second order hyperbolic equation

$$c^2(x)u_{xx} - u_{tt} = 0 \tag{1.28}$$

and the corresponding system

$$v_t = u_x, \quad u_t = c^2(x)v_x. \tag{1.29}$$

In spite of the apparent equivalence of a single PDE and a corresponding system of PDE's it does not necessarily follow that their respective invariance groups of point transformation are the same. It could happen that the group of point transformations leaving invariant the system is larger than that leaving invariant the single equation; also the converse could be true. This is indeed the case for the single equation (1.28) and the corresponding system (1.29). For example if $c(x) = (Ax+B)^2$, then (1.28) is invariant under an infinite Lie group of point transformations, whereas the Lie group of point transformations leaving invariant (1.29) has only four parameters; if $c(x) = \sqrt{A + Be^{kx}}$, then the Lie group of (1.28) has two parameters and that of (1.29) has four parameters.

Consequently it follows that invariant (similarity) solutions of a system of PDE's lead to noninvariant solutions of a corresponding equivalent single PDE and vice versa.

It is important to note that under the hodograph transformation (the interchange of dependent and independent variables), system (1.29) is equivalent to the nonlinear system

$$v_t = u_x, \quad u_t = c^2(v)v_x. \tag{1.30}$$

Consequently, if $\{u(x,t), v(x,t)\}$ solve (1.30) then $v(x,t)$ solves

$$(c^2(v)v_x)_x - v_{tt} = 0,$$

and introducing the potential $\phi(x,t)$, where $(u,v) = (\phi_t, \phi_x)$, the system (1.30) reduces to

$$c^2(\phi_x)\phi_{xx} - \phi_{tt} = 0.$$

2 Solitons and Group Invariance

A connection between soliton equations and group invariance was conjectured by Ablowitz, Ramani and Segur (1978). They hypothesized that any completely integrable equation may be reduced by appropriate group invariants, sometimes called similarity variables, to one of a canonical set of Painlevé type (see Ince, 1956) ordinary differential equations. These group invariants arise primarily because the soliton equations admit the translation group in all variables. In this section, important soliton equations, namely the Korteweg-de Vries and its generalizations, e.g., the Kadomtsev-Petviashvili, and $2 + 1$ linked equations, are subjected to group anlaysis.

2.1 The Korteweg-de Vries Equation

The Korteweg-de Vries (KdV) equation, originally derived in the context of the propagation of long water waves along a channel (Korteweg and de Vries, 1895), arises in a variety of physical models involving finite amplitude waves (Miles, 1981; Newell, 1985; Rogers and Shadwick, 1982; Rogers and Ames, 1989). Indeed, a large class of nonlinear evolution equations may be reduced to consideration of this important canonical form (Su and Gardner, 1968 and Taniuti and Wei, 1968). The Kadomtev-Petviashvili (KP) equation is also of considerable physical importance. It arises in the study of long gravity waves in a single layer or multilayered shallow fluids wherein waves propagate predominantly in one direction with a small orthogonal perturbation (Kadomtsev and Petviashvili, 1970 and Satsuma and Ablowitz, 1979).

The KdV equation

$$u_t + uu_x + u_{xxx} = 0 \qquad (2.1)$$

has been shown to be invariant under the four parameter class of Lie transformations (Shen and Ames, 1974)

$$
\begin{aligned}
x' &= x + \epsilon \left[\frac{\gamma x}{3} + \alpha t + \beta \right] \\
t' &= t + \epsilon[\gamma t + \delta] \\
u' &= u + \epsilon[(-2\gamma u/3) + \alpha]
\end{aligned}
\qquad (2.2)
$$

with generators of the associated Lie algebra given by

$$
\begin{aligned}
\Gamma_1 &= \frac{\partial}{\partial x}, \qquad \Gamma_2 = \frac{\partial}{\partial t} \quad \text{(Translations in } x \text{ and } t\text{)}, \\
\Gamma_3 &= t\frac{\partial}{\partial x} + \frac{\partial}{\partial u} \quad \text{(Galilean ``boost'')}, \\
\Gamma_4 &= \frac{1}{3}x\frac{\partial}{\partial x} + t\frac{\partial}{\partial t} - \frac{2}{3}u\frac{\partial}{\partial u} \quad \text{(Scaling or dilation)}.
\end{aligned}
\qquad (2.3)
$$

If λ is a real number and $u = f(x,t)$ is a solution of (2.1), then from Γ_1, so is $u_1 = f(x - \lambda, t)$. From Γ_2, so is $u_2 = f(x, t - \lambda)$ a solution. From Γ_3 a solution is also found to be $u_3 = f(x - \lambda t, t)$, while Γ_4 gives rise to the solution $u_4 = e^{-2\lambda/3} f(e^{-\lambda/3}x, e^{-\lambda}t)$. The study of many soliton equations is facilitated by combining Γ_1 and Γ_2 as $\Gamma = \lambda\Gamma_1 + \Gamma_2$ with solutions of the form $u = f(x - \lambda t)$. The quantity $w = x - \lambda t$ is the group invariant of the subgroup formed by

Γ_1 and Γ_2. Upon substituting $u = f(x - \lambda t)$ into (2.1) there results the ordinary differential equation

$$f''' + ff' - \lambda f' = 0. \tag{2.4}$$

This constitutes the so-called symmetry reduction of (2.1) under the translation subgroup. Zabusky (1967) based his study of solitons on equation (2.4) without knowledge of the symmetry group.

Although the list of symmetries for the KdV equation is small it turns out that group invariance plays a key role in the construction of multi-soliton solutions via Bäcklund transformations. Thus, in particular the celebrated auto-Bäcklund transformation (Rogers and Shadwick, 1982) for the KdV equation in fact represents a conjugation of a Bianchi transformation (Eisenhart, 1972) and the Lie symmetry corresponding to Galilean invariance (Dodd et al., 1982). Such combinations have also been described for the cubic Schrödinger equation by Stendel (1980) and for a variety of nonlinear integral equations by Huang (1982, 1983). For the general approach to the conjugation of Lie symmetries with Bianchi-type symmetries see Rogers and Shadwick (1982).

2.2   The Kadomtsev-Petviashvili Equation

One two dimensional version of the KdV equation is the K-P equation

$$\left(u_t + \frac{3}{2}uu_x + \frac{1}{4}u_{xxx}\right)_x + \frac{3}{4}u_{yy} = 0. \tag{2.5}$$

The invariance group for this equation was first obtained by Ames and Nucci (1985) manually and later calculated by David et al. (1986) using MACSYMA and by Schwarz (1988) using REDUCE-SPDE. A versatile *interative* computer algebra package called NUSY (Nucci, 1990) computed Lie symmetries by the classical and nonclassical methods. The generators for the KP equation involve three arbitrary functions $f(t)$, $g(t)$ and $h(t)$ in $C^\infty(\mathbb{R})$ and are given by

$$\Gamma_1 = f(t)\frac{\partial}{\partial t} + \left[\frac{1}{3}xf'(t) - \frac{2}{9}y^2f''(t)\right]\frac{\partial}{\partial x} + \frac{2}{3}yf'(t)\frac{\partial}{\partial y}$$

$$+ \left[\frac{2}{3}uf'(t) + \frac{2}{9}xf''(t) - \frac{4}{27}y^2 f'''(t)\right] \frac{\partial}{\partial u},$$

$$\Gamma_2 = h(t)\frac{\partial}{\partial x} + \frac{2}{3}h'(t)\frac{\partial}{\partial u}, \qquad (2.6)$$

$$\Gamma_3 = -\frac{2}{3}yg'(t)\frac{\partial}{\partial x} + g(t)\frac{\partial}{\partial y} - \frac{4}{9}yg''(t)\frac{\partial}{\partial u}.$$

If f, g and h are assumed to be constants, then the generators reduce to those associated with translation in the independent variables t, x and y respectively. If f, g, and h are taken as linear in t, then a six parameter group is obtained with the generators

$$\Gamma_1 = \frac{\partial}{\partial t}, \qquad \Gamma_2 = \frac{\partial}{\partial x}, \qquad \Gamma_3 = \frac{\partial}{\partial y},$$

$$\Gamma_4 = 2y\frac{\partial}{\partial x} - 3t\frac{\partial}{\partial y},$$

$$\Gamma_5 = 3t\frac{\partial}{\partial x} + 2\frac{\partial}{\partial u},$$

$$\Gamma_6 = 3t\frac{\partial}{\partial t} + x\frac{\partial}{\partial x} + 2y\frac{\partial}{\partial y} - 2u\frac{\partial}{\partial u}. \qquad (2.7)$$

Other subalgebras have been investigated in David et al. (1986). In particular, the subalgebra L_π of the K-P algebra obtained by restricting f, g, and h in the generators (2.7) to be Laurent polynomials in t is shown to be embedded in an affine loop algebra. In the same paper the finite invariance group is constructed and three symmetry reductions are detailed. For example if $f(t) \neq 0$ one solution is found to be

$$u(t, x, y) = f^{-2/3}q(\xi, \eta) + 2f'x/9f + 4(2gf' - 3fg')y/27f^2$$
$$+ 4(2f'^2 - 3ff'')y^2/81f^2 + 2g^2/9f^2 + 2h/3f$$

where $q(\xi, \eta)$ satisfies the Boussinesq equation

$$q_{\eta\eta} + (q^2)_{\xi\xi} + \frac{1}{3}q_{\xi\xi\xi} = 0 \qquad (2.8)$$

and the symmetry reduction variables ξ and η are

$$\xi = [x + (2gy/3f) + (2f'y^2/9f)]f^{-1/3}$$

$$-\int_0^t [2g^2(s)f^{-7/3}(s)/3 + h(s)f^{-4/3}(s)]ds$$

$$\eta = yf^{-2/3} - \int_0^t g(s)f^{-5/3}(s)ds.$$

2.3 The Potential K-P Equation

David, Levi and Winternitz (1986), using MACSYMA, have computed the invariance group for the potential Kadomtsev-Petviashvili equation (PKP)

$$[u_t + 3(u_x)^2/4 + u_{xxx}/4]_x + 3u_{yy}/4 = 0. \qquad (2.9)$$

Operation on (2.9) with $\partial/\partial x$ shows that $w = u_x$ satisfies the KP equation (2.5). The symmetry algebra, while again infinite dimensional, now depends upon five arbitrary functions f, g, h, k, ℓ, two more than the symmetry algebra for the KP equation. The general element of the symmetry algebra is

$$V = X(f) + Y(g) + Z(h) + W(k) + U(\ell) \qquad (2.10)$$

where

$$
\begin{aligned}
X(f) &= f(t)\frac{\partial}{\partial t} + \frac{2}{3}yf'(t)\frac{\partial}{\partial y} + \frac{1}{3}\left[f'(t)x - \frac{2}{3}f''(t)y^2\right]\frac{\partial}{\partial x} \\
&\quad - \left[\frac{1}{3}uf'(t) - \frac{1}{9}x^2 f''(t) + \frac{4}{27}xy^2 f'''(t) - \frac{4}{243}y^4 f'''(t)\right]\frac{\partial}{\partial u}, \\
Y(g) &= g(t)\frac{\partial}{\partial y} - \frac{2}{3}g'(t)y\frac{\partial}{\partial x} - \frac{4}{9}y\left[xg''(t) - \frac{2}{9}y^2 g'''(t)\right]\frac{\partial}{\partial u}, \\
Z(h) &= h(t)\frac{\partial}{\partial x} + \frac{2}{3}\left[h'(t)x - \frac{2}{3}h''(t)y^2\right]\frac{\partial}{\partial u}, \\
W(k) &= k(t)\frac{\partial}{\partial u}, \\
U(\ell) &= \ell(t)\frac{\partial}{\partial u}. \qquad (2.11)
\end{aligned}
$$

The finite group invariances corresponding to the above vector fields show increasing complexity beginning with $W(k)$ and $U(\ell)$.

They correspond to the invariance of the PKP equation under the transformation

$$\tilde{u} = u + \lambda(\ell(t) + k(t)y),$$
$$\tilde{x} = x, \qquad \tilde{y} = y, \qquad \tilde{t} = t,$$

whence, if $u(x, y, t)$ is a solution of the PKP equation, then so also is

$$\tilde{u}(\tilde{x}, \tilde{y}, \tilde{t}) = u(\tilde{x}, \tilde{y}, \tilde{t}) + \lambda[\ell(\tilde{t}) + k(\tilde{t})\tilde{y}].$$

The group invariances corresponding to $Z(h)$ and $Y(g)$ show, in turn, that if $u(x, y, t)$ is a solution of the PKP equation, then so also are

$$\tilde{u}(\tilde{x}, \tilde{y}, \tilde{t}) = u[\tilde{x} - \lambda h(\tilde{t}), \tilde{y}, \tilde{t}] + \frac{2}{3}\lambda h'\tilde{x} - \frac{4}{3}\lambda h''\tilde{y}^2 - \frac{1}{3}\lambda^2 hh',$$

and

$$
\begin{aligned}
\tilde{u}(\tilde{x}, \tilde{y}, \tilde{t}) = {} & u\left[\tilde{x} + \frac{2}{3}\lambda g'\tilde{y} - \frac{1}{3}\lambda^2 gg', \tilde{y} - \lambda g, \tilde{t}\right] \\
& - \frac{2}{9}\lambda g''(2\tilde{y} - \lambda g)\tilde{x} + \frac{8}{81}\lambda g'''\tilde{y}^3 \\
& - \frac{4}{27}\lambda^2(g'g'' + gg''')\tilde{y}^2 + \frac{4}{27}\lambda^3 g\left(g'g'' + \frac{2}{3}gg'''\right)\tilde{y} \\
& - \frac{1}{27}\lambda^4 g^2\left(g'g'' + \frac{2}{3}gg'''\right),
\end{aligned}
$$

with

$$h = h(t) = h(\tilde{t}), \qquad g = g(t) = g(\tilde{t}),$$

Finally, corresponding to the vector field $X(f)$, it is found that if $u(x, y, t)$ is a solution of the PKP equation, then so also is

$$
\begin{aligned}
\tilde{u}(\tilde{x}, \tilde{y}, \tilde{t}) = {} & e^{\tilde{h}/3} u(x, y, t) - \frac{1}{9}\tilde{h}'\tilde{x}^2 + \frac{4}{81}(3\tilde{h}'' - \tilde{h}'^2)\tilde{x}\tilde{y}^2 \\
& - \frac{4}{2187}(g\tilde{h}''' - g\tilde{h}'\tilde{h}'' + \tilde{h}'^3),
\end{aligned}
$$

where

$$\tilde{h} = \tilde{h}(\tilde{t}) = \ln\left[\frac{f(t)}{f(\tilde{t})}\right],$$

$$x = e^{\tilde{h}/3}\left(\tilde{x} - \frac{2}{9}\tilde{h}'\tilde{y}^2\right), \qquad y = e^{2\tilde{h}/3}\tilde{y}, \qquad t = \Phi^{-1}[\Phi(\tilde{t}) - \lambda],$$

and

$$\Phi(t) = \int_0^t \frac{dz}{f(z)}$$

The specializations $h = 1$ and $h = t$ in equations (2.11) correspond, in turn, to translational invariance and Galilean invariance in the x direction. The specializations $g = 1$ and $g = t$ in equations (2.11) give translational invariance in the y direction and invariance under quasi-rotations. For $f(t) = 1$ or $f(t) = t$ in equations (2.11), we obtain invariance under time translation or space-time scaling, respectively.

2.4   The Potential Modified KP Equations

Lastly, a group analysis has been performed by Schwarz (1988) for the potential modified Kadomtsev-Petviashvili equation

$$u_{xt} = u_{xxxx} - 6u_x^2 u_{xx} + 3u_{yy} - 6u_y u_{xx}$$

obtained by setting $v = u_x$ in the modified Kadomtsev-Petviashvili equation

$$v_t = v_{xxx} - 6v^2 v_x + 3\int_a^x v_{yy}dx - 6v_x\int_a^x v_y dx$$

The latter equation is linked to the K-P equation by the Miura type transformation (Rogers and Shadwick (1982))

$$w = v_x + v^2 + \int_a^x v_y dx.$$

Again, the generators form an infinite-dimensional Lie algebra with four arbitrary functions of time, f, g, h, and k,

$$
\begin{aligned}
\Gamma_1 &= f(t)\frac{\partial}{\partial t} + \left[\frac{1}{3}xf'(t) + \frac{1}{18}y^2 f''(t)\right]\frac{\partial}{\partial x} + \frac{2}{3}yf'(t)\frac{\partial}{\partial y} \\
&\quad + \left[\frac{1}{18}xyf''(t) + \frac{1}{324}y^3 f'''(t)\right]\frac{\partial}{\partial u}, \\
\Gamma_2 &= \frac{1}{6}yg'(t)\frac{\partial}{\partial x} + g(t)\frac{\partial}{\partial y} + \left[\frac{1}{12}xg'(t) + y^2 g''(t)\right]\frac{\partial}{\partial u}, \\
\Gamma_3 &= h(t)\frac{\partial}{\partial x} + \frac{1}{6}yh'(t)\frac{\partial}{\partial u}, \\
\Gamma_4 &= k(t)\frac{\partial}{\partial u}.
\end{aligned}
$$

2.5 Related Equations

The already mentioned Bousinesq equation (2.8), together with five related Bousinesq-type equations have been analyzed by Clarkson (1986). Korobeinikov (1983) and Zakharov and Korobeinikov (1980) have generated the Lie algebra for a generalized K-dV-Burgers' equation

$$u_t + \frac{j}{2}\frac{u}{t} + \gamma u^m u_x - \mu u_{xx} + \beta u_{xxx} = 0,$$

in which $j = 0, 1, 2$ and γ, μ, β are real constants. Levi and Winternitz (1987) study a generalized cylindrical K-P equation

$$[u_t + 6uu_x + u_{xxx} + (u/2t) + (H(t) + yG(t))u_x + F(t)u_y]_x + \frac{\alpha}{4t^2}u_{yy} = 0$$

with F, G, H arbitrary and $\alpha = \pm 1$. Dorizzi et al. (1986) finds the infinite symmetry group for the second member of a K-P hierarchy, called the Jumbo-Miwa equation

$$u_{xxxy} + 3u_{xy}u_x + 3u_y u_{xx} + 2u_{yt} - 3u_{xz} = 0.$$

The reader can find all of these groups tabulated in Rogers and Ames (1989).

Symmetries of a wide class of partial differential equations is the subject of a book edited by Vinogradov (1989). The equations called the Kadomtsev-Pogutse equations are of interest in this section. Their symmetry group is to be found on page 23 of this volume. Related work is also discussed in a comprehensive book by Konopelchenko (1987).

3   Fluid Mechanics and Group Invariance

Similarity solutions, as fluid mechanicians called invariant solutions, were developed early for the boundary layer equations (see Schlichting, 1960, for details and history). They were developed, mostly, by ad hoc use of the scaling and spiral subgroups. The first calculation of the Lie symmetry group for the two dimensional boundary layer equations appeared in 1961. A corresponding calculation in gas dynamics was published in 1962 (Ovsiannikov). For the Navier Stokes equations Poochnachev (1960) was the first to obtain its symmetry group for two dimensional flows. The motivator in many of these studies was L. V. Ovsiannikov. Since the aforementioned pioneering studies a considerable literature has been generated. It is described below in several subsections.

3.1   The Boundary Layer Equations

Extensive studies of the boundary layer equations, beginning with (see Ovsiannokov, 1982), have been published. Vyryshagina (1978) studied the three dimensional incompressible boundary layer system

$$
\begin{aligned}
u_x &+ v_y + w_z = 0 \\
u_t &+ uu_x + vu_y + wu_z = -\rho^{-1}p_x + \nu u_{yy} \\
w_t &+ uw_x + vw_y + ww_z = -\rho^{-1}p_z + \nu w_{yy} \\
p_y &= 0
\end{aligned}
\tag{3.1}
$$

with density (ρ) constant. The algebra contains an arbitrary function, $\phi(x, z, t)$, and so is infinite dimensional with generators

$$
\Gamma_1 = \frac{\partial}{\partial t}, \quad \Gamma_2 = \frac{\partial}{\partial x}, \quad \Gamma_3 = \frac{\partial}{\partial z}, \quad \Gamma_4 = \frac{\partial}{\partial p} \quad \text{(Translations)}
$$

$$\Gamma_5 = z\frac{\partial}{\partial x} - x\frac{\partial}{\partial z} + w\frac{\partial}{\partial u} - u\frac{\partial}{\partial w} \qquad \text{(Rotation)}$$

$$\Gamma_6 = x\frac{\partial}{\partial x} + z\frac{\partial}{\partial z} + u\frac{\partial}{\partial u} + w\frac{\partial}{\partial w} + 2p\frac{\partial}{\partial p} \qquad \text{(Scaling)}$$

$$\Gamma_7 = 2t\frac{\partial}{\partial t} + y\frac{\partial}{\partial y} - 2u\frac{\partial}{\partial u} - v\frac{\partial}{\partial v} - 2w\frac{\partial}{\partial w} - 4p\frac{\partial}{\partial p} \qquad \text{(Scaling)}$$

$$\Gamma_8 = t\frac{\partial}{\partial x} + \frac{\partial}{\partial u} \qquad \text{(Galilean Boost)}$$

$$\Gamma_9 = t\frac{\partial}{\partial z} + \frac{\partial}{\partial w} \qquad \text{(Galilean Boost)}$$

$$\Gamma_\phi = \phi(x,y,z,t)\frac{\partial}{\partial y} + (\phi_t + u\phi_x + w\phi_z)\frac{\partial}{\partial v}$$

It is interesting that there are two scaling (dilation) generators and two Galilean boosts. Similar results are observed for some gas dynamic equations.

A generalization to a three dimensional steady boundary layer system over an arbitrary surface is also available (Ovsiannikov, 1982).

In the two dimensional case the symmetry groups have been derived for the unsteady and steady cases of the following:

i) Incompressible

ii) Prescribed pressure

Additionally one finds in two dimensions the steady problem with moment stresses. These groups are found in Ovsiannikov (1982) and Rogers and Ames (1989).

3.2   The Navier Stokes Equations

The Navier-Stokes equations

$$\begin{aligned}
u_t &+ uu_x + vu_y + wu_z = -p_x + \mu\nabla^2 u \\
v_t &+ uv_x + vv_y + wv_z = -p_y + \mu\nabla^2 v \\
w_t &+ uw_x + vw_y + ww_z = -p_z + \mu\nabla^2 w
\end{aligned} \qquad (3.2)$$

are fundamental in fluid mechanics. Apparently, the first derivation of the Lie symmetry group was by Poochnachev (1960) in the two dimensional case and by Bytev (1972) in the general case. Bytev published his result in a journal outside the fluid and applied mathematics arena. As a result the work was redone in the western world by Lloyd (1981) and Boisvert (1982) (see also Boisvert et al., 1983).

The symmetry group for the Navier-Stokes equations, containing four arbitrary functions of time (f, g, h and j), is

$$\alpha : \Gamma_1 = \frac{\partial}{\partial t}$$

$$\beta : \Gamma_2 = 2t\frac{\partial}{\partial t} + x\frac{\partial}{\partial x} + y\frac{\partial}{\partial y} + z\frac{\partial}{\partial z} - u\frac{\partial}{\partial u} - v\frac{\partial}{\partial v} - w\frac{\partial}{\partial w} - 2p\frac{\partial}{\partial p}$$

$$\gamma : \Gamma_3 = -y\frac{\partial}{\partial x} + x\frac{\partial}{\partial y} - v\frac{\partial}{\partial u} + u\frac{\partial}{\partial v}$$

$$\lambda : \Gamma_4 = -z\frac{\partial}{\partial x} + x\frac{\partial}{\partial z} - w\frac{\partial}{\partial u} + u\frac{\partial}{\partial w}$$

$$\sigma : \Gamma_5 = -z\frac{\partial}{\partial y} + y\frac{\partial}{\partial z} - w\frac{\partial}{\partial v} + v\frac{\partial}{\partial w}$$

$$\Gamma_6(f) = f(t)\frac{\partial}{\partial x} + f'(t)\frac{\partial}{\partial u} - xf''(t)\frac{\partial}{\partial p}$$

$$\Gamma_7(g) = g(t)\frac{\partial}{\partial y} + g'(t)\frac{\partial}{\partial v} - yg''(t)\frac{\partial}{\partial p}$$

$$\Gamma_8(h) = h(t)\frac{\partial}{\partial z} + h'(t)\frac{\partial}{\partial w} - zh''(t)\frac{\partial}{\partial p}$$

$$\Gamma_9(j) = j(t)\frac{\partial}{\partial p} \tag{3.3}$$

The generators of translation in the space variables are included in Γ_6, Γ_7 and Γ_8 when f, g, and h are constant, consequently they are not listed separately.

The finite group invariances, corresponding to the above generators, have been set down in Olver (1986). They are given below

(i) Time Translations:

$$G_1 : (\mathbf{x}^*, t^*, \mathbf{q}^*, p^*) = (\mathbf{x}, t + \epsilon, \mathbf{q}, p),$$
$$\mathbf{x} = (x, y, z).$$

(ii) Scaling Transformations:

$$G_2 : (\mathbf{x}^*, t^*, \mathbf{q}^*, p^*) = (\lambda\mathbf{x}, \lambda t, \mathbf{q}, p),$$
$$G_3 : (\mathbf{x}^*, t^*\mathbf{q}^*, p^*) = (\mathbf{x}, \lambda t, \lambda^{-1}\mathbf{q}, \lambda^{-2}p).$$

Here, $\lambda = e^\epsilon$ is a multiplicative group parameter.

(iii) Transformation to an Arbitrary Moving Coordinate System:

$$G_\alpha : (\mathbf{x}^*, t^*, \mathbf{q}^*, p^*) = (\mathbf{x} + \epsilon\alpha, t, \mathbf{q} + \epsilon\alpha', p - \epsilon\mathbf{x} \cdot \alpha'' - \frac{1}{2}\epsilon^2\alpha \cdot \alpha'').$$

Here $\alpha = (\alpha(t), \beta(t), \gamma(t))$ and G_α is generated by the linear combination

$$\Gamma_\alpha = \Gamma_4 + \Gamma_5 + \Gamma_6.$$

(iv) Simultaneous Rotation in Space and the Velocity Vector Field:

$$SO(3) : (\mathbf{x}^*, t^*, \mathbf{q}^*, p^*) = (R\mathbf{x}, t, R\mathbf{q}, p).$$

Here R is an arbitrary 3×3 orthogonal matrix. $SO(3)$ is generated by a linear combination of Γ_7, Γ_8 and Γ_9.

(v) Pressure Changes:

$$G_{10} : (\mathbf{x}^*, t^*, \mathbf{q}^*, p^*) = (\mathbf{x}, t, \mathbf{q}, p + \epsilon\delta(t))$$

The action of the various above subgroups on a solution set $\{\mathbf{q}(\mathbf{x}, t), p(\mathbf{x}, t)\}$ of the Navier-Stokes system is to generate new solutions as set down below:

$$G_1 : \{\mathbf{q}(\mathbf{x}, t - \epsilon), p(\mathbf{x}, t - \epsilon)\},$$
$$G_2 : \{\mathbf{q}(\lambda^{-1}\mathbf{x}, \lambda^{-1}t), p(\lambda^{-1}\mathbf{x}, \lambda^{-1}t)\},$$
$$G_3 : \{\lambda^{-1}\mathbf{q}(\mathbf{x}, \lambda^{-1}t), \lambda^{-2}p(\mathbf{x}, \lambda^{-1}t)\},$$
$$G_\alpha : \{\mathbf{q}(\mathbf{x} - \epsilon\alpha, t) + \epsilon\alpha', p(\mathbf{x} - \epsilon\alpha, t) - \epsilon\mathbf{x}\cdot\alpha'' - \frac{1}{2}\epsilon^2\alpha \cdot \alpha''\},$$
$$SO(3) : \{R\mathbf{q}(R^{-1}\mathbf{x}, t), p(R^{-1}\mathbf{x}, t)\},$$
$$G_{10} : \{\mathbf{q}(\mathbf{x}, t), p(\mathbf{x}, t) + \epsilon\delta(t)\}.$$

Subgroups of the symmetry group may be employed to generate solutions of the Navier-Stokes system by symmetry reduction (Bosivert, 1982 and Rogers and Ames, 1989).

To save space we shall indicate several applications of the subgroups of (3.3) in two space dimensions (x, y) and time t. Eliminating all terms in z and w leaves the operators $\Gamma_1, \Gamma_2, \Gamma_3, \Gamma_6, \Gamma_7$, and Γ_9 for the two-dimensional system. With $\alpha = 1$ and $\beta = \gamma = 0$ (deleting Γ_2 and Γ_3), the Lie operator Γ becomes

$$\Gamma = \frac{\partial}{\partial t} + f(t)\frac{\partial}{\partial x} + g(t)\frac{\partial}{\partial y} + f'(t)\frac{\partial}{\partial u} + g'(t)\frac{\partial}{\partial v}$$
$$+ [j(t) - xf''(t) - yg''(t)]\frac{\partial}{\partial p} \qquad (3.4)$$

The first order equation $\Gamma I = 0$ has the characteristics (invariants)

$$\bar{x} = x - F(t) \qquad \bar{y} = y - G(t)$$

where $F' = f$, $G' = g$, while

$$
\begin{aligned}
u &= \bar{u}(\bar{x}, \bar{y}) + f(t) \\
v &= \bar{v}(\bar{x}, \bar{y}) + g(t) \\
p &= \bar{p}(\bar{x}, \bar{y}) - xf'(t) - yg'(t) + k(t)
\end{aligned}
\qquad (3.5)
$$

where

$$k(t) = \frac{1}{2}f^2 + \frac{1}{2}g^2 + \int j(t)dt$$

The functions, $\bar{u}, \bar{v}$, and $\bar{p}$ satisfy the *steady* Navier-Stokes equations

$$
\begin{aligned}
\bar{u}\bar{u}_{\bar{x}} + \bar{v}\bar{u}_{\bar{y}} &= -\bar{p}_{\bar{x}} + \mu[\bar{u}_{\bar{x}\bar{x}} + \bar{u}_{\bar{y}\bar{y}}] \\
\bar{u}\bar{v}_{\bar{x}} + \bar{v}\bar{v}_{\bar{y}} &= -\bar{p}_{\bar{y}} + \mu[\bar{v}_{\bar{x}\bar{x}} + \bar{v}_{\bar{y}\bar{y}}] \\
\bar{u}_{\bar{x}} + \bar{v}_{\bar{y}} &= 0
\end{aligned}
\qquad (3.6)
$$

As is true in other systems in fluid mechanics the generator of pure time translation, Γ_1, exists. Hence it is possible to "factor out" the explicit time dependence, as indicated above. This is a useful result since any solution of the steady equation, (3.6), can be transformed by means of equations (3.5), into a time dependent

solution containing three arbitrary functions of time. A similar result
holds in three dimensions.

Many applications of group analysis for the determination of solutions by symmetry reduction are available (Bosivert, 1982, Rogers
and Ames, 1989). We illustrate one surprising one here.

Dropping the bars on equations (3.6) and using the stream function form with $v = \psi_x$ and $u = -\psi_y$, there results

$$\psi_x \psi_{yyy} + \psi_x \psi_{xxy} - \psi_y \psi_{xyy} - \psi_y \psi_{xxx}$$
$$- \mu(\psi_{xxxx} + 2\psi_{xxyy} + \psi_{yyyy}) = 0 \qquad (3.7)$$

The rotation group is obtained from (3.3) by setting α and β equal to
zero and including only Γ_3 (in stream function form). The invariants
are $\eta = x^2 + y^2$, $\psi = f(\eta)$, and when these are substituted into (3.7)
one obtains the *linear equation*

$$2f'' + 4f''' + \eta^2 f'''' = 0$$

which is independent of μ! This is a surprising result. The general
solution, first obtained by Hamel (using ad hoc methods), is

$$f(\eta) = -c_1 \ln \eta + c_2 \eta \ln \eta + c_3 \eta + c_4$$

where the c_i are arbitrary constants. The associated time-dependent
solutions are

$$
\begin{aligned}
u &= f(t) + (y - G(t))[c_1 \eta^{-1} - c_5 \ln \eta - c_6] \\
v &= g(t) + (x - f(t))[-c_1 \eta^{-1} + c_5 \ln \eta + c_6] \\
p &= -xf'(t) - yg'(t) + k(t) - \frac{1}{2}c_1^2 \eta^{-1} \\
&\quad + (c_5^2 - c_5 c_6 + \frac{1}{2}c_6^2)\eta - c_1 c_6 \ln \eta + (c_5 c_6 - c_5^2)\eta \ln \eta \\
&\quad - \frac{1}{2}c_1 c_5 (\ln \eta)^2 + \frac{1}{2}c_5^2 \eta(\ln \eta)^2 - 4\mu c_5 \tan^{-1}\left[\frac{x - F(t)}{y - G(t)}\right]
\end{aligned}
$$

where $\eta = [x - F(t)]^2 + [y - G(t)]^2$.

Gusyatnikova and Yumaguzhin (1989) revisited the Navier-Stokes
equations and they derive the above result when potential extrinsic

force is employed. A complete answer for nonpotential extrinsic force is also found as is the space of all conservation laws for the Navier-Stokes equations.

Some symmetry group generation has taken place in other coordinate systems. For example Kapitanski (1978, 1980) does the group analysis of the Euler and Navier-Stokes equations in the presence of rotational symmetry. Nucci (1987) finds the algebra for a nonsteady axisymmetric incompressible viscous system. In related areas the collisonless plasma equations have been investigated by Taranov (1978), a nonsteady dissipative magneto-hydrodynamic system by Nucci (1984), a steady non-dissipative magneto-hydrodynamic system by Rogers and Winternitz (1988) and the Zakharov equations

$$i\psi_t + \psi_{xx} - n\psi = 0, \qquad n_t + u_x = 0$$

$$u_t + n_x + (|\psi|^2)_x = 0$$

by Verbovetsky (1989). This system describes the interaction of acoustic and Langmuir plasma waves.

3.3 Gas Dynamics Equations

Various problems in gas dynamics, and other systems modeled with equations that resemble the equations of gases, have benefited by the availability of the symmetry group. Ovsiannikov (1962) obtained the generators for the equations

$$
\begin{aligned}
\rho_t + \underline{v} \cdot \nabla \rho + \rho \, \text{div} \, \underline{v} &= 0 \\
\rho[\underline{v}_t + \underline{v} \cdot \nabla \underline{v}] + \nabla p &= 0 \\
p_t + \underline{v} \cdot \nabla p + \gamma p \, \text{div} \, \underline{v} &= 0 \\
\underline{v} = (v_1, \ldots, v_n) &
\end{aligned}
\tag{3.8}
$$

of motion for a polytropic gas.

Case (i): Arbitrary γ

In this case, the system admits an invariance group dependent upon $n(n+3)/2 + 4$ parameters. The generators are

$$\Gamma_0 = \frac{\partial}{\partial t}, \qquad \Gamma_i = \frac{\partial}{\partial x_i},$$

$$\Gamma_{ij} = x_j\frac{\partial}{\partial x_i} - x_i\frac{\partial}{\partial x_j} + v_j\frac{\partial}{\partial v_i} - v_i\frac{\partial}{\partial v_j}, \quad i \neq j,$$

$$G_i = t\frac{\partial}{\partial x_i} + \frac{\partial}{\partial v_i},$$

$$H_1 = t\frac{\partial}{\partial t} + \sum_i x_i\frac{\partial}{\partial x_i},$$

$$H_2 = 2t\frac{\partial}{\partial t} + \sum_i\left[x_i\frac{\partial}{\partial x_i} - v_i\frac{\partial}{\partial v_i}\right] + 2\rho\frac{\partial}{\partial \rho},$$

$$H_3 = \rho\frac{\partial}{\partial \rho} + p\frac{\partial}{\partial p},$$

$$i,j = 1,2,\ldots,n,$$

(in physical applications $n = 1, 2$, or 3).

Case (ii):
$$\gamma = \frac{n+2}{n}.$$

In this case, the set of generators of Case (i) is agumented by

$$H_4 = t^2\frac{\partial}{\partial t} + \sum_i\left[tx_i\frac{\partial}{\partial x_i} + (x_i - tv_i)\frac{\partial}{\partial v_i}\right] - nt\rho\frac{\partial}{\partial \rho} - (n+2)tp\frac{\partial}{\partial p}.$$

Other early studies were undertaken by (see Ovsiannikov, 1982, p. 138).

When one adjoins to equation (3.8) the irrotational condition

$$\underline{v} = \nabla\phi$$

and a constant specific entropy condition then the symmetry group is changed. The generator for this non-steady potential homentropic flow equation

$$\phi_{tt} + 2\nabla\phi\cdot\nabla\phi_t + \nabla\phi\cdot(\nabla\phi\cdot\nabla)\nabla\phi + (\gamma-1)[\phi_t + \frac{1}{2}|\nabla\phi|^2]\Delta\phi = 0 \quad (3.9)$$

were found by Ibragimov (1983) to be

Case (i):
                              Arbitrary γ.

The generators are

$$
\begin{aligned}
\Gamma_0 &= \frac{\partial}{\partial t}, \quad \Gamma_i = \frac{\partial}{\partial x_i}, \\
\Gamma_{ij} &= x_j \frac{\partial}{\partial x_i} - x_i \frac{\partial}{\partial x_j}, \quad i \neq j, \\
\Gamma_{n+1} &= \frac{\partial}{\partial \phi}, \\
G_i &= t \frac{\partial}{\partial x_i} + x_i \frac{\partial}{\partial \phi}, \\
H_1 &= t \frac{\partial}{\partial t} + \sum_i x_i \frac{\partial}{\partial x_i} + \phi \frac{\partial}{\partial \phi}, \\
H_2 &= \frac{2\gamma + n(\gamma - 1)}{\gamma + 1} t \frac{\partial}{\partial t} + \sum_i x_i \frac{\partial}{\partial x_i} + \frac{2 - n(\gamma - 1)}{\gamma + 1} \phi \frac{\partial}{\partial \phi},
\end{aligned}
$$

Case (ii):

$$
\gamma = \frac{n+2}{n}.
$$

In this case, the set of generators of Case (i) is augmented by

$$
H_3 = t^2 \frac{\partial}{\partial t} + \sum_i t x_i \frac{\partial}{\partial x_i} + \frac{1}{2} |\mathbf{x}|^2 \frac{\partial}{\partial \phi}.
$$

Various specializations and modifications of equations (3.8) and (3.9) have been studied. The one, two and three dimensional Lin-Tsien equations of transonic gas dynamics are particular examples. Ibragimov (1967, 1983) investigated the three dimensional Lin-Tsien equation

$$
2u_{tx} + u_x u_{xx} - u_{yy} - u_{zz} = 0
$$

and found the generators to be

$$
\begin{aligned}
\Gamma_1 &= \frac{\partial}{\partial t} \\
\Gamma_2 &= 5t \frac{\partial}{\partial t} + x \frac{\partial}{\partial x} + 3y \frac{\partial}{\partial y} + 3z \frac{\partial}{\partial z} - 3u \frac{\partial}{\partial u}
\end{aligned}
$$

$$\Gamma_3 = t\frac{\partial}{\partial t} - x\frac{\partial}{\partial x} - 3u\frac{\partial}{\partial u},$$

$$\Gamma_4 = z\frac{\partial}{\partial y} - y\frac{\partial}{\partial z}$$

$$\Gamma_5 = \frac{5}{2}t^2\frac{\partial}{\partial t} + \left[tx + \frac{3}{2}(y^2 + z^2)\right]\frac{\partial}{\partial x}$$
$$+3ty\frac{\partial}{\partial y} + 3tz\frac{\partial}{\partial z} + (x^2 - 3tu)\frac{\partial}{\partial u}$$

$$\Gamma_\gamma = \gamma(t)\frac{\partial}{\partial x}$$

$$\Gamma_\phi = \phi(t)\frac{\partial}{\partial u}$$

$$\Gamma_\alpha = \dot\alpha y\frac{\partial}{\partial x} + \alpha(t)\frac{\partial}{\partial y} + \left[3xy\ddot\alpha + (y^2 + z^2)y\ddot\alpha + \frac{2}{3}\ddot\alpha y^3\right]\frac{\partial}{\partial u}$$

$$\Gamma_\beta = \dot\beta z\frac{\partial}{\partial x} + \beta(t)\frac{\partial}{\partial z} + \left[3xz\ddot\beta + (y^2 + z^2)z\ddot\beta + \frac{2}{3}\dddot\beta z^3\right]\frac{\partial}{\partial u}$$

where α, β, γ are arbitrary functions of time and $\phi_{xx} + \phi_{yy} = 0$. In the second article Ibragimov found the symmetry group for $u_x u_{xx} + u_{yy} = 0$. The two dimensional case

$$2\phi_{tx} + \phi_x\phi_{xx} - \phi_{yy} = 0$$

has been analyzed by Ovsiannikov (1982) while Fejes (1985) has found the group for the more general equation

$$Au_{tt} + 2Bu_{xx} + Cu_{xx} + Du_x u_{xx} - u_{yy} = 0$$

The group structure for gas dynamic models of a two component plasma has been found by Smoodskii and Taranov (1975). A gas dynamic model of pellet fusion was group analyzed by Ervin et al. (1984). An interesting connection to the Riccati chain was uncovered.

Logan and Perez (1980) have determined the class of symmetry group solutions to a one-dimensional time dependent problem in shock gas dynamics with a chemical reaction taking place behind the shock. The Lie group for the finite motion of a rotation shallow liquid contained in a rigid basin is found by Levi et al. (1987). Titev (1989) calculates the symmetry group and conservation laws for the shallow water equations with an axisymmetric bottom profile.

4   Solid Mechanics and Symmetry Groups

A number of known solutions for deformation problems in nonlinear elasticity are associated with invariance of the governing system under stretching, rotation, or translation (Klingbeil and Shield, 1960; Wesolowski, 1969) that is by ad hoc use of subgroups of the Galilei-Newton group. Group methods have been applied in solid mechanics by Hill (1972, 1982), Olver (1986) and Ames-Ames (1985).

There is increasing use of these methods because of the appearance of fundamental books in the past few years.

4.1   Neo-Hookean Materials

Group invariant deformation of a neo-Hookean material has been studied by Hill (1982) in both two and three dimensions. We give some of the results in two dimensions below.

The plane deformation

$$x = x(X,Y), \qquad y = y(X,Y), \qquad z = Z \tag{4.1}$$

of a homogeneous, isotropic, incompressible, hyperelastic material of neo-Hookean type is governed by

$$\frac{\partial(x,y)}{\partial(X,Y)} = 1, \qquad \text{incompressibility,} \tag{4.2}$$

and

$$\left. \begin{array}{l} \frac{\partial p}{\partial x} = \mu \nabla^2 x, \\[2mm] \frac{\partial p}{\partial y} = \mu \nabla^2 y, \end{array} \right\} , \text{equilibrium,} \tag{4.3}$$

where (X,Y,Z) and (x,y,z) designate, in turn, material and spatial coordinates, p is the pressure and μ is the constant shear modulus. ∇^2 denotes the two-dimensional Laplacian given by

$$\nabla^2 = \frac{\partial^2}{\partial X^2} + \frac{\partial^2}{\partial Y^2}.$$

Invariance of the nonlinear elastostatic system (4.1)-(4.3) is sought under the Lie group with infinitesimal form

$$
\begin{aligned}
X^* &= X + \epsilon U(X, Y, x, y, p) + O(\epsilon^2), \\
Y^* &= Y + \epsilon V(X, Y, x, y, p) + O(\epsilon^2), \\
x^* &= x + \epsilon u(X, Y, x, y, p) + O(\epsilon^2), \\
y^* &= y + \epsilon v(X, Y, x, y, p) + O(\epsilon^2), \\
p^* &= p + \epsilon s(X, Y, x, y, p) + O(\epsilon^2).
\end{aligned}
$$

The determining equations are readily obtained via MACSYMA and may be solved to yield

$$
\begin{aligned}
X^* &= X + \epsilon(a_0 X + a_1 Y + a_2) + O(\epsilon^2), \\
Y^* &= Y + \epsilon(-a_1 X + a_0 Y + a_3) + O(\epsilon^2), \\
x^* &= x + \epsilon(a_0 x + a_4 y + a_5) + O(\epsilon^2), \\
y^* &= y + \epsilon(-a_4 x + a_0 y + a_6) + O(\epsilon^2), \\
p^* &= p + a_7 \epsilon + O(\epsilon^2),
\end{aligned}
$$

where a_i, $i = 0, 1, \ldots, 7$ are arbitrary constants. Thus, the invariance group is 8-parameter. This group may be embedded in a broader class of symmetry transformations whenever change in the stretch is introduced (Levi and Rogers, 1989).

The generators of the Lie algebra are given by

$$
\Gamma_1 = X\frac{\partial}{\partial X} + Y\frac{\partial}{\partial Y} + x\frac{\partial}{\partial x} + y\frac{\partial}{\partial y}, \quad \Gamma_2 = Y\frac{\partial}{\partial X} - X\frac{\partial}{\partial Y},
$$

$$
\Gamma_3 = \frac{\partial}{\partial X}, \qquad\qquad\qquad \Gamma_4 = \frac{\partial}{\partial Y},
$$

$$
\Gamma_5 = y\frac{\partial}{\partial x} - x\frac{\partial}{\partial y}, \qquad\qquad \Gamma_6 = \frac{\partial}{\partial x},
$$

$$
\Gamma_7 = \frac{\partial}{\partial y}, \qquad\qquad\qquad \Gamma_8 = \frac{\partial}{\partial p}.
$$

Here, the generator Γ_1 corresponds to invariance under stretching, Γ_2 and Γ_5 reflect invariance under rotation of the material and spatial coordinates, while the remaining generators correspond to invariance under translation of the respective variables.

Upon an appropriate choice of origins of the coordinate systems, the translation constants a_2, a_3, a_5 and a_6 can be reduced to zero. The group invariant equation

$$U\frac{\partial I}{\partial X} + V\frac{\partial I}{\partial Y} + u\frac{\partial I}{\partial x} + v\frac{\partial I}{\partial y} + s\frac{\partial I}{\partial p} = 0$$

then has characteristic equations

$$\frac{dX}{d\sigma} = a_0 X + a_1 Y, \qquad \frac{dY}{d\sigma} = -a_1 X + a_0 Y,$$
$$\frac{dx}{d\sigma} = a_0 x + a_4 y, \qquad \frac{dy}{d\sigma} = -a_4 x + a_0 y, \qquad (4.4)$$
$$\frac{dp}{d\sigma} = a_7,$$

where σ is a *characteristic parameter*. In terms of the cylindrical polar coordinates (R, Θ) and (r, θ), defined by

$$R = (X^2 + Y^2)^{1/2}, \qquad \Theta = \tan^{-1}(Y/X),$$
$$r = (x^2 + y^2)^{1/2}, \qquad \theta = \tan^{-1}(y/x),$$

the system (4.4) reduces to

$$\frac{dR}{d\sigma} = a_0 R, \qquad \frac{d\Theta}{d\sigma} = -a_1,$$
$$\frac{dr}{d\sigma} = a_0 r, \qquad \frac{d\theta}{d\sigma} = -a_4,$$
$$\frac{dp}{d\sigma} = a_7,$$

whence, on integration

$$r = Rf(\xi), \qquad \theta = \left(\frac{a_4}{a_1}\right)\Theta + g(\xi),$$

$$p = \left(\frac{a_7}{a_0}\right)\ln r + h(\xi), \qquad (4.5)$$

where

$$\xi = a_0\Theta + a_1 \ln R, \qquad a_0, a_1 \neq 0,$$

is the similarity variable and f, g, h are determined by the symmetry-reduced system obtained by substitution of the expressions (4.5) into the governing elastostatic equations.

It is noted that the class of finite deformations corresponding to the special case $a_4 = 0$ was considered by Klingbeil and Shield (1966). Moreover, the important solution of Ericksen's problem given in Klingbeil and Shield (1966) is also a special case of the deformation (4.5).

Integration of the initial value problems

$$\frac{dX^*}{d\epsilon} = a_0 X^* + a_1 Y^*, \qquad \frac{dY^*}{d\epsilon} = -a_1 X^* + a_0 Y^*,$$

$$X^*|_{\epsilon=0} = X, \qquad Y^*|_{\epsilon=0} = Y,$$

$$\frac{dx^*}{d\epsilon} = a_0 x^* + a_4 y^*, \qquad \frac{dy^*}{d\epsilon} = -a_4 x^* + a_0 y^*,$$

$$x^*|_{\epsilon=0} = x, \qquad y^*|_{\epsilon=0} = y,$$

produces the finite form of the invariance group, namely,

$$\begin{pmatrix} X^* \\ Y^* \end{pmatrix} = e^{a_0 \epsilon} \begin{pmatrix} \cos(a_1\epsilon) & \sin(a_1\epsilon) \\ -\sin(a_1\epsilon) & \cos(a_1\epsilon) \end{pmatrix} \begin{pmatrix} X \\ Y \end{pmatrix}$$

$$\begin{pmatrix} x^* \\ y^* \end{pmatrix} = e^{a_0 \epsilon} \begin{pmatrix} \cos(a_4\epsilon) & \sin(a_4\epsilon) \\ -\sin(a_4\epsilon) & \cos(a_4\epsilon) \end{pmatrix} \begin{pmatrix} x \\ y \end{pmatrix}$$

4.2   von-Karman Elastic Plate Equations

The von Karman equations provide an approximate model for the large deflection of elastic plates under the combined action of a uniform, perpendicular, lateral load and a tensile force in the middle plane of the plate. They constitute the nonlinear coupled system

$$\begin{aligned} \nabla^4 F &= (w_{xy})^2 - w_{xx} w_{yy}, \\ \nabla^4 w &= -w_{tt} + F_{yy} w_{xx} + F_{xx} w_{yy} - 2F_{xy} w_{xy}, \end{aligned} \qquad (4.6)$$

where w is the dimensionless displacement of the plate about its equilibrium position, F is the dimensionless stress function and

$$\nabla^4 = \frac{\partial^4}{\partial x^4} + 2\frac{\partial^4}{\partial x^2 \partial y^2} + \frac{\partial^4}{\partial y^4},$$

is the biharmonic operator.

The stresses $\sigma_x, \sigma_y, \tau_{xy}$ in the middle surface are given in terms of the stress function F by the relations

$$\sigma_x = F_{yy}, \quad \sigma_y = F_{xx}, \quad \tau_{xy} = -F_{xy}.$$

The symmetry group, for equations (4.6), has been obtained by hand calculation in Ames and Ames (1985) and by SPDE (Schwarz, 1984). The generators of the Lie algebra are

$$
\begin{aligned}
\Gamma_1 &= \frac{\partial}{\partial t}, \quad \Gamma_2 = \frac{\partial}{\partial x}, \quad \Gamma_3 = \frac{\partial}{\partial y}, \quad \Gamma_4 = \frac{\partial}{\partial w}, \\
\Gamma_5 &= y\frac{\partial}{\partial x} - x\frac{\partial}{\partial y}, \quad \Gamma_6 = t\frac{\partial}{\partial w}, \quad \Gamma_7 = y\frac{\partial}{\partial w} \\
\Gamma_8 &= x\frac{\partial}{\partial w}, \quad \Gamma_9 = 2t\frac{\partial}{\partial t} + x\frac{\partial}{\partial x} + y\frac{\partial}{\partial y}, \qquad\qquad (4.7) \\
\Gamma_{10} &= tx\frac{\partial}{\partial w}, \quad \Gamma_{11} = ty\frac{\partial}{\partial w}, \\
\Gamma_{12} &= f(t)\frac{\partial}{\partial F}, \quad \Gamma_{13} = xg(t)\frac{\partial}{\partial F}, \quad \Gamma_{14} = yh(t)\frac{\partial}{\partial F}.
\end{aligned}
$$

The presence of the arbitrary functions f, g and h of t in the generators Γ_{12}, Γ_{13} and Γ_{14} respectively indicates that the Lie algebra is infinite dimensional. The first four generators are translations while Γ_5 is rotation and Γ_9 is scaling.

In the case of static equilibrium, so that equations (4.6) are time-independent, certain simple symmetry reductions corresponding to invariance under rotation and translation subgroups are immediate. Thus, corresponding to the rotation subgroup associated with Γ_5, we obtain the finite invariance group

$$
\begin{aligned}
x^* &= x\cos\epsilon + y\sin\epsilon, \quad y^* = -x\sin\epsilon + y\cos\epsilon, \\
w^* &= w, \quad F^* = F,
\end{aligned}
$$

with invariants

$$F = \phi(\eta), \quad w = \psi(\eta), \quad \eta = x^2 + y^2. \qquad (4.8)$$

Introduction of (4.8) into the von Karman system leads to the reduction

$$4[\eta^2\phi'']'' = -[\eta(\psi')^2]',$$
$$2[\eta^2\psi'']'' = [\eta\psi'\phi']'.$$

Whence on integration

$$4[\eta^2\phi'']' = -\eta[\psi']^2 + \lambda,$$
$$2[\eta^2\psi'']' = \eta\psi'\phi' + \mu$$

where λ and μ are arbitrary constants.

4.3 Other Studies

In addition to the body of theoretical work by Olver (1986, et seq.) we should mention the early work of Lynskii (1966) on the one dimensional motion of a visco-plastic material. Lynskii found the symmetry group for the equation

$$u_t = f(u_x)u_{xx}.$$

The analysis includes several special cases for $f(u_x)$.

Generalizations have been undertaken by Ames (1988) (see Section 1).

5 Symmetry Groups and Computer Algebra Packages

Computer algebra systems now available on many machines make it feasible to perform a variety of analytical procedures automatically. In particular, the construction of symmetry groups, the application of the Painlevé test and generation of multi-soliton solutions via permutability theorems are all amenable to such symbolic computation.

The most important general purpose computer algebra systems currently available are MACSYMA (Math Lab Group MIT), REDUCE (A. C. Hearn, Rand Corporation), MAPLE (B. Char, University of Waterloo, Canada), Mu Math (D.R. Stoutemyer, The

Software-house, Honolulu, Hawaii), SMP (S. Wolfram) and SCRATCH-PAD I and II (R. D. Jenks and D. Yun, IBM Watson Laboratories).

A general introduction to applications of computer algebra methods in MACSYMA is given by Rand (1984). Schwarz (1987a) has undertaken completely automatic symmetry group calculations via computer algebra in REDUCE. Thus, in Version 3.3 of the REDUCE User's Manual, there is a description of SPDE (Symmetries of Partial Differential Equations). A package also exists for the calculation of group symmetries of ordinary differential equations (SODE). Details of the SPDE and SODE programs are given by Schwarz (1987b, 1988).

Champagne and Winternitz (1985) have developed a computer algebra package for symmetry calculations using MACSYMA. Kersten (1987, 1989), using REDUCE, has introduced semi-automatic routines. See also Gragert (1989). Nucci (1990) has had problems with automatic packages and has produced an interactive package called NUSY.

In Russia, early computer algebra packages developed were CINO and PASSIV (see Ovsiannikov, 1978, p. 571). The application of computer algebra to the calculation of generalized symmtries of contact and Lie-Bäcklund type has been discussed by Eliseev, Fedorova and Kornyak, 1985 and Fedorova and Kornyak, 1986. Many of the recent ideas and applications of computer algebra are described in the Proceedings of the *Conference on Computer Algebra and Its Applications in Theoretical Physics,* held in Dubna, USSR in 1985 (Covorun, Chairman, 1987). See also Hussin (1990) for the Proceedings of a Canadian Mathematical Society Seminar.

Bibliography

Ablowitz, M. J., Ramani, A., and Segur, H. (1978). *Nonlinear evolution equations and ordinary differential equations of Painlevé type,* Lett. Nuòvo Cimento, 23, 333–338.

Ames, K. A. and Ames, W. F. (1985). *Analysis of the von Kármán equations by group methods,* Int. J. Nonlinear Mechanics, 20, 201–209.

Ames, Karen A. (1989). *Group properties of equations for wave propagation in various media,* Int. J. Non-Linear Mechanics, 24, 29–39.

Ames, W. F. (1965). *Nonlinear Partial Differential Equations in Engineering,* Vol. I, Academic Press, New York.

Ames, W. F. (1968). *Nonlinear Ordinary Differential Equations in Transport Processes,* Academic Press, New York.

Ames, W. F. (1972). *Nonlinear Partial Differential Equations in Engineering,* Vol. II, Academic Press, New York.

Ames, W. F. (1977). *Numerical Methods for Partial Differential Equations,* 2nd Ed. Academic Press, New York.

Ames, W. F. and Donato, A. (1988). *On the evolution of weak discontinuities in a state characterized by invariant solutions.* Int. J. Non-Linear Mech. 23, 167–174.

Ames, W. F. and Hölzmuller, A. (1991, in press). *Group analysis of a class of third order partial differential equations.*

Ames, W. F. and Nucci, M. C. (1985). *Analysis of fluid equations by group methods,* J. Eng. Math. 20, 181–187, Corrigendum.

Ames, W. F. and Suliciu, I. (1982). *Some exact solutions for wave propagation in viscoelastic, viscoplastic and electrical transmission lines,* Int. J. Nonlinear Mech. 17, 223–230.

Ames, W. F., Lee, S. Y. and Zaiser, J. M. (1968). *Nonlinear vibration of a travelling threadline,* Int. J. Nonlinear Mech. 3, 449–469.

Ames, W. F., Lohner, R. J. and Adams, E. (1981). *Group properties of $u_{tt} = [f(u)u_x]_x$,* Int. J. Nonlinear Mech. 16, 439–447.

Ames, W. F., Donato, A. and Nucci, M. C. (1989). *Analysis of the thread-line equations,* Nonlinear Wave Motion, Pitman, London, 1–10.

Anderson, I. and Olver, P. J. (Eds.) (in press). *Symmetry Methods in Differential Equations: Proceedings of the 1987 Utah State University Conference.*

Anderson, R. L. and Ibragimov, N. H. (1979). *Lie-Bäcklund Transformations in Applications,* SIAM, Philadelphia.

Barenblatt, G. I. (1979) *Similiarity, Self-Similarity and Intermediate Asymptotics,* Consultants Bureau, New York.

Beiglbock, W., Böhm, A. and Takasugi, E. (Eds.) (1979). *Group Theoretical Methods in Physics, Proceedings of the 1978 Austin Conference,* Springer Lecture Notes in Physics 94, Springer-Verlag, New York.

Bluman, G. W. and Cole, J. D. (1974). *Similarity Methods for Differential Equations.* Springer-Verlag, New York.

Bluman, G. and Kumei, S. (1987). *On invariance properties of the wave equation.* J. Math. Phys. 28, 307–318.

Bluman, G. and Kumei, S. (1989). *Symmetries and Differential Equations.* Springer-Verlag, New York-Berlin.

Boisvert, R. E. (1982). *Group analysis of the Navier-Stokes equations.* Ph.D. Dissertation,Georgia Institute of Technology, Atlanta, Georgia, USA.

Boisvert, R. E., Ames, W. F. and Srivastava, U. N. (1983). *Group properties and new solutions of the Navier-Stokes equations.* J. Eng. Math. 17, 203–221.

Bytev, V. O. (1972). *Group properties of the equations of Navier-Stokes.* Numerical Meth. Mech. Fluids 3 (no. 5), Novosibirsk, V.Ts.So, Acad. Sciences USSR, 13–17.

Champagne, B. and Winternitz, P. (1985). *A MACSYMA program for calculating the symmetry group of a system of differential equations.* Centre de Recherches Mathématiques, Université de Montréal, Report No. 1278.

Clarkson, P. A. (1986). *The Painlevé property, a modified Boussinesq equation and a modified Kadomtsev-Petviashvili equation.* Physica D19, 447–450.

Cohen, A. (1911). *An Introduction to the Lie Theory of One-Parameter Groups with Applications to the Solution of Differential Equations.* D. C. Heath, Boston. Reprint by Stechert, New York, (1931).

David, D., Kamran, N., Levi, D. and Winternitz, P. (1986). *Symmetry reduction for the Kadomtsev-Petviashvili equation using a loop algebra,* J. Math. Phys. 27, 1225–1237.

David, D., Levi, D., and Winternitz, P. (1986). *Bäcklund transformations and the infinite-dimensional symmetry group of the Kadomtsev-Petviashvili equation,* Phys. Lett. A118, 390–394.

Dodd, R. K., Eilbeck, J., Gibbon, J. D. and Morris, H. C. (1982). *Solitons and Nonlinear Waves.* Academic Press, New York.

Donato, A. and Oliveri, F. (1988). *On nonlinear plane vibration of a moving threadline.* Zeit. für Ang. Math. u. Phys., 39, 367–375.

Dorizzi, B., Grammaticos, B., Ramani, A., and Winternitz, P. (1986). *Are all the equations of the Kadomtsev-Petviashvili hierarchy integrable?,* J. Math. Phys. 27, 2848–2852.

Eisenhart, L. P. (1972). *A Treatise on the Differential Geometry of Curves and Surfaces.* Dover, New York.

Ervin, V. J., Ames, W. F., and Adams, E. (1984). *Nonlinear waves in pellet fusion,* in Wave Phenomena: Modern Theory and Applications (C. Rogers and T. B. Moodie, eds.), pp. 199–210, North Holland, Amsterdam.

Fejes, L. (1985). *Similarity Solutions of the Transonic Equations.* Diplom. thesis, Univ. Kassel, Germany.

Gragert, P. K. H. (1989). *Lie algebra computations.* Acta Appl. Math. 16, 231–242.

Gusyatnikova, V. N. and Yumaguzhin, V. A. (1989). *Symmetries and conservation laws of Navier Stokes equations.* Acta Appl. Math. 15, 65–81.

Hansen, A. G. (1964). *Similarity Analyses of Boundary Value Problems in Engineering.* Prentice Hall, Englewood Cliffs, New Jersey.

Hill, J. M. (1972). *Some Partial Solutions of Finite Elasticity.* Ph.D. Thesis, University of Queensland, Brisbane, Australia.

Hill, J. M. (1982a). *On static similarity deformations for isotropic materials,* Q. Appl. Math. 40, 287–291.

Hill, J. M. (1982b). *Solution of Differential Equations by Means of One-Parameter Groups.* Pitman, Boston.

Huang, Xun-Cheng (1982). *Relations connecting scale transformation and Bäcklund transformation for the cylindrical Korteweg-de Vries equation,* J. Phys. a. Math. Gen. 15, 7–21.

Huang, Xun-Cheng (1983). *Decompositions of the Bäcklund transformations for the generalized Klein Gordon and other nonlinear evolution equations,* Acta Mathematica Scientia 3, 95–101.

Hussin, V. (Ed.) (1990). *Lie Theory, Differential Equations and Representation Theory,* Univ. Montreal, Montreal.

Ibragimov, N. H. (1967). *Group Properties of Some Differential Equations* (in Russian). NAUKA, Novosibirsk.

Ibragimov, N. H. (1983). *Groups of Transformations in Mathematical Physics.* NAUKA, Moscow. English Ed., Reidel Pub. Co., Dordrecht, Netherlands (1985).

Ibragimov, N. H. and Ovsiannikov, L. V., eds. (1978). *Group Theoretical Methods in Mechanics,* Proceedings of the Joint IUTAM/IMU Symposium at Novosibirsk.

Ince, E. L. (1956). *Ordinary Differential Equations.* Dover, New York.

Jumbo, M. and Miwa, T., eds. (1983). *Proceedings RIMS Symposium on Nonlinear Integrable Systems, Kyoto 1981.* World Science Publ. Co., Singapore.

Kadomtsev, B. and Petviashvili, V. I. (1970). *On the stability of solitary waves in weakly dispersing media,* Sov. Phys. Dokl. 15, 539–541.

Kapitanski, L. V. (1978). *Group analysis of the Euler and Navier-Stokes equations in the presence of rotational symmetry and new exact solutions of these equations,* Sov. Phys. Dokl. 23, 896–898.

Kapitanski, L. V. (1980). *Group-theoretic analysis of the Navier-Stokes equations in the rotationally symmetric case and some new exact solutions,* Sov. J. Math. 21, 314–327.

Kersten, P. H. M. (1987). *Infinitesimal symmetries: A computational approach,* C.W.I. Tract Vol. 34.

Kersten, P. H. M. (1989). *Software to compute infinitesimal symmetries of exterior differential systems, with applications.* Acta Appl. Math. 16, 207–229.

Klingbeil, W. W. and Shield, R. T. (1966). *On a class of solutions in plane finite elasticity,* Zeit. angew. Math. Phys. 17, 489–511.

Konopelchenko, B. G. (1987). *Nonlinear Integrable Equations,* Springer Lecture Notes in Physics 270, Springer-Verlag, New York.

Korobeinikov, V. P. (1983). *Some exact solutions of Korteweg-de Vries-Burgers' equation for plane, cylindrical and spherical waves,* in Nonlinear Deformation Waves, (U. Nigul and J. Engelbrecht, eds.), pp. 149–154. IUTAM Symposium, Tallinin, 1982. Springer-Verlag, Berlin.

Korteweg, D. J. and de Vries, G. (1895). *On the change of form of long waves advancing in a rectangular channel, and on a new type of long stationary wave,* Philos. Mag. 39, 422-443.

Levi, D. and Rogers, C. (1989). *Group invariance of a nonlinear elasto-static system: Incorporation of stretch.* Preprint, Dept. Applied Mathematics, University of Waterloo, Waterloo, Ontario.

Levi, D. and Winternitz, P. (1987). *The cylindrical Kadomtsev-Petviashvili equation; its Kac-Moody-Virasoro algebra and relation in KP equation,* Centre de Recherches Mathématiques, Université de Montréal, Report No. 1499.

Levi, D., Nucci, M. C., Rogers, C., and Winternitz, P. (1987). *The Motion of a Rotating Shallow Liquid in a Rigid Container: Applications of Lie Group Analysis.* Preprint, Dept. Applied Mathematics, University of Waterloo, Waterloo, Ontario.

Lie, S. (1881). *Diskussion der Differentialgleichung $\partial^2/z\partial x\partial y = F(z)$.* Arch. für Math. 6, 112–124.

Lie, S. (1888a). *Theorie der Transformationgruppen I.* Leipzig. 2nd Ed., Chelsea, New York (1970).

Lie, S. (1888b). *Klassifikation und Integration von gewöhnlichen Differentialgleichungen zwischen x, y die eine Gruppe von Transformationen gestatten,* Math. Ann. 32, 213–281.

Lie, S. (1890). *Theorie der Transformationgruppen II.* Leipzig. 2nd Ed., Chelsea, New York (1970).

Lie, S. (1891). *Differentialgleichungen.* Leipzig. Reprinted by Chelsea, New York (1967).

Lie, S. (1893). *Theorie der Transformationgruppen III.* Leipzig. 2nd Ed., Chelsea, New York (1970).

Lloyd, S. P. (1981). *The infinitesimal group of the Navier-Stokes equations,* Acata Mechanica 38, 85–98.

Logan, J. D. and Peréz, J. J. (1980). *Similarity solutions for reactive shock hydrodynamics,* SIAM J. Appl. Math. 39, 512–527.

Lynskii, E. V. (1966). *A group analysis of the equations of motion for a viscoplastic medium,* J. Math. Mech. 5, 116–125 (in Russian).

Miles, J. W. (1981). *The Korteweg-de Vries equations: A historical essay,* J. Fluid Mech. 106, 131–147.

Na, T. Y. (1979). *Computational Methods in Engineering Boundary Value Problems.* Academic Press, New York.

Newell, A. C. (1985). *Solitons in Mathematics and Physics.* SIAM, Philadelphia.

Nucci, M. C. (1984). *Group analysis for MHD equations,* Atti. Sem. Mat. Fis. Univ. Modena 33, 21–34.

Nucci, M. C. (1987). *Group analysis for unsteady incompressible viscous flow (kinematic approach),* J. Phys. A: Math. Gen. 20, 5053–5059.

Nucci, M. C. (1990) *Interactive REDUCE programs for calculating classical, non-classical and Lie-Bäcklund symmetries of differential equations.* School of Mathematics Report # 062090-051, Georgia Institute of Technology, Atlanta, Ga.

Olver, P. J. (1986). *Applications of Lie Groups to Differential Equations.* Springer-Verlag, New York.

Ovsiannikov, L. V. (1962). *Gruppovye Svoystva Differentsialny Utraveni.* USSR Academy of Sciences Novosibirsk. (*Group Properties of Differential Equations,* (G. Bluman, trans. [1967]), unpublished).

Ovsiannikov, L. V. (1978). *Gruppovĭ analiz differentsial'nykh uravenĭi.* NAUKA, Moscow. *Group Analysis of Differential Equations* (English Edition, W. F. Ames, ed.). Academic Press, Boston, (1982).

Page, J. M. (1897). *Ordinary Differential Equations with an Introduction to Lie's Theory of the Group of One Parameter.* Macmillan, New York.

Peters, J. E. and Ames, W. F. (1990). *Group properties of the non-linear dynamic equations of elastic strings.* Int. J. Nonlinear Mech. 25, 107–115.

Poochnachev, V. V. (1960). *Group properties of the equations of Navier-Stokes in the plane,* J. Appl. Mech. and Tech. Phys. I, 83–90 (in Russian).

Pucci, E. and Salvatori, M. C. (1986). *Group properties of a class of semilinear hyperbolic equations,* Int. J. Nonlinear Mech. 21, 147–155.

Rand, R. H. (1984). *Computer Algebra in Applied Mathematics: An Introduction to MACSYMA.* Pitman, New York.

Richards, P. C. (1987). *Group Analysis of Equations Arising in Ocean Acoustics.* Ph.D. Dissertation, Georgia Institute of Technology, Atlanta, Georgia, USA.

Rogers, C. and Ames, W. F. (1989). *Nonlinear Boundary Value Problems in Science and Engineering.* Academic Press, Boston.

Rogers, C. and Shadwick, W. F. (1982). *Bäcklund Transformations and Their Applications.* Academic Press, New York.

Rogers, C. and Winternitz, P. (1988). *Group Analysis of a Magneto-hydrodynamic System.* Preprint, Dept. Applied Mathematics, University of Waterloo, Waterloo, Ontario.

Satsuma, J. and Ablowitz, M. J. (1979). *Two dimensional lumps in nonlinear dispersive systems,* J. Math. Phys. 20, 1496–1501.

Sattinger, D. H. and Weaver, O. L. (1986). *Lie Groups and Algebras with Applications to Physics, Geometry and Mechanics.* Springer-Verlag, New York.

Schlichting, H. (1960). *Boundary Layer Theory,* 4th Ed. McGraw-Hill, New York.

Schwarz, F. (1984). *Lie symmetries of the von Kármán equations,* Comp. Phys. Commun. 31, 113–114.

Schwarz, F. (1987a). *The Package SPDE for Determining Symmetries of Partial Differential Equations: User's Manual.* Distributed with REDUCE 3.3.

Schwarz, F. (1987b). *Programming with abstract data types: The symmetry packages AIDE and SPDE in SCRATCHPAD,* Int. Symp. *Trends in Computer Algebra,* Bad Neuenahr, West Germany (1987). (Schwarz's address: GMD, Institut F1, Postfach 1240, 5205 St. Augustin, West Germany).

Schwarz, F. (1988). *Symmetries of differential equations: From Sophus Lie to computer algebra,* SIAM Review 30, 450–481.

Seshadri, R. and Na, T. Y. (1985). *Group Invariance in Engineering Boundary Value Problems.* Springer-Verlag, New York.

Sharomet, N. O. (1989). *Symmetries, invariant solutions, and conservation laws of the nonlinear acoustic equation.* Acta Appl. Math. 15, 83–120.

Shen, H. and Ames, W. F. (1974). *On invariant solutions of the Korteweg-de Vries equation.* Phys. Lett. A49, 313–314.

Smoodskii, A. A. and Taranov, V. B. (1975). *On symmetry motions of two component plasmas,* J. Tech. Physics 45, 158–161 (in Russian).

Steudel, H. (1980). *A relation connecting scale transformation, Galilean transformation and Bäcklund transformation for the nonlinear Schrödinger equation,* Physica D. Nonlinear Phenomena 1, 420–421.

Su, C. H. and Gardner, C. S. (1968). *Korteweg-de Vries equation and its generalizations: III Derivation of the Korteweg-de Vries and Burgers' equation,* J. Math. Phys. 9, 1204–1209.

Şuhubi, E. S. and Bakkaloğlu, A. (1991). *Group properties and similarity solutions for a quasilinear wave equation in the plane.* In press, Int. J. Nonlinear Mech.

Suliciu, I., Lee, S. Y. and Ames, W. F. (1973). *Nonlinear traveling waves for a class of rate-type materials.* J. Math. Anal. Appl. 42, 313–322.

Taniuti, T. and Wei, C. C. (1968). *Reductive perturbation method in nonlinear wave propagation I.* J. Phys. Soc. Japan. 24, 941–496.

Taranov, V. B. (1978). *Group analysis of collisionless plasma equations,* in Group Theoretical Methods in Mechanics (N. H. Ibragimov and L. V. Ovsiannikov, eds.), pp. 277–284. USSR Academy of Sciences, Novosibirsk.

Torrisi, M. and Valenti, A. (1985). *Group properties and invariant solutions for infinitesimal transformations of a nonlinear wave equation,* Int. J. Nonlinear Mech. 20, 135–144.

Verbovetsky, A. M. (1989). *Local nonintegrability of long-short wave interaction equations.* Acta Appl. Math. 15, 121–136.

Vinogradov, A. M. (Ed.) (1989). *Symmetries of Partial Differential Equations.* Kluwer Academic Publishers, Dordrecht.

Wesolowski, Z. (1969). *Finite deformations of an elastic wedge and cone,* (in Polish), Mech. Teor. i Stos. 7, 195–204.

Whitham, G. B. (1974). *Linear and Nonlinear Waves.* Wiley, New York.

Winternitz, P. (1983). *Lie groups and solutions of nonlinear differential equations,* in Nonlinear Phenomena, Lecture Notes in Physics 189, Springer-Verlag, New York.

Wolf, K. B., Ed. (1983). *Nonlinear Phenomena,* Lecture Notes in Physics 189, Springer-Verlag, New York.

Zabrusky, N. J. (1967). *A synergetic approach to problems of nonlinear dispersive wave propagation and interaction.* In Nonlinear Partial Differential Equations (W. F. Ames, ed.). Academic Press, New York, 223–256.

Zakharov, N. S. and Korobeinikov, V. P. (1980). *Group analysis of the generalized Korteweg-de Vries-Burgers' equation,* J. Appl. Math. Mech. 44, 668-671.

GEOMETRY OF THE MELNIKOV VECTOR

Shui-Nee Chow

Center for Dynamical Systems and Nonlinear Studies

School of Mathematics

Georgia Institute of Technology

Atlanta, GA 30332

and

Masahiro Yamashita

Department of Mathematics and Statistics

Wright State University

Dayton, OH 45435

§1. INTRODUCTION

The notion of a homoclinic point was introducted by Poincaré [17]. To recall this concept, consider a diffeomorphism in $\mathbb{R}^2$ with a hyperbolic fixed point p. A point q is called a homoclinic point of p if q is in the intersection of the stable and unstable manifold of p. The point q is called a transversal homoclinic point of p if the intersection of the stable and unstable manifolds is transversal, i.e. the tangent spaces at q to the stable and unstable manifolds span the whole space. We note that if one homoclinic point exists, there must be infinitely many homoclinic points.

Poincaré already observed that the existence of homoclinic points implies complexity of the dynamics of the diffemorphism. Later G.D. Birkhoff [3] proved that every transversal homoclinic point of a a two-dimensional diffeomorphism is accumulated by periodic orbits. The results by Smale [19], now called the Smale-Birkhoff theorem, extend the Birkhoff's results both in two dimensional and to higher dimensional cases and assert that the existence of a transversal homoclinic point implies the existence of an invariant Cantor set in which the periodic orbits are dense. See also Moser [14]. Moreover, Newhouse [15] has proved that there is a much more complicated dynamical behavior associated with a homoclinic tangency. Thus the dynamics of diffeomorphisms with transversal or tangential homoclinic points are fairly well understood.

However to apply the above abstract theories for diffeomorphisms to a system of differential equations, we need to know the existence of a homoclinic point of a diffeomorphism induced by this system. More precisely, since we shall deal with an autonomous system with a time-period perturbation, the above diffeomorphism appears as a time-one map, called a Poincaré map, induced by the flow of the system.

Our problem is the following: an autonomous system of ordinary differential equations with a time-periodic perturbation is given and we assume that the unperturbed autonomous sytem has two hyperbolic critical points (not necessarily distinct) and a homoclinic or heteroclinic orbit connecting them. Find computable conditions under which the Poincaré map induced by the perturbed system has a transversal homoclinic point. See §2 for more precise definitions of these notions and for a rigorous formulation of the problem.

Poincaré [17], Melnikov [12] and Arnold [2] developed such conditions for two-dimensional analytic Hamiltonian systems and it is now called the Poincaré Melnikov Arnold method or simply the Melnikov method. The Melnikov theory has been studied by several authors, e.g. Chow, Hale and Mallet-Paret [4], Holmes [9], and Palmer [16], and generalizations to higher dimensional cases have also been studied,e.g., Holmes and Marsden [10] and Gruendler [6]. The key of these theories is the use of the Melnikov function which measures the splitting distance between the perturbed stable and unstable manifolds.

One of the purposes of the present notes is to clarify the geometry of the Melnikov function (now should be called the Melnikov vector) in higher dimensional cases and to extend the previous theories for the two-dimensional case to higher dimensional cases.

Our theory is based on the theory of exponential dichotomy. Palmer [16] showed that the linear variational system along the homoclinic orbit of the unperturbed autonomous system has exponential dichotomies on half-lines. Using this fact we shall derive explicit expressions of the local stable and unstable manifolds of the perturbed system in §3. Then Fredholm's alternative, given in Chow, Hale and Mallet-Paret [4] for the two-dimensional case, in Palmer [16] in higher dimensional cases and explained in §4, is used to derive the Melnikov vector in §5 and we examine conditions for a transversal homoclinic point. In §6 we introduce a notion of an index of a homoclinic or heteroclinic orbit which is useful to classify the cases that can occur in higher dimensional cases. We discuss a relation between the dimension of the Melnikov vector and the index of the homoclinic or heteroclinic orbit. Numerical aspect of the Melnikov vector is discussed in §7. In §8 we consider several special cases in which the Melnikov vectors take simpler forms, and also we discuss the tangency of the stable and unstable manifolds. We apply these general theories to Hamiltonian systems in §9. In §10 we extend our theory to the case of a heteroclinic orbit to invariant tori and as a by-product we derive a formula which guarantees the transversal intersection of the stable and unstable manifolds of a two-dimensional system with a quasi-periodic perturbation. See also Meyer and Sell [13] and Wiggins [20]. Several interesting examples are discussed in §11 and also we discuss a serious limitation of the Melnikov method by using an example for which the Melnikov method does not work.

§2. FORMULATION OF THE PROBLEM

Consider a system of differential equations

$$(2.1) \qquad\qquad \dot{x} = f(x)$$

and its perturbed system

$$(2.2) \qquad\qquad \dot{x} = f(x) + \epsilon g(t, x)$$

where $x \epsilon \mathbb{R}^n$, $t \epsilon \mathbb{R}$, $\epsilon \in \mathbb{R}$ and $|\epsilon| << 1$. The vector fields f and g are assumed to be sufficiently smooth and bounded on bounded sets. The vector field g is periodic in t with the least period $T(> 0)$.

Assume that system (2.1) has two hyperbolic critical points x_+ and x_- (not necessarily distinct). Also assume that there is an orbit $\gamma(t), t \in \mathbb{R}$, of system (2.1) which connects the critical points x_+ and x_-. That is,

$$\gamma(t) \to x_\pm \text{ as } t \to \pm\infty.$$

If $x_+ = x_-$, the orbit γ is called a homoclinic orbit. Otherwise γ is called a heteroclinic orbit.

Let $x(t; x_0), x_0 \in \mathbb{R}^n$, be the solution of system (2.1) with the initial data $x(0; x_0) = x_0$. The stable manifold $W^s(x_+)$ of the hyperbolic critical point x_+ of system (2.1) is defined by

$$W^s(x_+) = \{x_0 \epsilon \mathbb{R}^n : x(t; x_0) \to x_+ \text{ as } t \to +\infty\},$$

and the unstable manifold $W^u(x_-)$ of the hyperbolic critical point x_- of the system (2.1) is defined by

$$W^u(x_-) = \{x_0 \in \mathbb{R}^n : x(t; x_0) \to x_- \text{ as } t \to -\infty\}.$$

Then we have

$$\gamma \subset W^s(x_+) \cap W^u(x_-)$$

from the above assumption.

Since the critical points $x_\pm$ are hyperbolic and system (2.2) is periodic in t, there exist unique T-periodic solutions $\bar{x}_\pm(t; \epsilon)$ of system (2.2) such that

$$\lim_{\epsilon \to 0} \bar{x}_\pm(t, \epsilon) = \bar{x}_\pm(t, 0) = x_\pm$$

uniformly in t. For details see Hale [7].

It will be shown in the next section that there exist sets $W^s_{loc}(\bar{x}_+, \epsilon)$ and $W^u_{loc}(\bar{x}_-, \epsilon)$ in $\mathbb{R}^n \times \{0\} \subset \mathbb{R}^n \times \mathbb{R}$, where $\mathbb{R}^n \times \mathbb{R}$ is the extended phase space of system (2.2), such that

$$W^s_{loc}(x_+, \epsilon) = \{(x_0, 0) \in \mathbb{R}^n \times \{0\} : |x(t; 0, x_0) - \bar{x}_\pm(t; \epsilon)| \to 0 \text{ as } t \to +\infty$$

$$\text{and } x_0 \text{ is in a sufficiently small neighborhood of } \gamma\}$$

and

$$W^u_{loc}(\bar{x}_-, \epsilon) = \{(x_0, 0) \in \mathbb{R}^n \times \{0\} : |x(t; 0, x_0) - \bar{x}_-(t; \epsilon)| \to 0 \text{ as } t \to -\infty$$

$$\text{and } x_0 \text{ is in a sufficiently small neighborhood of } \gamma\},$$

where $x(t; \tau, x_0)$ is the solution of system (2.2) with $x(\tau; \tau, x_0) = x_0, x_0 \in \mathbb{R}^n$.

If we define the time dependent stable and unstable manifolds, $\bar{W}^s(\bar{x}_+; \epsilon)$ and $\bar{W}^u(\bar{x}_-, \epsilon)$, of system (2.2) by

$$\bar{W}^s(\bar{x}_+, \epsilon) = \{(x_0, \tau) \in \mathbb{R}^n \times \mathbb{R} : |x(t; \tau, x_0) - \bar{x}_+(t; \epsilon)| \to 0 \text{ as } t \to +\infty\}$$

and

$$\bar{W}^u(\bar{x}_-, \epsilon) = \{(x_0, \tau) \in \mathbb{R}^n \times \mathbb{R} : |x(t; \tau, x_0) - \bar{x}_-(t; \epsilon)| \to 0 \text{ as } t \to -\infty\},$$

then $W_{loc}^s(\bar{x}_+, \epsilon)$ and $W_{loc}^u(\bar{x}_-, \epsilon)$ are the local cross sections at $t = 0$ of $\bar{W}^s(\bar{x}_+, \epsilon)$ and $\bar{W}^u(\bar{x}_-, \epsilon)$ respectively. That is,

$$W_{loc}^s(\bar{x}_+, \epsilon) \subset \bar{W}^s(\bar{x}_+, \epsilon) \cap (\mathbb{R}^n \times \{0\})$$

and

$$W_{loc}^u(\bar{x}_-, \epsilon) \subset \bar{W}^u(\bar{x}_-, \epsilon) \cap (\mathbb{R}^n \times \{0\}).$$

Since system (2.2) is periodic in t, its extended phase space can be regarded as $\mathbb{R}^n \times S^1$, where S^1 is the unit circle, and so $W_{loc}^s(\bar{x}_+, \epsilon)$ and $W_{loc}^u(\bar{x}_-, \epsilon)$ are the local stable and unstable manifolds of hyperholic critical points $\bar{x}_\pm \equiv \bar{x}_\pm(0; \epsilon)$ of the Poincaré map $\Pi_\epsilon : \mathbb{R}^n \to \mathbb{R}^n$, which is defined by the flow of system (2.2) as follows:

$$\Pi_\epsilon(x_0) = x(T; 0, x_0), \quad x_0 \in \mathbb{R}^n.$$

Now we state our problem.

Problem I. When does system (2.2) have an orbit $x(t)$, $t \in \mathbb{R}$, so that $x(t) \to \bar{x}_\pm(t; \epsilon)$ as $t \to \pm\infty$?

Following the above argument, it is clear that Problem I is equivalent to

Problem I'. When do $W_{loc}^s(\bar{x}_+, \epsilon)$ and $W_{loc}^u(\bar{x}_-, \epsilon)$, defined above, intersect each other?

Then the next natural question would be

Problem II. When do $W_{loc}^s(\bar{x}_+, \epsilon)$ and $W_{loc}^u(\bar{x}_+, \epsilon)$ intersect transversally?

Here the transversal intersection means that tangent spaces to $W_{loc}^s(\bar{x}_+, \epsilon)$ and to $W_{loc}^u(\bar{x}_-, \epsilon)$ at a point of intersection span the whole space $\mathbb{R}^n$.

§3. THE STABLE AND UNSTABLE MANIFOLDS

Since we would like to describe the local cross sections $W_{loc}^s(\bar{x}_+, \epsilon)$ and $W_{loc}^u(\bar{x}_-, \epsilon)$ of the time dependent stable and unstable manifolds $\bar{W}^s(\bar{x}_+, \epsilon)$ and $\bar{W}^u(\bar{x}_-, \epsilon)$ of system (2.2) as parts of the 'perturbed manifolds' of $W^s(x_+)$ and $W^u(x_-)$ of system (2.1) respectively along the orbit $\gamma(t)$, we let, for a fixed $\alpha \in \mathbb{R}$,

$$x(t) = \gamma(t + \alpha) + \epsilon z(t + \alpha).$$

Then system (2.2) becomes

(3.1) $$\dot{z} = A(t)z + g(t - \alpha, \gamma(t)) + h(t, z, \alpha, \epsilon)$$

where $A(t) = Df(\gamma(t))$ and

$$h(t, z, \alpha, \epsilon) = \frac{1}{\epsilon}\{f(\gamma(t) + \epsilon z(t)) - f(\gamma(t))$$

$$- \epsilon Df(\gamma(t))z(t) + \epsilon g(t - \alpha, \gamma(t) + \epsilon z(t))$$

$$- \epsilon g(t - \alpha, \gamma(t))\}.$$

We note that $|h(t, z, \alpha, \epsilon)| = O(|\epsilon|)$ uniformly in t, z and α.

Since $\bar{x}_+(0; \epsilon)$ is hyperbolic, $\gamma(\alpha) + \epsilon\xi \in W_{loc}^s(\bar{x}_+, \epsilon)$ if and only if the solution $z(t; \alpha, \xi)$ of system (3.1) with $z(\alpha; \alpha, \xi) = \xi$ is bounded on the time interval $[\alpha, \infty)$. Thus we have, by changing $\alpha \in \mathbb{R}$, that

$$W_{loc}^s(\bar{x}_+, \epsilon) = \bigcup_{\alpha \in \mathbb{R}} \{\gamma(\alpha) + \epsilon\xi^s : \text{ the solution}$$

$$z(t; \alpha, \xi^s) \text{ of system (3.1) is bounded on } [\alpha, \infty)\}.$$

Similarly we have

$$W_{loc}^u(\bar{x}_-, \epsilon) = \bigcup_{\alpha \in \mathbb{R}} \{\gamma(\alpha) + \xi^u : \text{ the solution } z(t; \alpha, \xi^u)$$
$$\text{of system (3.1) is bounded on } (-\infty, \alpha]\}.$$

We remark that $\alpha \in \mathbb{R}$ works as a 'sweeping' parameter along the orbit γ. See Figure 1.

Now as the orbit γ is assumed to be a homoclinic or heteroclinic orbit to hyperbolic fixed points, the linearized system

(3.2) $$\dot{z} = A(t)z$$

of system (2.1) along the orbit γ has exponential dichotomies on $[\alpha, \infty)$ and on $(-\infty, \alpha]$. (See Palmer [16].) Recall that system (3.2) has an exponential dichotomy on $[\alpha, \infty)$ if there exists a projection $P(\alpha) : \mathbb{R}^n \to \mathbb{R}^n$, $K \geq 1$ and $a > 0$ such that

$$|\Phi(t, \alpha)P(\alpha)\Phi(\alpha, s)| \leq Ke^{-a(t-s)}, \alpha \leq s \leq, t,$$

and

$$|\Phi(t, \alpha)(I - P(\alpha))\Phi(\alpha, s)| \leq Ke^{-a(s-t)}, \alpha, \leq t \leq s,$$

where $\Phi(t, \alpha)$ is the transition matrix of system (3.2). Similarly system (3.2) has an exponential dichotomy on $(-\infty, \alpha]$ if there exists a projection $Q(\alpha) : \mathbb{R}^n \to \mathbb{R}^n$, $L \geq 1$ and $b > 0$ such that

$$|\Phi(t, \alpha)Q(\alpha)\Phi(\alpha, s)| \leq Le^{-b(t-s)}, s \leq t \leq \alpha,$$

and

$$|\Phi(t, \alpha)(I - Q)(\alpha))\Phi(\alpha, s)| \leq Le^{-b(s-t)}, t \leq s \leq \alpha.$$

Now fix $\alpha \in \mathbb{R}$. Then from the variation of constants formula, the solution

$z(t; \alpha, \xi)$ of system (3.1) satisfies

$$z(t; \alpha, \xi) = \Phi(t, \alpha)P(\alpha)\xi + \int_\alpha^t \Phi(t, \tau)S(\tau)\{g(\tau - \alpha, \gamma(\tau))$$
$$+ h(\tau, z(\tau; \alpha, \xi), \alpha, \epsilon)\}d\tau + \Phi(t, \alpha)(I - P(\alpha))\xi$$
$$+ \int_\alpha^t \Phi(t, \tau)(I - S(\tau))\{g(\tau - \alpha, \gamma(\tau))$$
$$+ h(\tau, z(\tau; \alpha, \xi), \alpha, \epsilon)\}d\tau$$

where $S(\tau)$ is defined by

$$S(\tau) = \Phi(\tau, \alpha)P(\alpha)\Phi(\alpha, \tau).$$

It is easy to show that $z(t; \alpha, \xi^s)$ is a bounded solution of system (3.1) on $[\alpha, \infty)$ if and only if $z(t; \alpha, \xi^s)$ satisfies the following integral equation:

$$z(t; \alpha, \xi^s) = \Phi(t, \alpha)P(\alpha)\xi^s$$

(3.3)
$$+ \Phi(t, \alpha)P(\alpha) \int_\alpha^t \Phi(\alpha, \tau)\{g(\tau - \alpha, \gamma(\tau))$$
$$+ h(\tau, z(\tau; \alpha, \xi^s), \alpha, \epsilon)\}d\tau$$
$$+ \Phi(t, \alpha)(I - P(\alpha)) \int_\infty^t \Phi(\alpha, \tau\{g(\tau - \alpha, \gamma(\tau))$$
$$+ h(\tau, z(\tau; \alpha, \xi^s), \alpha, \epsilon)\}d\tau.$$

Here we used

$$\Phi(t, \tau)S(\tau) = \Phi(t, \alpha)P(\alpha)\Phi(\alpha, \tau).$$

Let $\eta^s = P(\alpha)\xi^s$. Then it can be shown by the contraction mapping principle that integral equation (3.3) has a unique solution $z(\eta^s)(t) \equiv z(t; \alpha, \xi^s(\eta^s))$ for $|\eta^s| << 1$ where $\xi^s = \xi^s(\eta^s)$ is a function of η^s. By letting $t = \alpha$, the function $\xi^s = \xi^s(\eta^s)$ is given by

$$\xi^s = \eta^s + (I - P(\alpha))\{\int_\infty^\alpha \Phi(\alpha, \tau)g(\tau - \alpha, \gamma(\tau))d\tau$$
$$+ \int_\infty^\alpha \Phi(\alpha, \tau)h(\tau, z(\eta^s)(\tau), \alpha, \epsilon)d\tau\}.$$

We remark that

$$\int_\infty^\alpha \Phi(\alpha, \tau)h(\tau, z(\eta^s)(\tau), \alpha, \epsilon)d\tau = O(|\epsilon|)$$

uniformly in η^s. Similarly $z(t; \alpha, \xi^u)$ is a bounded solution of system (3.1) on $(-\infty, \alpha]$ if and only if

$$\xi^u = \eta^u + Q(\alpha)\{\int_{-\infty}^{\alpha} \Phi(\alpha, \tau)g(\tau - \alpha, \gamma(\tau)d\tau$$
$$+ \int_{-\infty}^{\alpha} \Phi(\alpha, \tau)h(\tau, z(\eta^u)(\tau), \alpha, \epsilon)d\tau\}$$

where $\eta^u \in \mathcal{R}(I - Q(\alpha))$, the range of $I - Q(\alpha)$, $|\eta^u| \ll 1$ and $z(\eta^u)(t) \equiv z(t; \alpha, \xi^u(\eta^u))$ is the unique solution of

$$
\begin{aligned}
z(t; \alpha, \xi^u) = {}& \Phi(t, \alpha)\eta^u \\
& + \Phi(t, \alpha)(I - Q(\alpha))\int_{\alpha}^{t} \Phi(\alpha, \tau)\{g(\tau - \alpha, \gamma(\tau)) \\
& + h(\tau, z(\tau; \alpha, \xi^u), \alpha, \epsilon)\}d\tau \\
& + \Phi(\tau, \alpha)Q(d)\int_{-\infty}^{t} \Phi(\alpha, \tau)\{g(T - \alpha, \gamma(\tau)) \\
& + h(\tau, z(\tau; \alpha, \xi^u), \alpha, \epsilon)\}d\tau.
\end{aligned}
$$

(3.4)

We also remark that

$$\int_{-\infty}^{\alpha} \Phi(\alpha, \tau)h(\tau, z(\eta^u)(\tau), \alpha, \epsilon)d\tau = O(|\epsilon|).$$

Thus we have shown that $W_{loc}^s(\bar{x}_+, \epsilon)$ and $W_{loc}^u(\bar{x}_-, \epsilon)$ have the following expressions as functions of α, η^s or η^u.

Proposition 3.1.

(i) The local cross section $W_{loc}^s(\bar{x}_+, \epsilon)$ at $t = 0$ of the time dependent stable manifold $\bar{W}^s(\bar{x}_+, \epsilon)$ of system (2.2) is given by the following:

(3.5)
$$W_{loc}^s(\bar{x}_+, \epsilon) = \bigcup_{\alpha \in \mathbb{R}} \{\gamma(\alpha) + \epsilon\xi^s(\alpha, \eta^s, \epsilon)\}$$

where

(3.6)
$$
\begin{aligned}
\xi^s(\alpha, \eta^s, \epsilon) = {}& \eta^s + (I - P(\alpha))\{\int_{\infty}^{\alpha} \Phi(\alpha, \tau)g(\tau - \alpha, \gamma(\tau))d\tau \\
& + \int_{\infty}^{\alpha} \Phi(\alpha, \tau)h(\tau, z(\eta^s)(\tau), \alpha, \epsilon)d\tau\}
\end{aligned}
$$

and $\eta^s \in \mathcal{R}(P(\alpha))$, $|\eta^s| \ll 1$ and $z(\eta^s)(t)$ is the solution of equation (3.3) with $\eta^s = P(\alpha)\xi^s$.

(ii) The local cross section $W^u_{loc}(\bar{x}_-, \epsilon)$ at $t = 0$ of the time dependent unstable manifold $\bar{W}^u(\bar{x}_-, \epsilon)$ of system (2.2) is given by the following:

$$(3.7) \qquad W^u_{loc}(\bar{x}_-, \epsilon) = \bigcup_{\alpha \in \mathbf{R}} \{\gamma(\alpha) + \epsilon \xi^u(\alpha, \eta^u \epsilon)\}$$

where

$$(3.8) \qquad \begin{aligned} \xi^u(\alpha, \eta^u, \epsilon) = \eta^u + Q(\alpha)\{ &\int_{-\infty}^{\alpha} \Phi(\alpha, \tau)g(\tau - \alpha, \gamma(\tau))d\tau \\ &+ \int_{-\infty}^{\alpha} \Phi(\alpha, \tau)h(\tau, z(\eta^u)(\tau), \alpha, \epsilon)d\tau \} \end{aligned}$$

and $\eta^u \in \mathcal{R}(I - Q(\alpha))$, $|\eta^u| \ll 1$ and $z\eta^u)(t)$ is the solution of equation (3.4). ∎

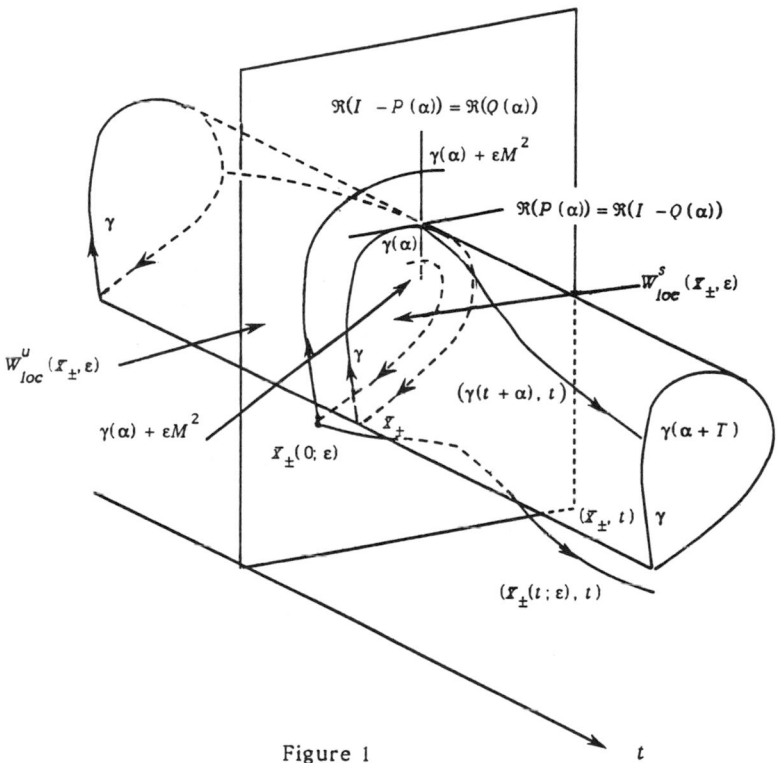

Figure 1

Remark 3.2. Notice that $\dot{\gamma}(\alpha) \in \mathcal{R}(P(\alpha)) \cap \mathcal{R}(I - Q(\alpha))$. Since we consider the cross sections of the time dependent stable and unstable manifolds in a vicinity of the orbit γ, it is sufficient, by the tubular neighborhood theorem, to consider coordinates in the normal bundle $\bigcup_{\alpha \in \mathbb{R}} T^{\perp}_{\gamma(\alpha)} \mathbb{R}^n$ of the submanifold γ, where $T^{\perp}_{\gamma(\alpha)} \mathbb{R}^n$ stands for the normal vector subspace, in the tangent space $T_{\gamma(\alpha)} \mathbb{R}^n$, to the one-dimensional vectorsubspace spanned by $\dot{\gamma}(\alpha)$, i.e., $T^{\perp}_{\gamma(\alpha)} \mathbb{R}^n \simeq T_{\gamma(\alpha)} \mathbb{R}^n / \mathrm{span}\{\dot{\gamma}(\alpha)\}$. Hence from now on, we assume, for $\eta^s \in T_{\gamma(\alpha)} \mathbb{R}^n$, $\eta^u \in T_{\gamma(\alpha)} \mathbb{R}^n$ and $\alpha \in \mathbb{R}$, that

$$\eta^s \in T^{\perp}_{\gamma(\alpha)} \mathbb{R}^n \cap \mathcal{R}P(\alpha)$$

and

$$\eta^u \in T^{\perp}_{\gamma(\alpha)} \mathbb{R}^n \cap \mathcal{R}(I - Q(\alpha)).$$

Remark 3.3. The 'stable subspace' $\mathcal{R}(P(\alpha))$ and the 'unstable subspace' $\mathcal{R}(I - Q(\alpha))$ are uniquely determined. However their complementary subspaces are not uniquely determined. See Coppel [5] for details. Thus we can assume that for each $\alpha \in \mathbb{R}$,

(3.9) $$\mathcal{R}(I - P(\alpha)) \subset T^{\perp}_{\gamma(\alpha)} \mathbb{R}^n, \quad \mathcal{R}Q(\alpha) \subset T^{\perp}_{\gamma(\alpha)} \mathbb{R}^n.$$

Under these assumptions, each point in $W^s_{loc}(\bar{x}_+, \epsilon)$ or $W^u_{loc}(\bar{x}_-, \epsilon)$ is uniquely expressed in terms of the coordinates α, η^s or η^u.

Remark 3.4. The higher order terms

$$(I - P(\alpha)) \int_{\infty}^{\alpha} \Phi(\alpha, \tau) h(\tau, z(\eta^s)(\tau), \alpha, \epsilon) d\tau \text{ and}$$

$$Q(\alpha) \int_{-\infty}^{\alpha} \Phi(\alpha, \tau) h(\tau, z(\eta^u)(\tau), \alpha, \epsilon) d\tau$$

in (3.6) and (3.8) are of order ϵ uniformly in α. Though these terms include the solutions $z(\eta^s)(\tau)$ and $z(\eta^u)(\tau)$ of equations (3.3) and (3.4), these solutions can be approximated in an arbitrarily high order of accuracy by an iterative scheme. In fact, second iteration is enough to obtain all information we need to determine the transversality of $W_{loc}^s(\bar{x}_+, \epsilon)$ and $W_{loc}^u(\bar{x}_-, \epsilon)$. (See §7).

§4. THE FREDHOLM'S ALTERNATIVE

In this section a linear version of problem I in §2 is considered to develop a geometrical idea to solve problem I in §2 in general cases. Consider a nonhomogeneous linear system

$$(4.1) \qquad\qquad \dot{z} = A(t)z + g(t), \ z \in \mathbb{R}$$

where $g(t)$ is bounded and continuous on $\mathbb{R}$, and suppose that the linear system

$$(4.2) \qquad\qquad \dot{z} = A(t)z, z \in \mathbb{R}^n$$

has exponential dichotomies on $[0, \infty)$ with projection P and on $(-\infty, 0]$ with projection Q. Consider the following

Problem: Find a condition under which system (4.1) has a bounded solution on $\mathbb{R}$. Let $\Phi(t, s)$ be the fundamental matrix of system (4.2). Then it is known that the solution $z(t; 0, \xi^s)$ of (4.1) is bounded on $[0, \infty)$ if and only if

$$(4.3) \qquad\qquad \xi^s = \eta^s + (I - P) \int_{\infty}^{0} \Phi(0, t)g(t)dt,$$

and the solution $z(t; 0, \xi^u)$ of (4.1) is bounded on $(-\infty, 0]$ if and only if

$$(4.4) \qquad\qquad \xi^u = \eta^u + Q \int_{-\infty}^{0} \Phi(0, t)g(t)dt,$$

where $\eta^s \in \mathcal{R}(P)$ and $\eta^u \in \mathcal{R}(I - Q)$.

Thus the set of initial data ξ^s(ξ^urespectively) which give bounded solutions on $[0, \infty)$($(-\infty, 0]$) constitutes the hyperplane which is a shift by the constant vector

$$(I - P) \int_{\infty}^{0} \Phi(0, t)g(t)dt(Q \int_{-\infty}^{0} \Phi(0, t)g(t)(dt)$$

of the unperturbed space $\mathcal{R}(P)$($\mathcal{R}(I - Q)$)).

Let

$$\vec{d} = Q \int_{-\infty}^{0} \Phi(0, t)g(t)dt - (I - P) \int_{\infty}^{0} \Phi(0, t)g(t)dt.$$

Then we have the following lemma which is the starting point of the paper.

Lemma 4.1. *System (4.1) has a bounded solution on* $\mathbb{R}$ *if and only if*

(4.5) $$\vec{d} \in \mathcal{R}(P) + \mathcal{R}(I - Q).$$

Proof. This is obvious because condition (4.5) is equivalent to say that two hyperplanes defined by (4.3) and (4.4) intersect. ∎

The geometrical statement in Lemma 4.1 can be expressed analytically by using bounded solutions of the adjoint system of (4.2):

(4.6) $$\dot{\phi} + A^*(t)\phi = 0$$

where $A^*(t)$ is the transpose of $A(t)$. Since system (4.2) has exponential dichotomies on $[0, \infty)$ with projection P and on $(-\infty, 0]$ with projection Q, its adjoint system (4.6) automatically has exponential dichotomies on $[0, \infty)$ with projection $I - P^*$ and on $(-\infty, 0]$ with projection $I - Q^*$, where P^* and Q^* are adjoint operators of P and Q respectively.

Notice that $\{\mathcal{R}(I - P^*) \cap \mathcal{R}(Q^*)\} = \{\mathcal{R}(P) + \mathcal{R}(I - Q)\}^{\perp}$ is the subspace of initial points at $t = 0$ of bounded solutions of the adjoint system (4.6). Therefore condition (4.5) is equivalent to saying that $\vec{d}$ is annihilated by $\phi(0)$ where $\phi(t)$ is any bounded solution of the adjoint system (4.6).

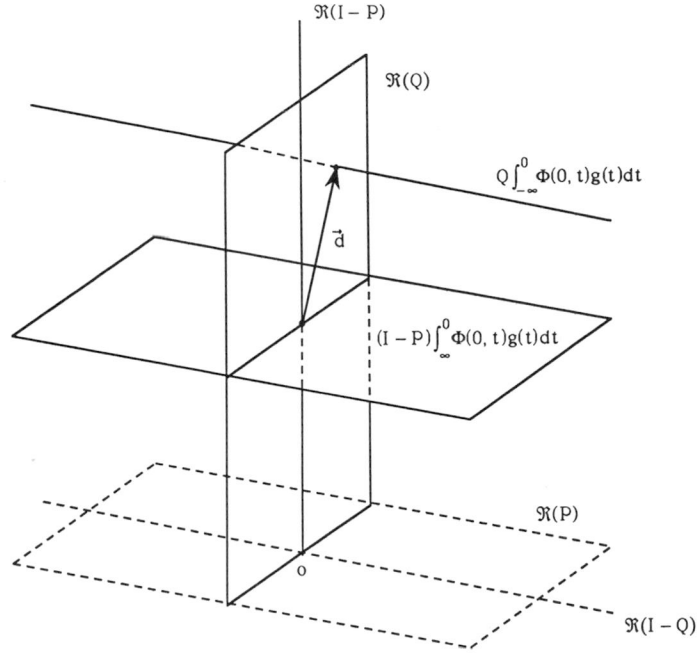

$$\Re(I-P)$$

$$\Re(Q)$$

$$Q\int_{-\infty}^{0}\Phi(0,t)g(t)dt$$

$$\vec{d}$$

$$(I-P)\int_{\infty}^{0}\Phi(0,t)g(t)dt$$

$$\Re(P)$$

$$\Re(I-Q)$$

o

Figure 2

Let $m = \dim\{\mathcal{R}(P) + \mathcal{R}(I-Q)\}^{\perp}$ and let $\{\phi_1(t),\dots,\phi_m(t)\}$ be a complete set of bounded solutions of system (4.6) which satisfies that

$$\{\mathcal{R}(P) + \mathcal{R}(I-Q)\}^{\perp} = \operatorname{span}\ \{\phi_1(0),\dots,\phi_m(0)\}.$$

Then we have the following analytical restatement of Lemma 4.1.

Lemma 4.2. $\vec{d} \in \mathcal{R}(P) + \mathcal{R}(I-Q)$ *if and only if*

$$(4.7) \qquad\qquad \int_{-\infty}^{\infty} \phi_i^*(t)g(t)dt = 0,\ i = 1,\dots,m,$$

where $\phi_i^*(t)$ is the transpose of $\phi_i(t)$.

Proof. $\vec{d} \in \mathcal{R}(P) + \mathcal{R}(I-Q)$ if and only if $\phi_i^*(0)\vec{d} = 0,\ i = 1,\dots,m.$

Since

$$Q^* \phi_i(0) = (I - P^*)\phi_i(0) = \phi_i(0)$$

and

$$\phi_i(t) = (\Phi^{-1}(t,0))^* \phi_i(0) = \Phi^*(0,t)\phi_i(0),$$

$$
\begin{aligned}
0 &= \phi_i^*(0)\vec{d} \\
&= \phi_i^*(0)Q \int_{-\infty}^0 \Phi(0,t)g(t)dt - \phi_i^*(0)(I-P)\int_\infty^0 \Phi(0,t)g(t)dt \\
&= \int_{-\infty}^0 \phi_i^*(0)\Phi(0,t)g(t)dt - \int_\infty^0 \phi_i^*(0)\Phi(0,t)g(t)dt \\
&= \int_{-\infty}^0 \phi_i^*(t)g(t)dt - \int_\infty^0 \phi_i^*(t)g(t)dt \\
&= \int_{-\infty}^\infty \phi_i^*(t)g(t)dt. \qquad \blacksquare
\end{aligned}
$$

We remark that Lemma 4.2 had been proved in Chow, Hale and Mallet-Paret [4] for the two-dimensional case and in Palmer [16] for the general case. However our proof is different and more geometrical. We shall apply in the next section the method of proof of Lemma 4.2 to the tangent space at each point of a homoclinic or heteroclinic orbit.

§5. INTERSECTION OF THE STABLE AND UNSTABLE MANIFOLDS

In this section we will prove our main result that gives computable conditions for the transversal intersection of the stable and unstable manifolds of system (2.2) given in Proposition 3.1. To this end we first derive the Melnikov vector which measures the 'distance' between $W_{loc}^s(\bar{x}_+, \epsilon)$ and $W_{loc}^u(\bar{x}_-, \epsilon)$. Define, for simplicity, the following quantities in the expressions in (3.6) and (3.8):

$$m^s(\alpha) = (I - P(\alpha)) \int_\infty^\alpha \Phi(\alpha, t) g(t - \alpha, \gamma(t)) dt,$$

$$\tilde{m}^s(\alpha, \eta^s, \epsilon) = (I - P(\alpha)) \int_\infty^\alpha \Phi(\alpha, t) h(t, z(\eta^s)(t), \alpha, \epsilon) dt,$$

$$m^u(\alpha) = Q(\alpha) \int_{-\infty}^\alpha \Phi(\alpha, t) g(t - \alpha, \gamma(t)) dt,$$

$$\tilde{m}^u(\alpha, \eta^u, \epsilon) = Q(\alpha) \int_{-\infty}^\alpha \Phi(\alpha, t) h(t, z(\eta^u)(t), \alpha, \epsilon) dt.$$

Then we have

$$\xi^s(\alpha, \eta^s, \epsilon) = \eta^s + m^s(\alpha) + \tilde{m}^s(\alpha, \eta^s, \epsilon)$$

and

$$\xi^u(\alpha, \eta^u, \epsilon) = \eta^u + m^u(\alpha) + \tilde{m}^u(\alpha, \eta^u, \epsilon).$$

Finally we define the distance vectors $\tilde{d}$ and $\vec{d}$ by

(5.1) $$\tilde{d}(\alpha, \eta^s, \eta^u, \epsilon) = \xi^u(\alpha, \eta^u, \epsilon) - \xi^s(\alpha, \eta^s, \epsilon)$$

and

$$\vec{d}(\alpha) = m^u(\alpha) - m^s(\alpha).$$

Recall that we are working on the normal bundle $\bigcup_{\alpha \in \mathbb{R}} T_{\gamma(\alpha)}^\perp \mathbb{R}^n$. Fix $\alpha \in \mathbb{R}$ and let $\eta^s, \eta^u \in T_{\gamma(\alpha)} \mathbb{R}^n$. Consider the following decomposition of $T_{\gamma(\alpha)}^\perp \mathbb{R}^n$:

$$T_{\gamma(\alpha)}^\perp \mathbb{R}^n = \{\mathcal{R}(I - Q(\alpha)) \cap \mathcal{R}P(\alpha) \cap T_{\gamma(a)}^\perp \mathbb{R}^n\}$$

$$\oplus \{\mathcal{R}(I - Q(\alpha)) \cap \mathcal{R}(I - P(\alpha)) \cap T_{\gamma(\alpha)}^\perp \mathbb{R}^n)\}$$

$$\oplus \{\mathcal{R}(Q(\alpha)) \cap \mathcal{R}P(\alpha) \cap T_{\gamma(\alpha)}^\perp \mathbb{R}^n\}$$

$$\oplus \{\mathcal{R}Q(\alpha) \cap \mathcal{R}(I - P(\alpha)) \cap T_{\gamma(\alpha)}^\perp \mathbb{R}^n\}.$$

According to this decomposition of $T_{\gamma(\alpha)}^\perp \mathbb{R}^n$, the vectors $\xi^u(\alpha, \eta^u, \epsilon)$ and $\xi^s(\alpha, \eta^s, \epsilon)$ are decomposed as follows:

$$\xi^u(\alpha, \eta^u, \epsilon) = (\eta_1^u, \eta_2^u, m_1^u + \tilde{m}_1^u, m_2^u + \tilde{m}_2^u)$$

Shui-Nee Chow and Masahiro Yamashita

and

$$\xi^s(\alpha, \eta^s, \epsilon) = (\eta_1^s, m_1^s + \tilde{m}_1^s, \eta_2^s, m_2^s + \tilde{m}_2^s).$$

See the diagram below.

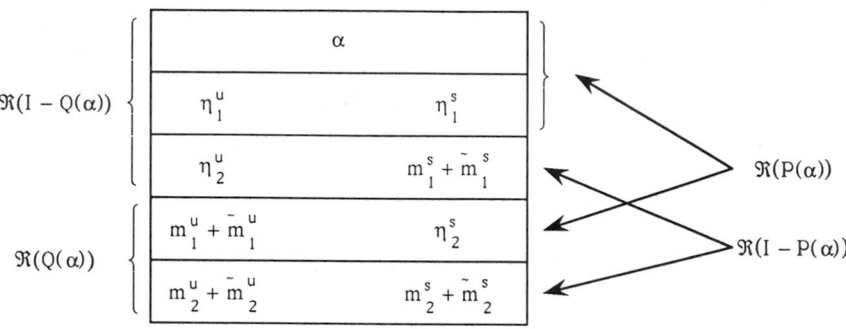

Diagram 1

To get familiar with the decomposition defined above we give an example in Figure 3. Here consider a homoclinic orbit γ in $\mathbb{R}^3$ and assume $\dim \mathcal{R}P(\alpha) = 2$ and $\dim \mathcal{R}(I - Q(\alpha)) = 1$.

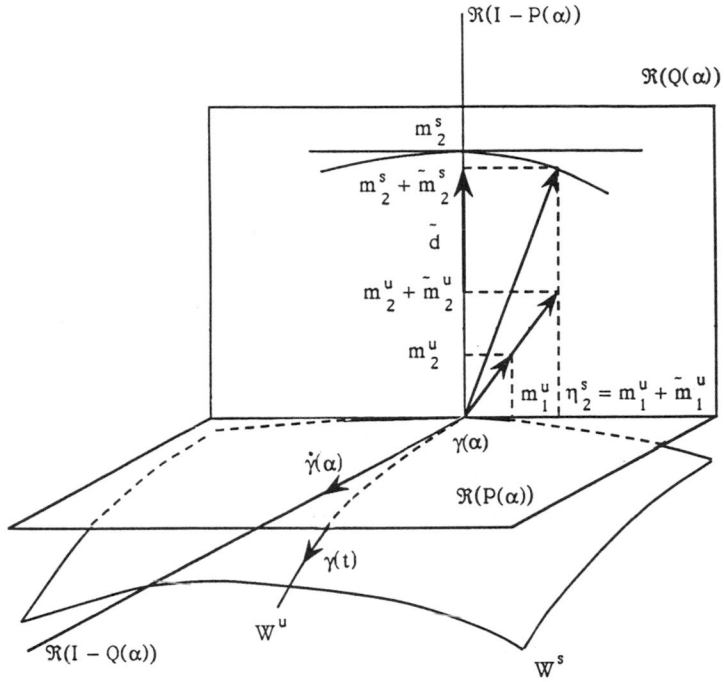

Figure 3

Now it is clear from (3.5), (3.7) and (5.1) that $W^s_{loc}(\bar{x}_+, \epsilon)$ and $W^u_{loc}(\bar{x}_-, \epsilon)$ intersect each other in the hyperplane $\gamma(\alpha) + T^{\perp}_{\gamma(\alpha)}\mathbb{R}^n$ if and only if $\tilde{d}(\alpha, \eta^s, \eta^u, \epsilon) = 0$ for some α, η^s and η^u. Since

(5.2)
$$\tilde{d}(\alpha, \eta^s, \eta^u, \epsilon) = (\eta^u_1 - \eta^s_1, \eta^u_2 - (m^s_1 + \tilde{m}^s_1), (m^u_1 + \tilde{m}^u_1)$$
$$- \eta^s_2, (m^u_2 + \tilde{m}^u_2) - (m^s_2 + \tilde{m}^s_2)),$$

$\tilde{d}(\alpha, \eta^s, \eta^u, \epsilon) = 0$ if and only if there exist $\alpha, \nu (= \eta^s_1 = \eta^u_1), \eta^s_2$ and η^u_2 such that the following three equations are satisfied.

(5.3)
$$\eta^u_2 - \{m^s_1(\alpha) + \tilde{m}^s_1(\alpha, \nu, \eta^s_2, \epsilon)\} = 0,$$

(5.4) $$\eta_2^s - \{m_1^u(\alpha) + \tilde{m}_1^u(\alpha, \nu, \eta_2^u, \epsilon)\} = 0,$$

(5.5) $$\{m_2^u(\alpha) + \tilde{m}_2^u(\alpha, \nu, \eta_2^u, \epsilon)\} - \{m_2^s(\alpha) + \tilde{m}_2^s(\alpha, \nu, \eta_2^s, \epsilon)\} = 0.$$

From (5.3) and (5.4),

$$F(\eta_2^u, \alpha, \nu) \equiv \eta_2^u - \{m_1^s(\alpha) + \tilde{m}_1^s(\alpha, \nu, m_1^u(\alpha) + \tilde{m}_1^u(\alpha, \nu, \eta_2^u, \epsilon), \epsilon)\} = 0.$$

We notice that

$$\frac{\partial}{\partial \eta_2^u} F(\eta_2^u, \alpha, \nu) = I - \frac{\partial \tilde{m}_1^s}{\partial \eta_2^s} \frac{\partial \tilde{m}_1^u}{\partial \eta_2^u}$$

is nonsingular for ϵ small enough because

$$\left| \frac{\partial \tilde{m}_1^s}{\partial \eta_2^s} \frac{\partial \tilde{m}_1^u}{\partial \eta_2^s} \right| = O(|\epsilon|^2).$$

Hence, it follows from the implicit mapping theorem that

(5.6) $$\eta_2^u = \eta_2^u(\alpha, \nu, \epsilon)$$

for $|\nu| << 1$. Similarly, we have

(5.7) $$\eta_2^s = \eta_2^s(\alpha, \nu, \epsilon)$$

for $|\nu| << 1$.

Therefore, by (5.5),

$$\tilde{d}(\alpha, \eta^s, \eta^u, \epsilon) = 0$$

if and only if

(5.8)
$$\{m_2^u(\alpha) + \tilde{m}_2^u(\alpha, \nu, \eta_2^u(\alpha, \nu, \epsilon), \epsilon)\}$$
$$- \{m_2^s(\alpha) + \tilde{m}_2^s(\alpha, \nu, \eta_2^s(\alpha, \nu, \epsilon), \epsilon)\} = 0.$$

To rewrite (5.8) in a more convenient form, we utilize bounded solutions of the adjoint system

(5.9) $$\dot{\phi} + A^*(t)\phi = 0$$

of system (3.2). Let

(5.10) $m = \dim\{\mathcal{R}(P(\alpha)) + \mathcal{R}(I - Q(\alpha))\}^{\perp} = \dim\{\mathcal{R}(I - P^*(\alpha)) \cap \mathcal{R}(Q^*(\alpha))\}$

and let $\{\phi_1(t), \ldots, \phi_m(t)\}$ be a complete set of bounded solutions of system (5.9) which satisfies

$$\{\mathcal{R}(P(\alpha)) + \mathcal{R}(I - Q(\alpha))\}^{\perp} = \text{span}\{\phi_1(\alpha), \ldots, \phi_m(\alpha)\}.$$

Then (5.8) is equivalent to the following equations:

(5.11)
$$\phi_i^*(\alpha)[\{m_2^u(\alpha) + \tilde{m}_2^u(\alpha, \nu, \eta_2^u(\alpha, \nu, \epsilon), \epsilon)\}$$
$$- \{m_2^s(\alpha) + \tilde{m}_2^s(\alpha, \nu, \eta_2^s(\alpha, \nu, \epsilon), \epsilon)\}] = 0, \ i = 1, \ldots, m,$$

where $\phi_i^*(\alpha)$ is the transpose of $\phi_i(\alpha)$. Since

$$\phi_i(t) = \Phi^*(\alpha, t)\phi_i(\alpha), \ i = 1, \ldots, m,$$

and

$$\phi_i^*(\alpha)Q(\alpha) = \phi_i^*(\alpha)(I - P(\alpha)) = \phi_i^*(\alpha), \ i = 1, \ldots, m,$$

(5.11) becomes

$$0 = \phi_i^*(\alpha)\{Q(\alpha)\int_{-\infty}^{\alpha} \Phi(\alpha, t)g(t - \alpha, \gamma(t))dt$$
$$+ Q(\alpha)\int_{-\infty}^{\alpha} \Phi(\alpha, t)h(t, z^u(\nu)(t), \alpha, \epsilon)dt$$
$$- (I - P(\alpha))\int_{\infty}^{\alpha} \Phi(\alpha, t)g(t - \alpha, \gamma(t)dt$$
$$- (I - P(\alpha))\int_{\infty}^{\alpha} \Phi(\alpha, t)h(t, z^s(\nu)(t), \alpha, \epsilon)dt\}$$
$$= \int_{-\infty}^{\infty} \phi_i^*(t)g(t - \alpha, \gamma(t))dt$$
$$+ \phi_i^*(\alpha)\{\int_{-\infty}^{\alpha} \Phi(\alpha, t)h(t, z^u(\nu)(t), \alpha, \epsilon)dt$$
$$+ \int_{\alpha}^{\infty} \Phi(\alpha, t)h(t, z^s(\nu)(t), \alpha, \epsilon)dt\},$$

$i = 1, \ldots, m$. Here $z^u(\nu)(t) \equiv z(t; \alpha, \xi^u(\nu, \eta_2^u(\alpha, \nu, \epsilon)))$ is the solution of (3.4) and η_2^u is given in (5.6). Similarly $z^s(\nu)(t) \equiv z(t; \alpha, \xi^s(\nu, \eta_2^s(\alpha, \nu, \epsilon)))$ is the solution of (3.3) and η_2^s is given in (5.7). Now it is reasonable to define the following quantities.

Definition 5.1. The Melnikov vector $M(\alpha, \nu, \epsilon)$ for system (2.2) is defined by

$$(5.12) \qquad M(\alpha, \nu, \epsilon) = (M_1(\alpha, \nu, \epsilon), \ldots, M_m(\alpha, \nu, \epsilon))$$

where

$$
\begin{aligned}
(5.13) \qquad M_i(\alpha, \nu, \epsilon) = & \int_{-\infty}^{\infty} \phi_i^*(t) g(t - \alpha, \gamma(t)) dt \\
& + \phi_i^*(\alpha) \{ \int_{-\infty}^{\alpha} \Phi(\alpha, t) h(t, z^u(\nu)(t), \alpha, \epsilon) dt \\
& + \int_{\alpha}^{\infty} \Phi(\alpha, t) h(t, z^s(\nu)(t), \alpha, \epsilon) dt \}, \quad i = 1, \ldots, m.
\end{aligned}
$$

The first approximation $\hat{M}(\alpha)$ of the Melnikov vector $M(\alpha)$ is defined by

$$(5.14) \qquad \hat{M}(\alpha) = (\hat{M}_1(\alpha), \ldots, \hat{M}_m(\alpha))$$

where

$$(5.15) \qquad \hat{M}_i(\alpha) = \int_{-\infty}^{\infty} \phi_i^*(t) g(t - \alpha, \gamma(t)) dt, \quad i = 1, \ldots, m.$$

Remark 5.2. $M(\alpha, \nu, \epsilon) = \hat{M}(\alpha) + O(|\epsilon|)$ uniformly in ν.

Remark 5.3. By (5.10) the dimension m of the Melnikov vector for system (2.2) is the same as the number of linearly independent bounded solutions of the adjoint system (5.9).

Remark 5.4. The above argument to derive the Melnikov vector is essentially the same as the Lyapunov-Schmidt reduction. However, we employed the above more elementary and geometrical argument which will be useful when we derive the condition for the transversal intersection.

Next we consider transversality of the intersection of the stable and unstable manifolds. Recall from Remark 3.2 that $W^u_{loc}(\bar{x}_-, \epsilon)$ and $W^s_{loc}(\bar{x}_+, \epsilon)$ are diffeomorphic to the graphs $F^u(\alpha, \eta^u_1, \eta^u_2)$ and $F^s(\alpha, \eta^s_1, \eta^s_2)$ respectively in a tabular neighborhood of γ which are given by

$$
F^u(\alpha, \eta^u_1, \eta^u_2) = \begin{pmatrix} \alpha \\ \eta^u_1 \\ \eta^u_2 \\ m^u_1(\alpha) + \tilde{m}^u_1(\alpha, \eta^u_1, \eta^u_2) \\ m^u_2(\alpha) + \tilde{m}^u_2(\alpha, \eta^u_1, \eta^u_2) \end{pmatrix} \begin{matrix} \in \mathbb{R} \\ \\ \in \mathrm{Range}(I - Q(\alpha)) \\ \\ \\ \in \mathrm{Range}\, Q(\alpha) \\ \\ \end{matrix}
$$

and

$$
F^s(\alpha, \eta^s_1, \eta^s_2) = \begin{pmatrix} \alpha \\ \eta^s_1 \\ m^s_1(\alpha) + \tilde{m}^s_1(\alpha, \eta^s_1, \eta^s_2) \\ \eta^s_2 \\ m^s_2(\alpha) + \tilde{m}^s_2(\alpha, \eta^s_1, \eta^s_2) \end{pmatrix} \begin{matrix} \in \mathbb{R} \\ \\ \in \mathrm{Range}\, P(\alpha) \\ \\ \in \mathrm{Range}(I - P(\alpha)) \\ \\ \end{matrix}
$$

It is clear that transversal intersection occurs when column vectors in the following matrices $D_{(\alpha, \eta^u_1, \eta^u_2)} F^u$ and $D_{\alpha, \eta^s_1, \eta^s_2} F^s$ at a point of intersection span the whole space

$\mathbb{R}^n$:

$$(5.16) \qquad D_{(\alpha,\eta_1^u,\eta_2^u)}F^u = \begin{pmatrix} 1 & 0 & 0 \\ 0 & I & 0 \\ 0 & 0 & I \\ \frac{\partial}{\partial \alpha}(m_1^u + \tilde{m}_1^u) & \frac{\partial}{\partial \eta_1^u}\tilde{m}_1^u & \frac{\partial}{\partial \eta_2^u}\tilde{m}_1^u \\ \frac{\partial}{\partial \alpha}(m_2^u + \tilde{m}_2^u) & \frac{\partial}{\partial \eta_1^u}\tilde{m}_2^u & \frac{\partial}{\partial \eta_2^u}\tilde{m}_2^u \end{pmatrix} \begin{matrix} (a) \\[12pt] (b) \\[12pt] (c) \end{matrix}$$

$$\qquad\qquad (1) \qquad\qquad (2) \qquad\qquad (3)$$

$$(5.17) \qquad D_{(\alpha,\eta_1^s,\eta_2^s)}F^s = \begin{pmatrix} 1 & 0 & 0 \\ 0 & I & 0 \\ \frac{\partial}{\partial \alpha}(m_1^s + \tilde{m}_1^s) & \frac{\partial}{\partial \eta_1^s}\tilde{m}_1^s & \frac{\partial}{\partial \eta_2^s}\tilde{m}_1^s \\ 0 & 0 & I \\ \frac{\partial}{\partial \alpha}(m_2^s + \tilde{m}_2^s) & \frac{\partial}{\partial \eta_1^s}\tilde{m}_2^s & \frac{\partial}{\partial \eta_2^s}\tilde{m}_2^s \end{pmatrix} \begin{matrix} \\[12pt] (d) \\[12pt] (e) \\[12pt] (f) \end{matrix}$$

$$\qquad\qquad (4) \qquad\qquad (5) \qquad\qquad (6)$$

Now we have the main result in this paper.

Theorem 5.5. *Assume that system (2.1) has two hyperbolic critical points x_+ and x_- (not necessarily distinct) and has an orbit γ connecting them: $\gamma(t) \to x_+$ as $t \to \infty$ and $\gamma(t) \to x_-$ as $t \to -\infty$. Let $k = \dim\{\mathcal{R}(I - Q(\alpha)) \cap \mathcal{R}(P(\alpha))\}$, $m = \dim\{\mathcal{R}(I - Q(\alpha)) + \mathcal{R}(P(\alpha))\}^\perp$ and let $\nu = (\nu_1, \ldots, \nu_{k-1}) \in \{\mathcal{R}(I - Q(\alpha)) \cap \mathcal{R}(P(\alpha))\} \cap T_{\gamma(\alpha)}^\perp \mathbb{R}^n$, where $P(\alpha)$ and $Q(\alpha)$ are respectively projections of exponential dichotomies on $(\alpha, \infty]$ and on $[-\infty, \alpha)$ of system (3.2) and satisfy the conditions in (3.9). Consider the perturbed* system (2.2) and define the Melnikov vector $M(\alpha, \nu, \epsilon)$ and its first approximation $\hat{M}(\alpha)$ by (6.30)-(6.33). Then

(i) the cross sections $W^s(\bar{x}_+, \epsilon)$ and $W^u(\bar{x}_-, \epsilon)$ at $t = 0$ of the time dependent stable and unstable manifolds of system (2.2) intersect each other if and only if $M(\alpha_0, \nu_0, \epsilon) = 0$ for some α_0 and ν_0, provided ϵ is sufficiently small.

In this case,

(ii) the intersection of $W^u_{loc}(\bar{x}_-, \epsilon)$ and $W^s_{loc}(\bar{x}_+, \epsilon)$ are transversal if there exist m nonzero column vectors in the $m \times k$ matrix

$$[\frac{d}{d\alpha}\hat{M}(\alpha_0) \frac{\partial}{\partial \nu} M(\alpha_0, \nu_0, \epsilon)].$$

Proof. (i) If $M(\alpha_0, \mu_0, \epsilon) = 0$ for some α_0 and ν_0, then it is obvious from the definition of the Melnikov vector that $W^s_{loc}(\bar{x}_+, \epsilon)$ and $W^u_{loc}(\bar{x}_-, \epsilon)$ intersect each other. Conversely, once $W^s(\bar{x}_+, \epsilon)$ and $W^u(x_-, \epsilon)$ intersect, then there is a bi-infinite sequence $\{p_i\}_{i=-\infty}^{\infty}$ of points of intersection which approaches $\bar{x}_+$ and $\bar{x}_-$ as $i \to +\infty$ and $i \to -\infty$ respectively. Hence for sufficiently large $|i|, p_i \in W^s_{loc}(\bar{x}_+, \epsilon) \cap W^u_{loc}(\bar{x}_-, \epsilon)$ which implies that $M(\alpha_0, \nu_0, \epsilon) = 0$ for some α_0 and ν_0.

(ii) Consider column vectors in matrices (5.16) and (5.17). Firstly, it is clear that all column vectors in blocks (3) and (6) are always linearly independent. Secondly, by (5.11), $M(\alpha_0, \nu_0, \epsilon) = 0$ implies that

$$|m_2^u(\alpha_0) - m_2^s(\alpha_0)| = |\tilde{m}_2^s(\alpha_0, \nu_0, \eta_2^s(\alpha_0, \epsilon), \epsilon)$$

$$- \tilde{m}_2^u(\alpha_0, \nu_0, \eta_2^u(\alpha_0, \nu_0, \epsilon), \epsilon)| = O(|\epsilon|).$$

So

$$\frac{\partial}{\partial \alpha}\hat{M}(\alpha_0) = \begin{bmatrix} \dot{\phi}_1^*(\alpha_0)(m_2^u(\alpha_0) - m_2^s(\alpha_0)) \\ \vdots \\ \dot{\phi}_m^*(\alpha_0)(m_2^u(\alpha_0) - m_2^s(\alpha_0)) \end{bmatrix} + \begin{bmatrix} \phi_1^*(\alpha_0)\frac{\partial}{\partial \alpha}(m_2^u(\alpha_0) - m_2^s(\alpha_0)) \\ \vdots \\ \phi_m^*(\alpha_0)\frac{\partial}{\partial \alpha}(m_2^u(\alpha_0) - m_2^s(\alpha_0)) \end{bmatrix}$$

$$= \begin{bmatrix} \phi_1^*(\alpha_0)\frac{\partial}{\partial\alpha}(m_2^u(\alpha_0)-m_2^s(\alpha_0)) \\ \vdots \\ \phi_m^*(\alpha_0)\frac{\partial}{\partial\alpha}(m_2^u(\alpha_0)-m_2^s(\alpha_0)) \end{bmatrix} + O(|\epsilon|).$$

Therefore if $\frac{\partial}{\partial\alpha}\hat{M}(\alpha_0) \neq 0$, then $\frac{\partial}{\partial\alpha}(m_2^u(\alpha_0)-m_2^s(\alpha_0)) \neq 0$ and hence column vectors (1) and (4) are linearly independent at the point of intersection. Finally

$$\frac{\partial}{\partial\nu_j}M(\alpha_0,\nu_0,\epsilon)$$

$$= \begin{bmatrix} \phi_1^*(\alpha_0)\frac{\partial}{\partial\nu_j}\{\tilde{m}_2^u(\alpha_0,\nu_0,\eta_2^u(\alpha_0,\nu_0,\epsilon)\epsilon)-\tilde{m}_2^s(\alpha_0,\nu_0,\eta_2^s(\alpha_0,\nu_0,\epsilon)\epsilon))\} \\ \vdots \\ \phi_m^*(\alpha_0)\frac{\partial}{\partial\nu_j}\{\tilde{m}_2^u(\alpha_0,\nu_0,\eta_2^u(\alpha_0,\nu_0,\epsilon)\epsilon)-\hat{m}_2^s(\alpha_0,\nu_0,\eta_2^s(\alpha_0,\nu_0,\epsilon)\epsilon))\} \end{bmatrix},$$

$$j=1,\ldots,k-1.$$

Therefore if $\frac{\partial}{\partial\nu_j}M(\alpha_0,\nu_0,\epsilon) \neq 0$, then $\frac{\partial}{\partial\nu_j}(\tilde{m}_2^u-\tilde{m}_2^s) \neq 0$ and hence j-th column vectors in blocks (2) and (5) are linearly independent at the point of intersection.

Now consider the following decomposition of $T_{\gamma(\alpha)}\mathbb{R}^n$.

(5.18)
$$T_{\gamma(\alpha)}\mathbb{R}^n = \{\mathcal{R}(I-Q(\alpha))\cap\mathcal{R}(P(\alpha))\} \oplus \{\mathcal{R}(I-Q(\alpha))\cap\mathcal{R}(I-P(\alpha))\}$$
$$\oplus \{\mathcal{R}(Q(\alpha))\cap\mathcal{R}(P(\alpha))\} \oplus \{\mathcal{R}(Q(\alpha))\cap\mathcal{R}(I-P(\alpha))\}$$

with the following dimensions:

$$\dim\{\mathcal{R}(I-Q(\alpha))\cap\mathcal{R}(P(\alpha))\} = k,$$
$$\dim\{\mathcal{R}(I-Q(\alpha))\cap\mathcal{R}(I-P(\alpha))\} = n-n_--k,$$
$$\dim\{\mathcal{R}(Q(\alpha))\cap\mathcal{R}(P(\alpha))\} = n_+-k,$$
$$\dim\{\mathcal{R}(Q(\alpha))\cap\mathcal{R}(I-P(\alpha))\} = m,$$

where $n_- = \dim W^s(x_-)$ and $n_+ = \dim W^s(x_+)$.

See diagram below.

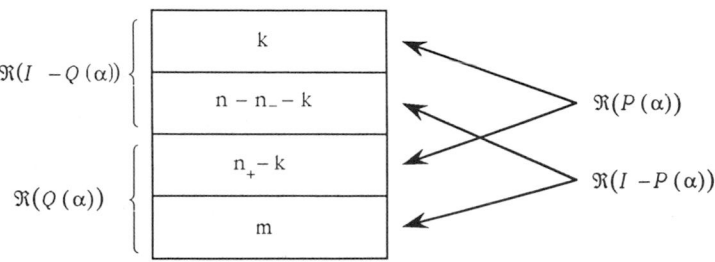

Diagram 2

From the above argument, $(n - n_- - k) + (n_+ - k) = (n - m - k)$ column vectors in blocks (3) and (6) are linearly independent. The condition of the theorem and the above argument imply that there are $(k-m)+2m = k+m$ column vectors in blocks (1), (2), (4) and (5) which are linearly independent and so we have $(n-m-k)+(k+m) = n$ linearly independent column vectors in matrices (5.16) and (5.17). ∎

Remark 5.6. It can easily be shown, by the implicit mapping theorem, that if $\hat{M}(\alpha_1) = 0$ and $\frac{\partial}{\partial \alpha}\hat{M}(\alpha_1) \neq 0$ for some α_1, then $M(\alpha, \nu, \epsilon) = 0$ for some α and ν.

Remark 5.7. The condition for transveral intersections in part (ii) of the theorem can not be, in general, the necessary and sufficient condition for transversal intersections since the Melnikov vector is the projection of the distance vector (5.2) to the subspace $\mathcal{R}(Q(\alpha)) \cap \mathcal{R}(I - P(\alpha))$. See the decomposition in (5.18). In other words, the Melnikov vector contributes only to the last components of column vectors, i.e. components (c) and (f) in matrices (5.16) and (5.17). See Example 3 in Section 11.

However, if

(5.19) $\mathcal{R}(P(\alpha)) = \mathcal{R}(I - Q(\alpha)),$

the components (a), (b), (d) and (e) of column vectors of matrices (5.16) and (5.17) are
missing and hence the Melnikov vector can be used to give a necessary and sufficient
condition for transversal intersection. Thus, we have the following

Corollary 5.8. Under the same assumption as in Theorem 5.5 and assumption
(5.19), the intersection of $W_{loc}^u(\bar{x}_-, \epsilon)$ and $W_{loc}^s(\bar{x}_+, \epsilon)$ is transversal if and only if
$M(\alpha_0, \nu_0, \epsilon) = 0$ for some α_0 and ν_0 and there exist m nonzero column vectors in the
$m \times k$ matrix

$$[\frac{d}{d\alpha}\hat{M}(\alpha_0)\frac{\partial}{\partial\nu}M(\alpha_0, \nu_0, \epsilon)].$$ ∎

We remark that condition (5.19) implies $m = n - k$.

§6.   THE SPLITTING INDEX OF γ AND THE DIMENSION OF THE MELNIKOV VECTOR

In this section we define an index of a homoclinic or heteroclinic orbit γ. Then
this index is used to classify the possible cases of the transversal intersection of the
stable and unstable manifolds. We also interpret this index as the Fredholm index of
a certain operator.

Consider system (2.1) and assume that $\gamma \subset W^u(x_-) \cap W^s(x_+)$ is a homoclinic or
heteroclinic orbit joining hyperbolic critical points $x_\pm$ of system (2.1).

Definition 6.1. The splitting index $\delta(\gamma)$ of a homoclinic or heteroclinic orbit γ is defined by

(6.1) $\delta(\gamma) = \dim W^s(x_-) - \dim W^s(x_+).$

Notice from (5.10) and diagram 1 in §5 that the dimension m of the Melnikov vector
is given by

$$m = n - [\{n - \dim W^s(x_-)\} + \dim W^s(x_+) - k].$$

Thus m, k and $\delta(\gamma)$ satisfy the following relation.

(6.2) $m = k + \delta(\gamma).$

The splitting index $\delta(\gamma)$ expresses the 'degeneracy' of γ in the following sense.
Here we distinguish the homoclinic and heteroclinic cases.

(i) Suppose that γ is a homoclinic orbit. Then the splitting index $\delta(\gamma)$ is always
zero. Thus by theorem 5.5(ii), all of k column vectors in $\left[\frac{\partial}{\partial \alpha} \cap M \frac{\partial}{\partial \nu} M\right]$ must be
nonzero to guarantee the transversal intersection. This situation is only the case in
the homoclinic case.

(ii) Suppose that γ is a heteroclinic orbit in $\mathbb{R}^n$. In this case we have three
different situations.

(ii-1) $\delta(\gamma) > 0$. Then $m = k + \delta(\gamma) > k$. Thus theorem 5.5(ii) implies that
there is no transversal intersection because the matrix $\left[\frac{\partial}{\partial \alpha} \hat{M} \frac{\partial}{\partial \nu} M\right]$ is of size $m \times k$.
A reason for this is that $\dim W^s(x_+) < \dim W^s(x_-)$ is equivalent to saying that
$\dim W^u(x_-) + \dim W^s(x_+) < n$.

(ii-2) $\delta(\gamma) = 0$. In this case we have the same situation as in the homoclinic case.

(iii-3) $\delta(\gamma) < 0$. Then $m = k + \delta(\gamma) < k$. Thus transversal intersection is possible.

In this way, we can classify the possibility and impossibility of transversal intersection
by using the splitting index $\delta(\gamma)$.

Though the splitting index $\delta(\gamma)$ is defined by local data, that is, the dimensions of stable manifolds of hyperbolic critical points, $\delta(\gamma)$ is global in nature since it can be used to distinguish homoclinic and heteroclinic orbits, and also used to classify heteroclinic orbits. Furthermore, the relation (6.2) shows how the dimension of the Melnikov vector depends on γ.

Next we shall show that the splitting index $\delta(\gamma)$ is the Fredholm index of a certain differential operator defined below.

Let $A(t) = Df(\gamma(t))$ and define a differential operator $L : BC^1(\mathbb{R}, \mathbb{R}^n) \to BC^0(\mathbb{R}, \mathbb{R}^n)$ by

$$(6.3) \qquad\qquad (Lz)(t) = \dot{z}(t) - A(t)z,$$

where $BC^1(\mathbb{R}, \mathbb{R}^n)$ is the space of bounded C^1 functions from $\mathbb{R}$ to $\mathbb{R}^n$ and $BC^0(\mathbb{R}, \mathbb{R}^n)$ is the space of bounded continuous functions from $\mathbb{R}$ to $\mathbb{R}^n$. As we shall show, L is a Fredholm operator. See also Palmer [16]. Recall that the Fredholm index of a linear operator L is defined by

$$\text{index } L = \dim(\ker L) - \text{codim}(\text{Range } L).$$

Proposition 6.2. The differential operator L defined in (6.3) is Fredholm and $\delta(\gamma) = -\text{index} \quad L$.

Proof. Define a bounded linear operator $A : BC^0(\mathbb{R}, \mathbb{R}^n) \to \mathbb{R}^m$ by

$$Ag = (\int_{-\infty}^{\infty} \phi_1^*(t)g(t)(dt, \ldots, \int_{-\infty}^{\infty} \phi_m^*(t)g(t)dt)$$

where $\phi_i, i = 1, \ldots, m$ are independent bounded solutions of the adjoint system $\dot{\phi} + A^*(t)\phi = 0$. Then Lemma 4.1 and 4.2 imply that Range L = Kernel A, which means that Range L is closed and

$$\text{codim}(\text{Range } L) = m,$$

and hence L is a Fredholm operator. Thus

$$\text{index } L = \dim(\ker L) - \text{codim}(\text{Range } L)$$

$$= k - m$$

$$= -\delta(\gamma). \qquad \blacksquare$$

Remark 6.3. The splitting index $\delta(\gamma)$ in (6.1) was defined in Sacker [18] in a more general setting and Proposition 6.2 was also proved there. However, his definition is for linear systems. Our definition of the splitting index is to relate a local information about eigenvalues of $Df(x_\pm)$ to a global information about a homoclinic or a heteroclinic orbit.

§7 Computation of Higher Order Terms

In the case of $\dim\{\mathcal{R}(P)(\alpha)) \cap \mathcal{R}(I - Q(\alpha))\} > 1$ for n-dimensional systems ($n \geq 3$), we need to know nonlinear terms in expression (5.13) of the Melnikov vector to examine the transversality condition. To this end, we consider again bounded solutions on $[\alpha, \infty)$ and on $(-\infty, \alpha]$ of system (3.1). We use the same assumption of exponential dichotomies as in §3. These bounded solutions are given as unique solutions of integral equations (3.3) and (3.4) respectively.

Let $\eta^s \in \mathcal{R}(P(\alpha)), |\eta^s| << 1$, and let $z(\eta^s)(t)$ be the unique solution of (3.3) which is guaranteed by the contraction mapping principle. That is, $z(\eta^s)(t)$ is the solution of the following integral equation:

$$(7.1) \qquad z(t) = \mathcal{T}_s(\eta^s)(g(t - \alpha, \gamma(t)) + h(t, z(t), \alpha, \epsilon)),$$

where the operator $T_s(\eta^s)$ is defined by

$$T_s(\eta^s)(g(t - \alpha, \gamma(t)) + h(t, z(t), \alpha, \epsilon))$$

$$= \Phi(t, \alpha)\eta^s + \Phi(t, \alpha)P(\alpha) \int_\alpha^t \Phi(\alpha, \tau)\{g(\tau - \alpha, \gamma(\tau)) + h(\tau, z(\tau), \alpha, \epsilon)\}d\tau$$

$$+ \Phi(t, \alpha)(I - P(\alpha)) \int_\infty^t \Phi(\alpha, \tau)\{g(\tau - \alpha, \gamma(\tau)) + h(\tau, z(\tau), \alpha, \epsilon)\}d\tau.$$

To approximate $z(\eta^s)(t)$, we use the following iteration scheme:

$$z_s^{(n+1)}(\eta^s)(t) = T_s(\eta^s)(g(t - \alpha, \gamma(t)) + h(t, z_s^{(n)}(\eta^s)(t), \alpha, \epsilon)).$$

Set $z_s^{(0)}(\eta^s)(t) \equiv 0$. Then

$$z_s^{(1)}(\eta^s)(t) = T_s(\eta^s)(g(t - \alpha, \gamma(t))$$

and

$$z_s^{(2)}(\eta^s)(t) = T_s(\eta^s)(g(t - \alpha, \gamma(t)) + h(t, T_s(\eta^s)(g(t - \alpha, \gamma(t)), \alpha, \epsilon)).$$

Notice that $T_s(\eta^s)(h(t, T_s(\eta^s)(g(t - \alpha, \gamma(t), \alpha, \epsilon)) = O(|\epsilon|)$ and hence

$$z_s^{(2)}(\eta^s)(t) = z_s^{(1)}(\eta^s)(t) + \epsilon \tilde{z}_s^{(1)}(\eta^s)(t)$$

for some function $\tilde{z}_s^{(1)}(\eta^s)(t)$. The true solution $z(\eta^s)(t)$ of (7.1) satisfies

$$z(\eta^s)(t) = z_s^{(2)}(\eta^s)(t) + O(|\epsilon|^2)$$

$$= z_s^{(1)}(\eta^s)(t) + \epsilon \tilde{z}_s^{(1)}(\eta^s)(t) + O(|\epsilon|^2), t \geq \alpha.$$

Apparently $z_s^{(1)}(\eta^s)(t)$ is the bounded solution on $[\alpha, \infty)$ of linear system $\dot{z} = A(t)z + g(t - \alpha, \gamma(t))$.

Similarly, define $T_u(\eta^u)$ by

$$T_u(\eta^u)(g(t - \alpha, \gamma(t) + h(t, z(t), \alpha, \epsilon))$$

$$= \Phi(t, \alpha)\eta^u + \Phi(t, \alpha)(I - Q(\alpha)) \int_\alpha^t \Phi(\alpha, \tau)\{g(\tau - \alpha, \gamma(\tau)) + h(\tau, z(\tau), \alpha, \epsilon)\}d\tau$$

$$+ \Phi(t, \alpha)Q(\alpha) \int_{-\infty}^t \Phi(\alpha, \tau)\{g(\tau - \alpha, \gamma(\tau)) + h(\tau, z(\tau), \alpha, \epsilon)\}d\tau,$$

where $\eta^u \in \mathcal{R}(I - Q(\alpha)), |\eta^u| << 1$ and let $z(\eta^u)(t)$ be the unique solution of (4.17). That is, $z(\eta^u)(t)$ is the unique solution of

$$(7.2) \qquad z(t) = \mathcal{T}_u(\eta^u)(g(t - \alpha, \gamma(t)) + h(t, z(t), \alpha, \epsilon)).$$

We use the following iteration scheme.

$$z_u^{(n+1)}(\eta^u)(t) = \mathcal{T}_u(\eta^u)(g(t - \alpha, \gamma(t)) + h(t, z_u^{(n)}(\eta^u)(t), \alpha, \epsilon)).$$

By setting $z_u^{(0)}(\eta^u)(t) \equiv 0$, we have

$$z_u^{(1)}(\eta^u)(t) = \mathcal{T}_u(\eta^u)(g(t - \alpha, \gamma(t)),$$

$$z_u^{(2)}(\eta^u)(t) = z_u^{(1)}(\eta^u)(t) + \epsilon \tilde{z}_u^{(1)}(\eta^u)(t),$$

and the true solution $z(\eta^u)(t)$ satisfies

$$z(\eta^u)(t) = z_u^{(1)}(\eta^u)(t) + \epsilon \tilde{z}_u^{(1)}(\eta^u)(t) + O(|\epsilon|^2), t \geq \alpha.$$

Notice that, by taking $\nu \equiv \eta^s = \eta^u \in \mathcal{R}(P(\alpha)) \cap \mathcal{R}(I - Q(\alpha)), z_s^{(1)}(\nu)(t)$ and $z_u^{(1)}(\nu)$ give the first approximation of the Melnikov vector, and $\tilde{z}_s^{(1)}(\nu)(t)$ and $\tilde{z}_u^{(1)}(\nu)(t)$ give the term of order ϵ or higher in the Melnikov vector. Thus we have derived the following expression of the Melnikov vector:

$$(7.3) \qquad \begin{aligned} M_i(\alpha, \nu) = \hat{M}_i(\alpha) + \epsilon \hat{\phi}_i^*(\alpha)\{ \int_{-\infty}^{\alpha} \Phi(\alpha, t) h(t, z_n^{(1)}(\nu)(t), \alpha, \epsilon) dt \\ + \int_{\alpha}^{\infty} \Phi(\alpha, t) h(t, z_s^{(1)}(\nu)(t), \alpha, \epsilon) dt\} + O(|\epsilon|^2). \end{aligned}$$

In this way we can compute the Melnikov vector in arbitraily high order of accuracy. It is also clear that it is sufficient to consider the first two terms in expression (7.3) for the transversality condition in Theorem 5.5.

§8 THE FIRST APPROXIMATION OF THE MELNIKOV VECTOR

In this section we consider some special cases for which the first approximation of the Melnikov vector gives sufficient information for transversal and tangential intersection of the stable and unstable manifolds. We consider system (2.1) and (2.2) under the same assumption as in §2.

Case (i). Suppose

$$(8.1) \qquad k = 1 \text{ and } \delta(\gamma) = 0.$$

where $k = \dim\{\mathcal{R}(I - Q(\alpha)) \cap \mathcal{R}(P(\alpha))\}$ and $\delta(\gamma)$ is defined in (6.1). This means that $\mathcal{R}(P(\alpha)) \cap \mathcal{R}(I - Q(\alpha)) = \text{span}\{\dot{\gamma}(\alpha)\}$ and $m = 1$.

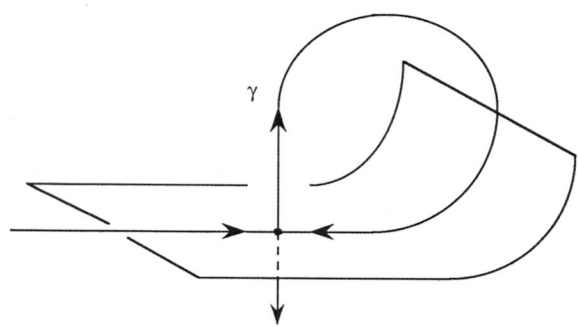

Figure 4

Note that the Melnikov function in this case is

$$(8.2) \qquad M(\alpha, \epsilon) = \hat{M}(\alpha) + O(|\epsilon|).$$

Proposition 8.1. Assume (8.1) and suppose that there exists $\alpha_0 \in \mathbb{R}$ such that

(8.3) $$\hat{M}(\alpha_0) = 0 \text{ and } \frac{d}{d\alpha}\hat{M}(\alpha_0) \neq 0.$$

Then $W_{loc}^u(x_-, \epsilon)$ and $W_{loc}^s(x_+, \epsilon)$ of system (2.2) have a point of transversal intersection.

Proof. By the implicit function theorem, condition (8.3) combining (8.2) implies $M(\alpha, \epsilon) = 0$ and $\frac{d}{d\alpha}M(\alpha, \epsilon) \neq 0$ for some α near α_0. Hence this proposition follows from Theorem 5.5. ∎

Note: Condition (8.3) can not be a necessary and sufficient condition for transversal intersection.

Apparently the two-dimensional case satisfies condition (8.1). In this special case, we have the following corollary.

Corollary 8.2. Suppose that system (2.1) and (2.2) are two-dimensional. Then $W_{loc}^s(x_+, \epsilon)$ and $W_{loc}^u(x_-, \epsilon)$ of system (2.2) have a point of transversal intersection if and only if there exists $\alpha_0 \in \mathbb{R}$ so that

(8.4) $$\hat{M}(\alpha_0) = 0 \text{ and } \frac{d}{d\alpha}\hat{M}(\alpha_0) \neq 0.$$

Proof. 'If' part is a special case of Proposition 8.1. Conversely, if transversal intersection exists, then by Corolloary 5.8, there exists α_1 such that $M(\alpha_1, \epsilon) = 0$ and $\frac{d}{d\alpha}M(\alpha_1, \epsilon) \neq 0$. Using (8.2), the conclusion follows from the implicit function theorem. ∎

Case (ii). Suppose that

(8.5) $\{\gamma(t,\nu) : t \in \mathbb{R}, \nu \in S \subset \mathbb{R}^{k-1}\} \subset W^u(x_-) \cap W^s(x_+)$

where S is an open subset of $\mathbb{R}^{k-1}$ and $\gamma(t,\nu)$ is a homoclinic or heteroclinic orbit connecting x_- and x_+ for each $\nu \in S$.

In other words, the 'homoclinic or heteroclinic manifold' is locally parametrized by $(t,\nu) \in \mathbb{R} \times S$. This case can occur when the system and its perturbation have some symmetry properties. See example 2 in §11.

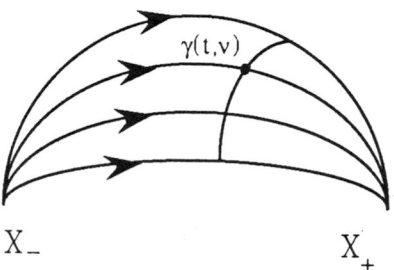

Figure 5

In this case the first approximation of the Melnikov vector has the form

$$\hat{M}_i(\alpha,\nu) = \int_{-\infty}^{\infty} \phi_i^*(t)g(t-\alpha,\gamma(t,\nu))dt, i = 1,\ldots,m.$$

Note that

$$M(\alpha, \nu, \epsilon) = \hat{M}(\alpha, \nu) + O(|\epsilon|).$$

Hence we have

Proposition 8.3. Assume (10.5) and suppose that there exist α_0 and ν_0 such that

(8.6) $$\hat{M}(\alpha_0, \nu_0) = 0$$

and

(8.7) $$\text{rank}\big[\frac{d}{d\alpha}\hat{M}(\alpha_0, \nu_0)\frac{\partial}{\partial\nu}\hat{M}(\alpha_0, \nu_0)\big] = m.$$

Then $W_{loc}^u(x_-, \epsilon)$ and $W_{loc}^s(x_+, \epsilon)$ of system (2.2) have a point of transversal intersection.

Proof. By the implicit mapping theorem, we have $M(\alpha_1, \nu_1, \epsilon) = 0$ and $\text{rank}\big[\frac{\partial}{\partial\nu}M(\alpha_1, \nu_1, \epsilon)\frac{\partial}{\partial\nu}M(\alpha_1, \nu_1, \epsilon)\big] = m$ for (α_1, ν_1) near (α_0, ν_0). Then the statement follows from Theorem 5.5. ∎

Remark 8.4. In the case $m = 1$, the rank condition (8.7) gives a necessary and sufficient condition for transversal intersection.

Next we turn to the tangency condition. Here a tangential intersection of the stable and unstable manifolds means that the tangent spaces of the stable and unstable manifolds at a point of intersection do not span the whole space. Our discussion of tangency is based on Corollary 5.8. Since Corollary 5.8 gives a necessary and sufficient condition for transversality, we consider the situations in which the condition in Corollary 5.8 is violated.

We consider the following system with parameters.

$$(8.8) \qquad\qquad \dot{x} = f(x) + \epsilon g(t, x, \mu)$$

where $x \in \mathbb{R}^n, \mu \in \mathbb{R}^N, \epsilon \ll 1, f$ and g are sufficiently smooth in all arguments, and g is periodic in t. Assume, as before, that the unperturbed system ($\epsilon = 0$) has a homoclinic or heteroclinic orbit $\gamma(t)$.

Recall, first of all, that $m = k + \delta(\gamma)$. Thus it is clear that if $\delta(\gamma) > 0$, then intersection is always tangential (see §6).

Assume $\delta(\gamma) \le 0$ and assume that

$$(8.9) \qquad\qquad \mathcal{R}(P(\alpha)) = \mathcal{R}(I - Q(\alpha))(= k).$$

We consider only several special cases here. Extension to more general cases is straightforward.

(i) Assume $m = 1$.

This case includes, e.g., $k = 1, \delta(\gamma) = 0$ in $\mathbb{R}^2$ and $k = 2, \delta(\gamma) = -1$ in $\mathbb{R}^3$. We also assume that $N \ge k$.

Proposition 8.5. Suppose that

$$(8.10) \qquad\qquad \hat{M}(\alpha_0, \mu_0) = \frac{\partial}{\partial \alpha} \hat{M}(\alpha_0, \mu_0) = 0,$$

$$(8.11) \qquad\qquad \frac{\partial^2}{\partial \alpha^2} \hat{M}(\alpha_0, \mu_0) \ne 0$$

and

$$(8.12) \qquad\qquad \frac{\partial}{\partial \mu} \hat{M}(\alpha_0, \mu_0) \text{ has rank } k.$$

Then there exists a point of tangential intersection for sufficiently small ϵ.

Proof. Define

$$F(\alpha, \nu, \mu, \epsilon) = (M(\alpha, \nu, \mu, \epsilon), \frac{\partial}{\partial \alpha}M(\alpha, \nu, \mu, \epsilon), \frac{\partial}{\partial \nu}M(\alpha, \nu, \mu, \epsilon)).$$

Note that $F : \mathbb{R}^{2k+1} \to \mathbb{R}^{k+1}$ and $F(\alpha_0, 0, \mu_0, 0) = 0$. Since conditions (8.10), (8,11) and (8.12) imply that the matrix

$$D_{(\alpha, \mu)}F(\alpha_0, 0, \mu_0, 0) = \begin{bmatrix} \frac{\partial}{\partial \alpha}\hat{M}(\alpha_0, \mu_0) & \frac{\partial}{\partial \mu}\hat{M}(\alpha_0, \mu_0) \\ \frac{\partial^2}{\partial \alpha^2}\hat{M}(\alpha_0, \mu_0) & \frac{\partial^2}{\partial \alpha \partial \mu}\hat{M}(\alpha_0, \mu_0) \end{bmatrix}$$

has rank $(k + 1)$, by the implicit mapping theorem there exist functions $\alpha(\nu, \epsilon)$ and $\mu(\nu, \epsilon)$ such that

$$F(\alpha(\nu, \epsilon), \nu, \mu(\nu, \epsilon)) = 0$$

for sufficiently small ν and ϵ. Hence, the condition in Corollary (5.8) is violated and the statement follows. ∎

See Wiggins and Holmes [21] for a similar result.

(ii) Assume that $m = 2$ and $k = 2$ (and hence $\delta(\gamma) = 0$). Assume also that $N \geq 3$.

Proposition 8.6. Suppose that

(8.13) $$\hat{M}(\alpha_0, \mu_0) = \frac{\partial}{\partial \alpha}\hat{M}(\alpha_0, \mu_0) = 0$$

and the matrix

(8.14) $$\begin{bmatrix} \frac{\partial}{\partial \alpha}\hat{M}(\alpha_0, \mu_0) & \frac{\partial}{\partial \mu}\hat{M}(\alpha_0, \mu_0) \\ \frac{\partial^2}{\partial \alpha^2}\hat{M}(\alpha_0, \mu_0) & \frac{\partial^2}{\partial \alpha \partial \mu}\hat{M}(\alpha_0, \mu_0) \end{bmatrix}$$

has rank 4.

Then there exists a point of tangential intersection for sufficiently small ϵ.

Proof. Define

$$F(\alpha, \nu, \mu, \epsilon) = (M(\alpha, \nu, \mu, \epsilon), \frac{\partial}{\partial \alpha} M(\alpha, \nu, \mu, \epsilon)).$$

Then the proof is identical to the one in Proposition 8.5 ∎

Next we consider a more special case.

(ii)$'$ Assume that $m = 2, k = 2$ and $N \geq 2$, and assume condition (8.5).

Proposition 8.7. Suppose that

$$(8.15) \qquad \hat{M}(\alpha_0, \nu_0, \mu_0) = \frac{\partial}{\partial \alpha} \hat{M}(\alpha_0, \nu_0, \mu_0) = 0$$

and the matrix

$$(8.16) \qquad \begin{bmatrix} \frac{\partial}{\partial \alpha} \hat{M} & \frac{\partial}{\partial \nu} \hat{M} & \frac{\partial}{\partial \mu} \hat{M} \\ \frac{\partial^2}{\partial \alpha^2} \hat{M} & \frac{\partial^2}{\partial \nu \partial \alpha} \hat{M} & \frac{\partial^2}{\partial \mu \partial \alpha} \hat{M} \end{bmatrix} \text{ or } \begin{bmatrix} \frac{\partial}{\partial \alpha} \hat{M} & \frac{\partial}{\partial \nu} \hat{M} & \frac{\partial}{\partial \mu} \hat{M} \\ \frac{\partial^2}{\partial \alpha \partial \nu} \hat{M} & \frac{\partial^2}{\partial \nu^2} \hat{M} & \frac{\partial^2}{\partial \mu \partial \nu} \hat{M} \end{bmatrix}$$

has rank 4 at (α_0, ν_0, μ_0).

Then there exists a point of tangential intersection for sufficiently small ϵ.

Proof. Similar to Proposition 8.6. ∎

§9 HAMILTONIAN SYSTEMS

In the case that the unperturbed system (2.1) is a completely integrable Hamiltonian system, the Melnikov vector takes particularly simple form even if a perturbation is non-Hamiltonian.

Assume that

$$(9.1) \qquad \dot{x} = X_H(x), \, x \in \mathbb{R}^{2n}$$

is completely integrable, and we consider its non-Hamiltonian and Hamiltonian perturbations

(9.2) $$\dot{x} = X_H(x) + \epsilon g(t, x),$$

and

(9.3) $$\dot{x} = X_H(x) + \epsilon X_G(t, x).$$

We shall derive the Melnikov vectors for systems (9.2) and (9.3).

We first recall some basic facts from Hamiltonian systems. Let $H \in C^{\infty}(\mathbb{R}^{2n})$.
Then the Hamiltonian vector field X_H with Hamiltonian H on $\mathbb{R}^{2n}$ is defined by

$$X_H(x) = J \nabla H(x)$$

where

$$J = \begin{bmatrix} 0 & I \\ -I & 0 \end{bmatrix} \text{ and } \nabla H(x) = \begin{bmatrix} \frac{\partial}{\partial x_1} H(x) \\ \vdots \\ \frac{\partial}{\partial x_{2n}} H(x) \end{bmatrix}.$$

Let $F_1, F_2 \in C^{\infty}(\mathbb{R}^{2n})$. The Poisson bracket $\{F_1, F_2\}$ of functions F_1 and F_2 is defined
by

$$\{F_1, F_2\}(x) = dF_1(x) X_{F_2}(x), \quad x \in \mathbb{R}^{2n}$$

where dF_1 is a differential 1-form on $\mathbb{R}^{2n}$.

One of the key facts on the Poisson bracket is the following:
$\{F_1, F_2\} = 0$ if and only if F_i is invariant under the flow of X_{F_j} where $(i, j) = (1, 2)$ or $(2, 1)$. See Arnold [1].

We suppose that system (9.1) has two hyperbolic critical points x_- and x_+ (not
necessarily distinct) joined by an orbit $\gamma(t)$ of system

(9.1) : $$\lim_{t \to \pm\infty} \gamma(t) = x_{\pm}.$$

Then the linearized system of (9.1) along the orbit γ is given by

(9.4) $\dot{x} = A(t)z, \ A(t) = DX_H(\gamma(t)).$

Since $A(t) = JD^2H(\gamma(t))$, $A(t)$ is infinitesimally symplectic for each $t \in \mathbb{R}$. Namely

(9.5) $A^*(t)J + JA(t) \equiv 0, \ t \in \mathbb{R}$

where $A^*(t)$ is the transpose of $A(t)$. Now let us define the adjoint system of (9.4) by

(9.6) $\dot{\phi} = -\phi A(t)$

where $\phi(t) \in T^*_{\gamma(t)}\mathbb{R}^{2n} \simeq (\mathbb{R}^{2n})^*$.

We note that $z(t)$ is a solution of (9.4) if and only if $\phi(t) = (J^{-1}z(t))^*$ is a solution of (9.6). This is clear from (9.5). We have the following

Proposition 9.1. Let $F \in C^\infty(\mathbb{R}^{2n})$. If $\{F, H\} = 0$, then $X_F(\gamma(t))$ is a solution of (9.4) and hence $dF(\gamma(t))$ is a solution of (9.6).

Proof. Let $\langle \cdot, \cdot \rangle$ be standard inner product on $\mathbb{R}^{2n}$. Then

$$\{F, H\} = \langle \nabla F, J \nabla H \rangle.$$

Since

$$0 = \nabla \langle \nabla F, J \nabla H \rangle$$
$$= D^2 F J \nabla H - D^2 H J \nabla F$$
$$= D^2 F X_H - D^2 H X_F,$$

we have

$$DX_F X_H = DX_H X_F.$$

Hence

$$\frac{d}{dt} X_F(\gamma(t)) = DX_F(\gamma(t)) X_H(\gamma(t))$$
$$= DX_H(\gamma(t)) X_F(\gamma(t))$$
$$= A(t) X_F(\gamma(t)).$$

Finally

$$(J^{-1}X_F(\gamma(t)))^* = dF(\gamma(t)). \qquad \blacksquare$$

One of the special situations of the Hamiltonian nature appears in the splitting index $\delta(\gamma)$ of γ. Since $DX_H(x_+) = \lim_{t\to\infty} A(t)$, $DX_H(x_+)$ is infinitesimally symplectic. Hence if $\gamma \in \sigma(DX_H(x_+))$, then $\bar{\lambda}, -\lambda, -\bar{\lambda} \in \sigma(DX_H(x_+))$, where $\sigma(DX_H(x_+))$ is the spectrum of $DX_H(x_+)$ and $\bar{\lambda}$ is the complex conjugate of λ. This symmetry property implies that both the stable and the unstable subspaces of $DX_H(x_+)$ have dimension n. Similarly, the stable and the unstable subspaces of $DX_H(x_-) = \lim_{t\to-\infty} A(t)$ have dimension n and hence we have

Proposition 9.2. Suppose that Hamiltonian system (9.1) has a homoclinic or heteroclinic orbit γ. Then $\delta(\gamma) = 0$. $\blacksquare$

Now we suppose that Hamiltonian system (9.1) is completely integrable. That is, there exist n C^∞-functions $F_1 = H, F_2, \ldots, F_n$ on $\mathbb{R}^{2n}$ which are in involution, namely $\{F_i, F_j\} = 0$ for $1 \le i, j \le n$, and dF_i, $i = 1, \ldots, n$, are linearly independent everywhere in $\mathbb{R}^{2n} - \{x_\pm\}$. We recall that $dH(x_\pm) = 0$ in our case.

Let $F_i(\gamma(t)) = f_i \in \mathbb{R}$, $i = 1, \ldots, n$, and define the set $M_f = \{x \in \mathbb{R}^{2n} - \{x_\pm\} : F_i(x) = f_i, i = 1, \ldots, n\}$. Then the Liuville integrability theorem (cf. Arnold [1]) asserts that M_f is an n-dimensional smooth manifold in $\mathbb{R}^{2n}$ which is invariant under the flow of each Hamiltonian vector field X_{F_i}, $i = 1, \ldots, n$. Therefore the components of the stable and unstable manifolds both of which contain the orbit γ coincide each other and it is contained in M_f. Furthermore, $X_{F_i}(\gamma(t))$, $i = 1, \ldots, n$, constitute a basis of $T_{\gamma(t)}M_f$.

For our purposes, it is enough to assume that there exist n first integrals $F_1 = H, F_2, \cdots, F_n$ which are locally defined in a neighborhood of γ. Now since system (9.4) has exponential dichotomies on $[\alpha, \infty)$ and on $(-\infty, \alpha]$ $\alpha \in \mathbb{R}$, we denote projections at $t = \alpha$ by $P(\alpha)$ and $Q(\alpha)$ respectively. Then we have the following

Proposition 9.3. Suppose that Hamiltonian system (9.1) is completely integrable. Then

(i) $\mathcal{R}(P(\alpha)) = \mathcal{R}(I - Q(\alpha)) = T_{\gamma(\alpha)} M_f$,

and

(ii) $\{dF_i(\gamma(t)); i = 1, \ldots, n\}$ forms a complete set of bounded solutions of the adjoint system (9.6).

Proof. These are clear from the above argument and proposition 9.1. ∎

Thus for completely integrable Hamiltonian systems in $\mathbb{R}^{2n}$, we have $k = m = n$ where $k = \dim\{\mathcal{R}(P(\alpha)) \cap \mathcal{R}(I - Q(\alpha))\}$ and $m = \dim\{\mathcal{R}(P(\alpha)) + \mathcal{R}(I - Q(\alpha))\}^{\perp}$.

Now we will give special forms for the Melnikov vector of systems (9.2) and (9.3). Let $x(t) = \gamma(t + \alpha) + \epsilon z(t + \alpha)$. Then systems (9.2) and (9.3) become

$$\text{(9.7)} \qquad \dot{z} = A(t)z + g(t - \alpha, \gamma(t)) + h(t, z, \alpha, \epsilon)$$

and

$$\text{(9.8)} \qquad \dot{z} = A(t)z + X_G(t - \alpha, \gamma(t)) + X_{\tilde{G}}(t, z, \alpha, \epsilon)$$

respectively. Here

$$A(t) = JD^2H(\gamma(t)),$$

$$h(t, z, \alpha, \epsilon) = \frac{1}{\epsilon}\{f(\gamma(t) + \epsilon z) - f(\gamma(t)) - \epsilon Df(\gamma(t))z$$

$$+ \epsilon g(t - \alpha, \gamma(t) + \epsilon z) - \epsilon g(t - \alpha, \gamma(t))\},$$

$$\tilde{G}(t, z, \alpha, \epsilon) = \frac{1}{\epsilon}\{H(\gamma(t) + \epsilon z) - H(\gamma(t)) - \epsilon DH(\gamma(t))z$$

$$+ \epsilon G(t - \alpha, \gamma(t) + \epsilon z) - \epsilon G(t - \alpha, \gamma(t))\}.$$

Theorem 9.4. *For system (9.2), the Melnikov vector $M(\alpha, \nu, \epsilon)$ and its first approximation $\hat{M}(\alpha)$ are given by*

(9.9) $$\hat{M}_i(\alpha) = \int_{-\infty}^{\infty} dF_i(\gamma(t))g(t - \alpha, \gamma(t))dt, \ i = 1, \ldots, n$$

and

(9.10)
$$M_i(\alpha, \nu, \epsilon) = \hat{M}_i(\alpha) + \int_{-\infty}^{\alpha} dF_i(\gamma(t))h(t, z^u(\nu)(t), \alpha, \epsilon)dt$$
$$+ \int_{\alpha}^{\infty} dF_i(\gamma(t))h(t, z^s(\nu)(t), \nu, \epsilon)dt, \ i = 1, \ldots, n$$

where $z^u(\nu)(t)$ and $z^s(\nu)(t)$ are bounded solutions on $(-\infty, \alpha]$ and on $[\alpha, \infty)$ respectively of system (9.7).

Proof. These are simple consequences of Proposition 9.3 (ii) and Definition 5.1 of the Melnikov vector. ∎

Corollary 9.5. *For system (9.3), the Melnikov vector $M(\alpha, \nu, \epsilon)$ and its first approximation $\hat{M}(\alpha)$ are given by*

(9.11) $$\hat{M}_i(\alpha) = \int_{-\infty}^{\infty} \{F_i, G(t - \alpha, \cdot)\}(\gamma(t))dt, \ i = 1, \ldots, n$$

and

$$M_i(\alpha, \nu, \epsilon) = \hat{M}_i(\alpha) + \int_{-\infty}^{\alpha} dF_i(\gamma(t))X_{\tilde{G}}(t, z^u(\nu)(t), \alpha, \epsilon)dt$$

(9.12)

$$+ \int_{\alpha}^{\infty} dF_i(\gamma(t))X_{\tilde{G}}(t, z^s(\mu)(t), \alpha, \epsilon)dt, \quad i = 1, \ldots, n$$

where $z^u(\nu)(t)$ and $z^s(\mu)(t)$ are bounded solutions on $(-\infty, \alpha]$ and on $[\alpha, \infty)$ respectively of system (9.8).

Proof. These are simple consequences of Proposition 9.4 and the definition of the Poisson bracket. ∎

The expression of the Melnikov vector in a more special case was given in Holmes and Marsden [10]. See also Wiggins [20].

Remark 9.6. The dimension of the Melnikov vector for a completely integrable Hamiltonian system is the same as the degree of freedom of that system.

Remark 9.7. Since the first approximation $\hat{M}_i(\alpha)$ can be written as

$$\hat{M}_i(\alpha) = \int_{-\infty}^{\infty} \{F_i, G(\tau, \cdot)\}(\gamma(t + \alpha))d\tau,$$

we have

$$\frac{d}{d\alpha}\hat{M}_i(\alpha) = \int_{-\infty}^{\infty} \frac{\partial}{\partial\alpha}\{F_i, G(\tau, \cdot)\}(\gamma(t + \alpha))d\tau$$

(9.13)

$$= \int_{-\infty}^{\infty} \{H, \{F_i, G(\tau, \cdot)\}(\tau(t + \alpha))d\tau$$

$$= \int_{-\infty}^{\infty} \{H, \{F_i, G(t - \alpha, \cdot)\}(\gamma(t))dt.$$

Here we used the following fact: for $f \in C^\infty(\mathbb{R}^{2n})$,

$$\frac{d}{dt}(\overleftarrow{F}_t f) = \{H, \overleftarrow{F}_t f\}$$

where F_t is the flow of X_H and $\overleftarrow{F}_t f$ is the pullback of f by F_t. More generally, we have

(9.14) $$\frac{d^k}{d\alpha^k}\hat{M}_i(\alpha) = \int_{-\infty}^{\infty} \underbrace{\{H, \{H, \cdots, \{H, \{F_i, G(t-\alpha, \cdot)\}\} \cdots\}\}}_{k-times}(\gamma(t))dt .$$

§10 A HETEROCLINIC ORBIT TO INVARIANT TORI

In this section we extend our theory developed in previous sections to the case of a heteroclinic orbit to invariant tori. The system we consider is

(10.1) $$\dot{x} = f(x), \ x \in \mathbb{R}^n$$

and its perturbed system

(10.2) $$\dot{x} = f(x) + \epsilon g(x),$$

where f and g are sufficiently smooth and are bounded on bounded sets. We assume that system (10.1) has two normally hyperbolic smooth invariant tori M^1 and M^2 on which the flow of f is quasi-periodic, and also that system (10.1) has a heteroclinic orbit γ to M^1 and M^2. That is, there exist orbits $\{x^i(t) : t \in \mathbb{R}\} \subset M^i$, $i = 1, 2$, such that

$$|\gamma(t) - x^1(t)| \to 0 \text{ as } t \to -\infty$$

and

$$|\tau(t) - x^2(t)| \to 0 \text{ as } t \to +\infty.$$

If $M^1 = M^2$, we have a homoclinic orbit to invariant torus M^1 as a special case. We remark that M^1 and M^2 could be any normally hyperbolic smooth invariant sets with no stationary points. However, we assume the above conditions for its simplicity and applications. We also remark that the normal hyperbolicity is nothing but the hyperbolicity to the normal direction.

To develop an analogous theory to the one in previous sections, we need the expressions of the stable and the unstable manifolds. For this reason we shall decompose system (10.2), in neighborhoods of invariant tori M^i, into the tangential and the normal components and apply the theory of exponential dichotomy to the normal components.

Let $\dim M^i = d_i$ ($i = 1, 2$) and assume for $i = 1, 2$ that M^i is given by the embedding

$$u^i : T^i \to M^i \subset \mathbb{R}^n$$

where $T^i = S^1 \times \ldots \times S^1$ (d_i-times) is a standard d_i-dimensional torus with coordinates $\theta^i = (\theta_1^i, \ldots, \theta_{d_i}^i)$. From now on we suppress i and assume that M stands for M^1. The case of M^2 is done exactly in the same fashion.

Let $T\mathbb{R}^n \mid M$ be the restriction of the tangent bundle $T\mathbb{R}^n$ to M. Then

$$(T\mathbb{R}^n \mid M)_x = T_x M \oplus T_x^\perp M, \; x \in M$$

where $T_x M$ and $T_x^\perp M$ are respectively the tangent space and the normal space to M at x.

Now by the tubular neighborhood theorem, there exist a neighborhood U of T in $T \times \mathbb{R}^{n-d}$, a neighborhood V of M in $T^\perp M$ and a linear map $N(\theta) : \mathbb{R}^{n-d} \to \mathbb{R}^n$ for each $\theta \in T$ such that the vector bundle map $F : U \subset T \times \mathbb{R}^{n-d} \to V \subset T^\perp M$ defined by

(10.3) $$F(\theta, z) = u(\theta) + \epsilon N(\theta) z$$

is a diffeomorphism. Here N depends on θ smoothly. Clearly, we have

(10.4) $$(Du(\theta))^* N(\theta) = 0$$

and

$$(10.5) \qquad\qquad N^*(\theta)N(\theta) = I^{n-d}$$

where $(Du(\theta))^*$ and $N^*(\theta)$ are transposes of $Du(\theta)$ and $N(\theta)$ respectively. By using F, we transform the vector field $f + \epsilon g$ on $\mathbb{R}^n$ to the vector field $\overleftarrow{F}(f + \epsilon g)$ on $T \times \mathbb{R}^{n-d}$ where $\overleftarrow{F}$ is the pullback by F. We set

$$\overleftarrow{F}(f + \epsilon g) = \sum_{j=1}^{d} Aj\frac{\partial}{\partial\theta_j} + \sum_{\ell=1}^{n-d} B_\ell\frac{\partial}{\partial z_\ell}.$$

That is,

$$f(u(\theta) + \epsilon N(\theta)z) + \epsilon g(u(\theta) + \epsilon N(\theta)z)$$

$$= [Du(\theta) + \epsilon\frac{\partial}{\partial\theta}N(\theta)z]\begin{bmatrix} A_1 \\ \vdots \\ A_d \end{bmatrix} + \epsilon N(\theta)\begin{bmatrix} B_1 \\ \vdots \\ B_{n-d} \end{bmatrix}.$$

By using (10.4) and (10.5) we have

$$\begin{bmatrix} A_1 \\ \vdots \\ A_d \end{bmatrix} = (Du(\theta))^* f(u(\theta)) + \epsilon(Du(\theta))^*[-\frac{\partial}{\partial\theta}(N(\theta)z)Du(\theta)f(u(\theta))$$

$$-Df(u(\theta))N(\theta)z - g(u(\theta))] + 0(\epsilon)$$

$$\equiv \omega + \epsilon\Theta(\theta, z) + O(|\epsilon|^2).$$

Here we assumed that $(Du(\theta))^* f(u(\theta)) = \omega$, $\omega = (\omega_1, \ldots, \omega_d)$ are rationally independent. Under this assumption we also have

$$\begin{bmatrix} B_1 \\ \vdots \\ B_{n-d} \end{bmatrix} = N^*(\theta)[Df(u(\theta))N(\theta)z + g(u(\theta)) + O(|\epsilon|) - \frac{\partial}{\partial\theta}N(\theta)z\begin{bmatrix} A_1 \\ \vdots \\ A_d \end{bmatrix}]$$

$$= N^*(\theta)Df(u(\theta))N(\theta)z - N^*(\theta)(\frac{\partial}{\partial\theta}N(\theta)z)\omega$$

$$+ N^*(\theta)g(u(\theta)) + O(|\epsilon|)$$

$$\equiv A(\theta)z + N^*(\theta)g(u(\theta)) + O(|\epsilon|).$$

Note that $A(\theta)$ is a linear mapping for each θ.

To obtain the differential equations in $T \times \mathbb{R}^{n-d}$, we need a scale change in time. This is because ω will be changed after perturbation. Thus we set

$$(10.6) \qquad x((1 + \epsilon\beta)t + \epsilon a) = F(\theta(t), z(t)),$$

where $\beta \in \mathbb{R}$ and $a \in \mathbb{R}$. By using (10.6) we have the following system in $T \times \mathbb{R}^{n-d}$.

$$\dot{\theta} = (1 + \epsilon\beta)[\omega + \epsilon\Theta(\theta, z) + O(|\epsilon|^2)] = \omega + \epsilon(\beta\omega + \Theta(\theta, z)) + O(|\epsilon|^2)$$

$$(10.7) \qquad \dot{z} = (1 + \epsilon\beta)[A(\theta)z + N^*(\theta)g(u(\theta)) + O(|\epsilon|)]$$

$$= A(\theta)z + N^*(\theta)g(u(\theta)) + O(|\epsilon|).$$

Now we choose $t^1 \ll -1$ so that $\gamma(t^1)$ is close enough to M. Then from (10.3), there exist unique $\alpha^1 \equiv \alpha^1(t^1) \in T$ and $w^1 = w^1(t^1) \in \mathbb{R}^{n-d}$ such that

$$(10.8) \qquad \gamma(t^1) = u(\alpha^1) + \epsilon N(\alpha^1)w^1.$$

Hereafter we use α and w instead of α^1 and w^1.

Consider system (10.7) and let $z = \bar{z}(\theta, t)$ where $\bar{z}$ is a bounded function for $t \in (-\infty, t^1]$ which will be determined later. Then the 'θ-equation' in (10.7) becomes

$$(10.9) \qquad \dot{\theta} = \omega + \epsilon(\beta\omega + \Theta(\theta, \bar{z}(\theta, t))) + O(|\epsilon|^2).$$

Let $\bar{\theta}(t) \equiv \bar{\theta}(t; t^1, \alpha; \bar{z})$ be the solution of (10.9) with $\bar{\theta}(t^1) = \alpha$. By using $\bar{\theta}(t)$ for θ, the 'Z-equation' in (10.7) becomes

$$(10.10) \qquad \dot{z} = A(\bar{\theta}(t))z + N^*(\bar{\theta}(t))g(u(\bar{\theta}(t))) + O(|\epsilon|)$$

with $z(t^1) = \bar{z}(\alpha, t^1)$. Since M is normally hyperbolic, the linear system $\dot{z} = A(\bar{\theta}(t))z$ has an exponential dichotomy on $(-\infty, t^1]$. Let $Q : \mathbb{R}^{n-m} \to \mathbb{R}^{n-m}$ be a projection of the exponential dichotomy. Then $z(t; t^1, \bar{z}(\alpha, t^1))$ is a bounded solution of (12.16) on $(-\infty, t^1]$ if and only if

$$z(t; t^1, \bar{z}(\alpha, t^1)) = \Phi(t, t^1)(I - Q)\bar{z}(\alpha, t^1)$$

$$+ \Phi(t, t^1)(I - Q)\int_{t^1}^{t} \Phi(t^1, s)\{N^*(\bar{\theta}(t))g(u(\theta(\bar{\theta}(t)))$$

$$+ O(|\epsilon|)\}ds + \Phi(t, t^1)Q\int_{-\infty}^{t} \Phi(t^1, s)\{N^*(\bar{\theta}(t))g(u(\bar{\theta}(t)))$$

$$+ O(|\epsilon|)\}ds,$$

where $\Phi(t,s)$ is the transition matrix of $\dot{z} = A(\bar{\theta}(t))z$. By letting $t = t^1$, we have

(10.11) $\bar{z}(\alpha, t^1) = \eta^u + Q \displaystyle\int_{-\infty}^{t^1} \Phi(t^1, s)\{N^*(\bar{\theta}(s; t^1, \alpha; \bar{z}))g(u(\bar{\theta}(s; t^1, \alpha; \bar{z}))) + O(|\epsilon|)\}ds$

where $\eta^u \in \mathcal{R}(I - Q)$. We can show, by the contraction mapping principle, that equation (10.11) has a unique solution $\bar{z}(\eta_u)(\alpha, t^1)$ for sufficiently small η_u. Thus (10.11) gives a local expression of the unstable manifold of system (10.7) near $\alpha \in T$ in the space $T \times \mathbb{R}^{n-d}$. (10.11) has the following expression in the 'original space' $\mathbb{R}^n$. (We use $i = 1$ this time.)

(10.12)
$\xi^u \equiv u^1(\alpha^1) + \epsilon N^1(\alpha^1)\bar{z}^1(\alpha^1, t^1)$

$= u^1(\alpha^1)$

$\quad + \epsilon[\bar{\eta}^u + N^1(\alpha^1)Q \displaystyle\int_{-\infty}^{t^1} \Phi(t^1, s)\{N^{1^*}(\bar{\theta}^1(s; t^1, \alpha^1; \bar{z}^1))g(u^1(\bar{\theta}^1(s; t^1, \alpha^1; \bar{z}^1)))$

$\quad + O(|\epsilon|)\}ds]$

where $\bar{\eta}^u = N^1(\alpha^1)\eta^u \in \mathcal{R}(N^1(\alpha^1)(I - Q))$.

Next we choose $t^2 \gg 1$ so that $\gamma(t^2)$ is close enough to M^2. Then there exist unique $\alpha^2 = \alpha^2(t^2) \in T^2$ and $w^2 = w^2(t^2) \in \mathbb{R}^{n-d_2}$ such that

(10.13) $\gamma(t^2) = u^2(\alpha^2) + \epsilon N^2(\alpha^2)w^2.$

In exactly the same fashion as before, we have the following local expression of the stable manifold of system (10.7) near $u^2(\alpha^2) \in M^2$ in the 'original space' $\mathbb{R}^n$.

(10.14)
$\xi^s \equiv u^2(\alpha^2) + \epsilon N^2(\alpha^2)\bar{z}^2(\alpha^2, t^2)$

$= u^2(\alpha^2)$

$\quad + \epsilon[\bar{\eta}^s + N^2(\alpha^2)(I - P) \displaystyle\int_{\infty}^{t^2} \Phi^2(t^2, s)\{N^{2^*}(\bar{\theta}^2(s; t^2, \alpha^2; \bar{z}^2))g(u^2(\bar{\theta}^2(s; t^2, \alpha^2, \bar{z}^2)))$

$\quad + O(|\epsilon|)\}ds]$

where $\bar{\eta}^s \in (N^2(\alpha^2)P)$ and P is a projection of the exponential dichotomy on $[t^2, +\infty)$.

To measure the distance between the sections of the stable and unstable manifolds given in (10.12) and (10.14), we shall 'carry' the section of the unstable manifold given in (10.12) to the hyperplane $\gamma(t^2) + \mathcal{R}(N^2(\alpha^2))$ by the flow of (10.2). Hence, we shall need the following expressions of the unstable and stable manifolds. From (10.8) and (10.13), we have

$$
\begin{aligned}
\xi^u = \gamma(t^1) &+ \epsilon N^1(\alpha^1)[-w^1 + \eta^u \\
&+ Q \int_\infty^{t^1} \Phi^1(t^1, s)\{N^{1*}(\bar{\theta}^1(s; t^1, \alpha^1; \bar{z}^1))g(u^1(\bar{\theta}^1(s; t^1, \alpha^1; \bar{z}^1))) \\
&+ O(|\epsilon|)\}ds]
\end{aligned}
$$

(10.15)

$$
\equiv \gamma(t^1) + \epsilon M^u(\eta^u)
$$

and

$$
\begin{aligned}
\xi^s = \gamma(t^2) &+ \epsilon N^2(\alpha^2)[-w^2 + \eta^s \\
&+ (I - P) \int_\infty^{t^2} \Phi^2(t^2, s)\{N^{2*}(\bar{\theta}^2(s; t^2, \alpha^2, \bar{z}^2))g(u^2(\bar{\theta}^2(s; t^2, \alpha^2; \bar{z}^2))) \\
&+ O(|\epsilon|)\}ds]
\end{aligned}
$$

(10.16)

$$
\equiv \gamma(t^2) + \epsilon M^s(\eta^s).
$$

Now we consider system (10.2) along the heteroclinic orbit γ. Let $\beta(t)$ and $a(t)$ be C^∞-bump functions such that

$$
\beta(t) \equiv \begin{cases} \beta^1 & \text{for } t \leq t^1 \\ \beta^2 & \text{for } t \geq t^2 \end{cases} \quad \text{and} \quad a(t) \equiv \begin{cases} a^1 & \text{for } t \leq t^1 \\ a^2 & \text{for } t \geq t^2 \end{cases},
$$

and let

$$
x((1 + \epsilon\beta(t))t + \epsilon a(t)) = \gamma(t) + \epsilon y(t).
$$

Then system (10.2) becomes

(10.17) $\dot{y} = B(t)y + g(\gamma(t)) + \{\beta(t) + \dot{a}(t) + \dot{\beta}(t)t\}f(\gamma(t)) + O(|\epsilon|)$

where $B(t) = Df(\gamma(t))$. Now let $X(t, s)$ be the transition matrix of $\dot{y} = B(t)y$ and let $y(\eta^u)(t)$ be the solution of (10.17) with the initial condition $y(\eta^u)(t^1) = M^u(\eta^u)$. Also define $\tau^2(\eta^u)$ so that

$$
y(\eta^u)(\tau^2(\eta^u)) \in \gamma(t^2) + \mathcal{R}(N^2(\alpha^2)).
$$

We note that $\tau^2(\eta^u) = t^2 + O(|\epsilon|)$ because of the continuous dependence of solutions on initial conditions. Therefore

(10.18)
$$
\begin{aligned}
y(\eta^u)(\tau^2(\eta^u)) =& X(\tau^2(\eta^u), t^1)M^u(\eta^u) \\
& + \int_{t^1}^{\tau^2(\eta^u)} X(\tau^2(\eta^u), s)\{g(\gamma(s)) + (\beta(s) + \dot{a}(s) \\
& + \dot{\beta}(s)s)f(\gamma(s)) + O(|\epsilon|)\}ds \\
=& X(t^2, t^1)M^u(\eta^u) + \int_{t^1}^{t^2} X(t^2, s)\{g(\gamma(s)) \\
& + (\beta(s) + \dot{a}(s) + \dot{\beta}(s)s)f(\gamma(s))\}ds + O(|\epsilon|).
\end{aligned}
$$

To measure the distance between $y(\eta^u)(\tau^2(\eta^u))$ and $M^s(\eta^s)$, we use linearly independent bounded solutions on $\mathbb{R}$ of the adjoint system

(10.19)
$$
\dot{\phi} + B^*(t)\phi = 0.
$$

Let $\{\phi_1, \ldots, \phi_m\}$ be a complete set of linearly independent bounded solutions on $\mathbb{R}$ of system (10.19) which satisfy

$$
\phi_i(t^2) \in \{\mathcal{R}(p) \cap \mathcal{R}(N^2(\alpha^2))\}^\perp, \quad i = 1, \ldots, m.
$$

Define

$$
M_i = \phi_i^*(t^2)[y(\eta^u)(\tau^2(\eta^u)) + \frac{1}{\epsilon}(\gamma(\tau^2(\eta^u)) - \gamma(t^2)) - M^s(\eta^s)], \quad i = 1, \ldots, m.
$$

By using (10.15), (10.16) and (10.18), M_i is computed as follows.

$$\phi_i^*(t^2)[y(\eta^u)(\tau^2(\eta^u)) + \frac{1}{\epsilon}(\gamma(\tau^2(\eta^u)) - \gamma(t^2))]$$

$$= \phi_i^*(t^2)[X(t^2, t^1)M^u(\eta^u) + \int_{t^1}^{t^2} X(t^2, s)\{g(\gamma(s))$$

$$+ \beta(s) + \dot{a}(s) + \dot{\beta}(s)s)f(\gamma(s))\}ds + \frac{1}{\epsilon}(\gamma(\tau^2(\eta^u))$$

$$- \gamma(t^2)) + O(|\epsilon|)]$$

$$= \phi_i^*(t^1)M^u(\eta^u) + \int_{t^1}^{t^2} \phi_i^*(s)\{g(\gamma(s)) + (\beta(s) + \dot{a}(s)$$

$$+ \dot{\beta}(s)s)f(\gamma(s))\}ds + O(|\epsilon|)$$

$$= \phi_i^*(t^1)M^u(\eta^u) + \int_{t^1}^{t^2} \phi_i^*(s)g(\gamma(s))ds + O(|\epsilon|)$$

$$= \phi_i^*(t^1)N^1(\alpha^1)[-w^1 + \eta^u$$

$$+ Q\int_{-\infty}^{t^1} \Phi^1(t^1, s)\{N^{1^*}(\bar{\theta}^1(s))g(u^1(\bar{\theta}^1(s))$$

$$+ O(|\epsilon|)\}ds] + \int_{t^1}^{t^2} \phi_i^*(s)g(\gamma(s))ds + O(|\epsilon|)$$

$$= \int_{\infty}^{t^1} \Psi_i^{1^*}(s)N^{1^*}(\bar{\theta}^1(s))g(u^1(\bar{\theta}^1(s)))ds + \int_{t^1}^{t^2} \phi_i^*(s)g(\gamma(s))ds + O(|\epsilon|)$$

$$= \int_{-\infty}^{t^1} \phi_i^*(s)g(u^1(\bar{\theta}^1(s)))ds + \int_{t^1}^{t^2} \phi_i^*(s)g(\gamma(s))ds + O(|\epsilon|)$$

$$= \int_{-\infty}^{t^1} \phi_i^*(s)g(\gamma(s))ds + \int_{t^1}^{t^2} \phi_i^*(s)g(\gamma(s)))ds + O(|\epsilon|)$$

$$= \int_{-\infty}^{t^2} \phi_i^*(s)g(\gamma(s))ds + O(|\epsilon|).$$

Here we let $\phi_i(t) = \Psi_i^1(t)N^{1*}(\bar{\theta}^1(t))$, $t \leq t^1$.

$$\phi_i^*(t^2)M^s(\eta^s) = \phi_i^*(t^2)N^2(\alpha^2)[-w^2 + \eta^s$$
$$+ (I-P)\int_\infty^{t^2} \Phi^2(t^2,s)\{N^{2*}(\bar{\theta}^2(s))g(u^2(\bar{\theta}^2(s))) + O(|\epsilon|)\}ds]$$
$$= \int_\infty^{t^2} \Psi_i^{2*}(s)N^{2*}(\bar{\theta}^2(s))g(u^2(\bar{\theta}^2(s)))ds + O(|\epsilon|)$$
$$= \int_\infty^{t^2} \phi_i^*(s)g(u^2(\bar{\theta}^2(s)))ds + O(|\epsilon|)$$
$$= \int_\infty^{t^2} \phi_i^*(s)g(\gamma(s))ds + O(|\epsilon|).$$

Here we let $\phi_i(t) = \Psi_i^2(t)N^{2*}(\bar{\theta}^2(t))$, $t \geq t^2$.

Thus we have

(10.20)
$$M_i = \int_{-\infty}^\infty \phi_i^*(s)g(\gamma(s))ds + O(|\epsilon|)$$

and we have proved the following

Theorem 10.1. *The first approximation of the Melnikov vector* $\hat{M} = (\hat{M}_1, \ldots, \hat{M}_m)$ *for system (10.2) is given by*

(10.21)
$$\hat{M}_i = \int_{-\infty}^\infty \phi_i^*(t)g(\gamma(t))dt, \ i = 1, \ldots, m,$$

where m *is the number of linearly independent bounded solutions of* $\dot{\phi} + [Df(\gamma(t))]^*\phi = 0$. ∎

Remark 10.2. Since the heteroclinic orbit γ is contained both in the unstable manifold $W^u(M^1)$ of M^1 and the stable manifold $W^s(M^2)$ of M^2, it is interesting to consider how these manifolds intersect each other along the heteroclinic orbit γ. Consider the case of a time-independent perturbation and define

$$k = \dim[T_{\gamma(t)}W^u(M^1) \cap T_{\gamma(t)}W^s(M^2)]$$

where $T_{\gamma(t)}W^w(M^1)$ and $T_{\gamma(t)}W^s(M^2)$ are tangent spaces at $\gamma(t)$ to $W^u(M^1)$ and $W^s(M^2)$ respectively. Then k and the dimension m of the Melnikov vector have the following relation.

(10.22) $$m = n - [\dim W^u(M^1) + \dim W^s(M^2) - k],$$

where

$$\dim W^u(M^1) = \dim \mathcal{R}(I - Q_1) + d_1$$

and

$$\dim W^s(M^2) = \dim \mathcal{R}(P_2) + d_2 .$$

Note that $\dim W^u(M^1) + \dim W^s(M^1) = n + d_1$.

If we define the splitting index $\delta(\gamma)$ of γ by

$$\delta(\gamma) = \dim W^s(M^1) - \dim W^s(M^2) ,$$

we have, from (12.40), the following relation

(10.23) $$m = k + \delta(\gamma) - d_1$$

which is a generalization of (6.2).

Now we go into a special case to which Theorem 10.1 can easily be applied. Consider a system with a quasi-periodic perturbation

(10.24) $$\dot{z} = f(z) + \epsilon g(z, \omega_1 t, \dots, \omega_d t), \ z \in \mathbb{R}^n$$

where g is periodic in each 't' argument and $\omega_1, \dots, \omega_d$ are rationally independent, see Meyer and Sell [13]. We assume that the unperturbed system $\dot{z} = f(z)$ has a homoclinic or heteroclinic orbit γ to hyperbolic critical point(s). System (10.24) is equivalent to the following system on the torus T^d.

(10.25)
$$\dot{z} = f(z) + \epsilon g(z, \theta)$$
$$\dot{\theta} = \omega$$

where $\theta = (\theta^1, \ldots, \theta^d), \omega = (\omega_1, \ldots, \omega_d)$ and $(z, \theta) \in \mathbb{R}^n \times T^d$. This is a special case of system (10.2) in the sense that the 'z-dynamics' of the unperturbed system of (10.25) is globally defined in the normal bundle of T^d. By using the homoclinic orbit γ, the homoclinic orbit $\bar{\gamma}$ of system (10.25) to the torus T^d is given by

$$\bar{\gamma}(t) = (\gamma(t), \omega_1(t), \, \omega_1 t + \theta_1, \ldots, \omega_d t + \theta_d)$$

where $\theta_i \in [0, 2\pi)$, $i = 1, \ldots, d$. This is because the 'z-dynamics' and 'θ-dynamics' of the unperturbed system of (10.25) are completely decoupled. By Theorem 10.1, we have the following corollary in this case.

Corollary 10.2. The first approximation of the Melnikov vector $\hat{M}(\theta_1, \ldots, \theta_d) = (\hat{M}_1(\theta_1, \ldots, \theta_d), \ldots, \hat{M}_m(\theta_1, \ldots, \theta_d))$for system (10.44) is given

$$(10.26) \qquad \hat{M}_i(\theta_1, \ldots, \theta_d) = \int_{-\infty}^{\infty} \phi_i^*(t) g(\gamma(t), \omega_1 t + \theta_1, \ldots, \omega_d t + \theta_d) dt,$$

$i = 1, \ldots, m$. Here $\{\phi_1, \ldots, \phi_m\}$ is a set of linearly independent bounded solutions of $\dot{\phi} + [Df(\gamma(t))]^* \phi = 0.$ ∎

As a special case, we shall prove the following proposition for two-dimensional systems. See also Meyer and Sell [13] and Wiggins [20].

Proposition 10.3. Consider system (10.24) with the same assumption as before and let $n = 2$ and $d \geq 2$. Then the stable and unstable manifolds of system (10.24) intersect transversally if and only if for the first approximation of the Melnikov function $\hat{M}(\theta_1, \ldots, \theta_d)$ defined in (10.26) ($i = 1$ in this case), there exist $(\bar{\theta}_1, \ldots, \bar{\theta}_d)$ such that

$$(10.27) \qquad \qquad \hat{M}(\bar{\theta}_1, \ldots, \bar{\theta}_d) = 0$$

and

$$(10.28) \qquad \qquad (\mathcal{L}_\omega \hat{M})(\bar{\theta}_1, \ldots, \bar{\theta}_d) \neq 0$$

where $\omega = (\omega_1, \ldots, \omega_d)$ and $\mathcal{L}_\omega \hat{M}$ is the Lie derivative of $\hat{M}$ with respect to ω.

Proof. Let $\theta_i(\alpha) = \bar{\theta}_i - \omega_i \alpha,\ i = 1, \ldots, d,\ \alpha \in \mathbb{R}$ and define

$$D(\alpha) = \hat{M}(\theta_1(\alpha), \ldots, \theta_d(\alpha)).$$

Then we have

(10.29)
$$
\begin{aligned}
D(\alpha) &= \int_{-\infty}^{\infty} \phi^*(t) g(\gamma(t), \omega_1(t - \alpha) + \bar{\theta}_1, \ldots, \omega_d(t - \alpha) + \bar{\theta}_d) dt \\
&= \int_{-\infty}^{\infty} \phi^*(t + \alpha) g(\gamma(t + \alpha), \omega_1 t + \bar{\theta}_1, \ldots, \omega_d t + \bar{\theta}_d) dt
\end{aligned}
$$

and

(10.30)
$$D(0) = \hat{M}(\bar{\theta}_1, \ldots, \bar{\theta}_d) = 0.$$

From (10.29) α works as a 'sweeping' parameter along γ. See Figure 6. Since $D(\alpha)$ changes along $\gamma(\alpha), \alpha \in \mathbb{R}$ and since the difference of the true distance of the stable and unstable manifolds and $D(s)$ is of order ϵ, the implicit mapping theorem implies, using (10.30), that the transversal intersection exists if and only if $D(0) = 0$ and $D'(0) \neq 0$. Finally

$$
\begin{aligned}
D'(0) &= \sum_{i=1}^{d} \frac{\partial \hat{M}}{\partial \theta_i}(\bar{\theta}_1, \ldots, \bar{\theta}_d) \frac{d\theta_i}{d\alpha}(0) \\
&= -\sum_{i=1}^{d} \omega_i \frac{\partial \hat{M}}{\partial \theta_i}(\bar{\theta}_1, \ldots, \bar{\theta}_d) = -(\mathcal{L}_\omega \hat{M})(\bar{\theta}_1, \ldots, \bar{\theta}_d). \qquad \blacksquare
\end{aligned}
$$

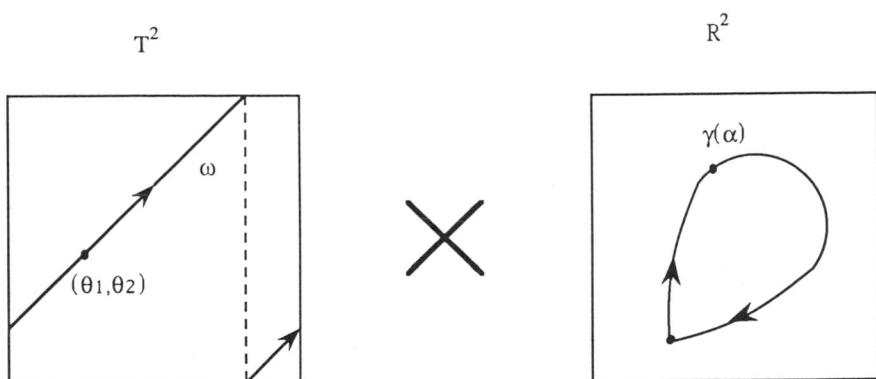

Figure 6

Remark 10.4. We note that the case of two-dimensional systems with periodic perturbations in §8 is a special case of this proposition. That is, $\frac{d}{d\alpha}\hat{M}(\alpha_0)$ in Corollary 8.2 is generalized to $(\mathcal{L}_\omega\hat{M})(\bar{\theta}_1,\ldots,\bar{\theta}_d)$.

§11. EXAMPLES

In this section we apply the methods developed in previous sections to three examples. We shall examine (1) a two-dimensional system which has transversal intersections, (2) a four-dimensional system which has both transversal and tangential intersections and (3) a system for which condition (ii) in Theorem 5.5 is not satisfied but transversal intersection exists. We also give an example to which the Melnikov method can not be applied to detect the intersection of the stable and unstable manifolds and discuss a limitation of the Melnikov method as a perturbation technique.

138   Shui-Nee Chow and Masahiro Yamashita

Example 1 (Chow, Hale and Mallet-Paret [4])

We consider the following second order equation

(11.1) $$\ddot{x} - x + \frac{3}{2}x^2 = \epsilon \cos t,$$

where ϵ is sufficiently small. That is,

(11.2) $$\frac{d}{dt}\begin{bmatrix} x \\ y \end{bmatrix} = f(x,y) + \epsilon g(t)$$
$$\equiv \begin{bmatrix} y \\ x - \frac{3}{2}x^2 \end{bmatrix} + \epsilon \begin{bmatrix} 0 \\ \cos t \end{bmatrix}.$$

We notice that the unperturbed system (i.e. $\epsilon = 0$) has a homoclinic orbit,

(11.3) $$\gamma(t) = \begin{bmatrix} P(t) \\ \dot{p}(t) \end{bmatrix} \equiv \begin{bmatrix} \sec h^2(t/2) \\ -4\sec h^2(t/2) & \tan h(t/2) \end{bmatrix}$$

to the origin. See the figure below.

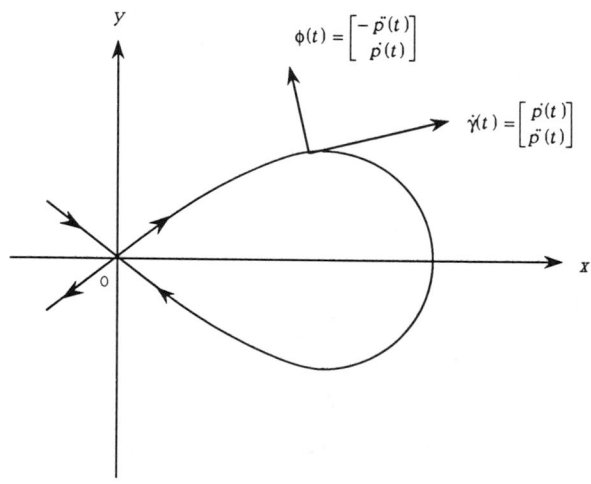

Figure 7

The linearized system along γ is

(11.4) $$\dot{z} = A(t)z$$

where

(11.5) $$A(t) = Df(\gamma(t)) = \begin{bmatrix} 0 & 1 \\ 1 - 3p(t) & 0 \end{bmatrix}.$$

The adjoint system $\dot{\phi} + A^*(t)\phi = 0$ has only one linearly independent bounded solution

(11.6) $$\phi(t) = \begin{bmatrix} -\ddot{p}(t) \\ \dot{p}(t) \end{bmatrix}$$

and hence the first approximation of the Melnikov function is

(11.7) $$\hat{M}(\alpha) = \int_{-\infty}^{\infty} \phi^*(t)g(t - \alpha)dt$$
$$= \int_{-\infty}^{\infty} \dot{p}(t)\cos(t - \alpha)dt$$
$$= -c\sin\alpha$$

where $c > 0$ is a constant.

Since

(11.8) $$\frac{d}{d\alpha}\hat{M}(n\pi) = (-1)^{n+1}c, \ n = 0, \pm 1, \pm 2, \ldots,$$

the perturbed stable and unstable manifolds always intersect transversally and so tangential intersections never occur. See Corollary 8.2.

Example 2 (Gruendler [6])

In this example we consider the following system of two second-order equations.

(11.9)
$$\ddot{x}_1 = x_1 - 2x_1(x_1^2 + x_2^2) + \epsilon\{-3\mu_1 x_1 - \mu_2 \dot{x}_1$$
$$+ \frac{2\mu_3}{1 + \omega^2}(3x_1^2 + x_2^2)\cos\omega t + \frac{4\mu_4}{1 + \omega^2}x_1 x_2 \cos\omega t\},$$
$$\ddot{x}_2 = x_2 - 2x_2(x_1^2 + x_2^2) + \epsilon\{-\mu_1 - x_2 - \mu_2 \dot{x}_2 + \frac{4\mu_3}{1 + \omega^2}x_1 x_2 \cos\omega t$$
$$+ \frac{2\mu_4}{1 + \omega^2}(x_1^2 + 3x_2^2)\cos\omega t\}.$$

Here μ_1, μ_2, μ_3 and μ_4 are parameters, and ϵ is assumed to be sufficiently small. We consider first the unperturbed system ($\epsilon = 0$). As easily seen, this unperturbed system is a Hamiltonian system. Let $\dot{x}_1 = x_3$ and $\dot{x}_2 = x_4$, and let $x = (x_1, x_2, x_3, x_4)$. Then the unperturbed system can be written in the form

$$(11.10) \qquad \dot{x} = X_H(x)$$

where the Hamiltonian function $H(x)$ is given by

$$(11.11) \qquad H(x_1, x_2, x_3, x_4) = -\frac{1}{2}(x_1^2 + x_2^2) + \frac{1}{2}(x_1^2 + x_2^2)^2 + \frac{1}{2}(x_3^2 + x_4^2).$$

Furthermore, system (11.10) has one more first integral

$$(11.12) \qquad F(x_1, x_2 x_3 x_4) = x_1 x_4 - x_2 x_3$$

which results from the conservation of the angular momentum.

Since

$$\{F, H\}(x) = dF(x)X_H(x)$$

$$(11.13) \qquad = [x_4 \ - x_3 \ - x_2 x_1] \begin{bmatrix} x_3 \\ x_4 \\ x_1 & -2x_1(x_1^2 + x_2^2) \\ x_2 & -2x_2(x_1^2 + x_2^2) \end{bmatrix}$$

$$= 0$$

and since $dF(x)$ and $dH(x)$ are linearly independent for any $x \in \mathbb{R}^4 \backslash \{0\}$ unperturbed system, (11.10) is a completely integrable system in $\mathbb{R}^4 \backslash \{0\}$. So we shall utilize this special structure (see Proposition 9.3(ii) and Theorem 9.4) to derive the Melnikov vector even though the perturbed system is not a Hamiltonian system.

Next we notice that the unperturbed system (11.10) has a homoclinic orbit $\gamma(t, 0)$ to the origin.

$$(11.14) \qquad \gamma(t, 0) = (p(t), 0, \dot{p}(t), 0)$$

where $p(t) = \operatorname{sech} t$.

In terms of the complete integrability, we know that the stable and the unstable manifolds of system (11.10), both of which have dimension two, must coincide along $\gamma(t, 0)$ and in fact, by using a symmetry property of X_H, this 'homoclinic manifold' can be expressed as

$$\gamma(t, \nu) = (p(t) \cos \nu, p(t) \sin \nu, \dot{p}(t) \cos \nu, \dot{p}(t) \sin \nu),$$

(11.15)

$$\nu \in [0, 2\pi), \ t \in \mathbb{R}.$$

That is, system (11.10) has a family of homoclinic orbits parametrized by ν. Thus system (11.10) is an example to case (ii) in Section 8.

Now we go back to the original perturbed system (8.8).

Let

$$g(t, x, \mu) = \left(0, 0, -3\mu_1 x_1 - \mu_2 x_3 + \frac{2\mu_3}{1 + \omega^2}(3x_1^2 + x_2^2) \cos \omega t + \frac{4\mu_4}{1 + \omega^2} x_1 x_2 \cos \omega t,\right.$$

$$\left. - \mu_1 x_2 - \mu_2 x_4 + \frac{4\mu_3}{1 + \omega^2} x_1 x_2 \cos \omega t + \frac{2\mu_4}{1 + \omega^2}(x_1^2 + 3x_2^2) \cos \omega t\right),$$

where $\mu = (\mu_1, \mu_2, \mu_3, \mu_4)$. Then system (11.9) has the form

(11.16) $\dot{x} = X_H(x) + \epsilon g(t, x, \mu).$

As we mentioned in case(ii) in Section 8, the first approximation of the Melnikov vector can be used to detect a point of transversal or tangential intersection of the perturbed stable and unstable manifolds of system (11.16). Furthermore, by virtue of Proposition 9.3(ii), bounded solutions of the adjoint system of the linearized system of system (11.10) along an orbit $\gamma(t, \nu)$ are given by

$$dH(\gamma(t, \nu)) = (-p(t) \cos \nu + 2p(t)^3 \cos \nu, -p(t) \sin \nu$$

(11.17)

$$+ 2p(t)^3 \sin \nu, \dot{p}(t) \cos \nu, \dot{p}(t) \sin \nu)$$

and

(11.18) $dF(\gamma(t, \nu)) = (\dot{p}(t) \sin \nu, -\dot{p}(t) \cos \nu, -p(t) \sin \nu, p(t) \cos \nu).$

It is easily shown that $dH(\gamma(t,\nu))$ and $dF(\gamma(t,\nu))$ are linearly independent.

We can now compute the first approximation of the Melnikov vector $\hat{M}(\alpha,\nu) = (\hat{M}_1(\alpha,\nu), \hat{M}_2(\alpha,\nu))$ as follows.

$$
\begin{aligned}
\hat{M}_1(\alpha,\nu) &= \int_{-\infty}^{\infty} dH(\gamma(t,\nu))g(t-\alpha,\gamma(t,\nu),\mu)dt \\
&= \int_{-\infty}^{\infty} \{-\mu_1(3\cos^2\nu + \sin^2\nu)p(t)\dot{p}(t) - \nu_2\dot{p}(t)^2 \\
&\quad + \frac{6}{1+\omega^2}(\nu_3\cos\nu + \mu_4\sin\nu)p(t)^2\dot{p}(t)\cos\omega(t-\alpha)\}dt \\
&= -\frac{2}{3}\mu_2 - \pi\omega\sec h(\frac{\pi\omega}{2})(-\mu_3\sin\nu + \mu_4\cos\omega\alpha) \\
\hat{M}_2(\alpha,\nu) &= \int_{-\infty}^{\infty} dF(\gamma(t,\nu))g(t-\alpha,\gamma(t,\nu),\mu)dt \\
&= 2\mu_1\sin 2\nu + c(-\mu_3\sin\nu + \mu_4\cos\nu)\cos\omega\alpha.
\end{aligned}
$$

(11.19)

Let $c = \pi\sec h(\frac{\pi\omega}{2})$. Then the Melnikov vector becomes

(11.20)
$$
\hat{M}(\alpha,\nu,\mu) = \begin{bmatrix} -\frac{2}{3}\mu_2 - c\omega(\mu_3\cos\nu + \mu_4\sin\nu)\sin\omega\alpha \\ 2\mu_1\sin 2\nu + c(-\mu_3\sin\nu + \mu_4\cos\nu)\cos\omega\alpha. \end{bmatrix}.
$$

To find points of intersection, consider, for example, the case $\nu = 0$. In this case the Melnikov vector becomes

(11.21)
$$
\hat{M}(\alpha,0;\mu) = \begin{bmatrix} -\frac{2}{3}\mu_2 - \quad c\omega\mu_3\sin\omega\alpha \\ c\mu_4\cos\omega\alpha \end{bmatrix}.
$$

Solving $\hat{M}(\alpha,0;\mu) = 0$, we have the following bifurcation set S in the parameter space $(\mu_1,\mu_2,\mu_3,\mu_4)$.

(11.22)
$$
S = A \cup B \cup C,
$$

where

$$
A = \{(\mu_1,\mu_2,\mu_3,\mu_4) : \mu_2 = \pm\frac{3}{2}c\omega\mu_3, \mu_4 \neq 0, \mu_1,\mu_2 \in \mathbb{R}\},
$$

$$
B = \{(\mu_1,\mu_2,\mu_3,\mu_4) : |\mu_2| < |\frac{3}{2}c\omega\mu_3|, \mu_4 = 0, \mu_1,\mu_3 \in \mathbb{R}\},
$$

$$
C = \{(\mu_1,\mu_2,\mu_3,\mu_4) : \mu_2 = \pm\frac{3}{2}c\omega\mu_3, \mu_4 = 0, \mu_1,\mu_3 \in \mathbb{R}\}.
$$

Next we examine the transversality and the tangency of intersection in the case $\nu = 0$. The derivatives of $\hat{M}$ are given by

(11.23)
$$\frac{\partial}{\partial \alpha} \hat{M}(\alpha, 0; \mu) = \begin{bmatrix} -c\omega^2 \mu_3 \cos \omega\alpha \\ -c\omega \mu_4 \sin \omega\alpha \end{bmatrix}$$

and

(11.24)
$$\frac{\partial}{\partial \nu} \hat{M}(\alpha, 0; \mu) = \begin{bmatrix} -c\omega \mu_4 \sin \omega\alpha \\ 4\mu_1 - c\mu_3 \cos \omega\alpha \end{bmatrix}.$$

Since, in this example, the stable and the unstable manifolds of the unperturbed system coincide and constitute a two dimensional manifold in $\mathbb{R}^4$, $\mathrm{rank}\left[\frac{\partial}{\partial \alpha}\hat{M}(\alpha, 0; \mu)\right.$ $\left.\frac{\partial}{\partial \nu}\hat{M}(\alpha, 0; \mu)\right] = 2$ implies a transversal intersection (see Proposition 8.3).

(i) Let $\mu \in A$. Since

$$\mathrm{rank}\left[\frac{\partial}{\partial \alpha}\hat{M}(\alpha, 0, \mu) \frac{\partial}{\partial \nu}\hat{M}(\alpha, 0, \mu)\right] = \mathrm{rank}\begin{bmatrix} 0 & \pm c\omega\mu_4 \\ \pm c\omega\mu_4 & 4\mu_1 \end{bmatrix} = 2,$$

intersection is always transversal.

(ii) Let $\mu \in B$. In this case we have

$$\left[\frac{\partial}{\partial \alpha}\hat{M}(\alpha, 0, \mu)\frac{\partial}{\partial \nu}\hat{M}(\alpha, 0, \mu)\right] = \begin{bmatrix} -c\omega^2 \mu_3 \cos \omega\alpha & 0 \\ 0 & 4\mu_1 - c\mu_3 \cos \omega\alpha \end{bmatrix}.$$

(ii-1) If $\mu_1 \neq \frac{1}{4}c\mu_3 \cos \omega\alpha$, then the rank of the above matrix is 2 and so we have a transversal intersection.

(ii-2) If $\mu_1 = \frac{1}{4}c\mu_3 \cos \omega\alpha$, then $\frac{\partial}{\partial \nu}\hat{M}(\alpha, 0, \mu) = 0$. By computing the rank of (8.16), we have a tangential intersection.

(iii) Let $\mu \in C$. In this case

$$\left[\frac{\partial}{\partial \alpha}\hat{M}(\alpha, 0, \mu)\frac{\partial}{\partial \nu}\hat{M}(\alpha, 0, \mu)\right] = \begin{bmatrix} 0 & 0 \\ 0 & 4\mu_1 \end{bmatrix}.$$

By computing the rank of (8.16), we have a tangential intersection f $\mu_1 \neq 0, \mu_3 \neq 0$.

Example 3.

   The aim of this example is to give an example in which condition (ii) of Theorem
5.5 is not satisfied but the stable and unstable manifolds intersect transversally. To
this end we modify the system in Example 2 slightly and consider the following system.

$$\dot{x}_1 = x_3$$

$$\dot{x}_2 = x_4$$

$$\dot{x}_3 = x_1 - 2x_1(x_1^2 + x_2^2) + \epsilon\{-3\mu_1 x_1 - \mu_2 x_3 + \frac{2\mu_3}{1+\omega^2}(3x_1^2 + x_2^2)(\cos\omega t$$

(11.25)
$$+ \frac{4\mu_4}{1+\omega^2} x_1 x_2 \cos\omega t)\}$$

$$\dot{x}_4 = x_2 - 2x_2(x_1^2 + x_2^2) + \epsilon\{-\mu_1 x_2 - \mu_2 x_4 + \frac{4\mu_3}{1+\omega^2} x_1 x_2 \cos\omega t$$

$$+ \frac{2\mu_4}{1+\omega^2}(x_1^2 + 3x_2^2) \cos\omega t\}$$

$$\dot{y} = y + \epsilon \cos\omega t$$

Notice that the unperturbed system ($\epsilon = 0$) has the following stable and unstable
manifolds.

(11.26) $W^u = (p(\alpha)\cos\nu,\ p(\alpha)\sin\nu,\ \dot{p}(\alpha)\cos\nu,\ \dot{p}(\alpha)\sin\nu, y)$,

(11.27) $W^s = (p(\alpha)\cos\nu,\ p(\alpha)\sin\nu,\ \dot{p}(\alpha)\cos\nu,\ \dot{p}(\alpha)\sin\nu, 0)$

where $p(t) = \operatorname{sech} t$ and $\alpha, y \in \mathbb{R}$, $\nu \in [0, 2\pi]$.

Note that dim $W^u = 3$ and dim $W^s = 2$, and the 'homoclinic manifold' is $W^u \cap W^s$.

From (11.17) and (11.18), it is clear that the adjoint system of the linearized system of
the unperturbed system has two linearly independent bounded solutions on $\mathbb{R}$ which
are given by

$$\phi^*(t) =(-p(t)\cos\nu + 2p(t)^3\cos\nu,\ -p(t)\sin\nu$$

(11.28)
$$+ 2p(t)^3\sin\nu,\ \dot{p}(t)\cos\nu,\ \dot{p}(t)\sin\nu, 0)$$

and

(11.29) $\quad \phi_2^*(t) = (\dot{p}(t)\sin\nu,\ -\dot{p}(t)\cos\nu,\ -p(t)\sin\nu,\ p(t)\cos\nu,\ 0).$

Hence the Nelnikov vector for system (11.25) is precisely the same as (11.20). Consider the case $\nu = 0$. From now on we assume that

(11.30) $\qquad\qquad\qquad \mu \in C \text{ and } \mu_1 \neq 0$

where C is defined in (11.22). Then we know that $\frac{\partial}{\partial\alpha}\hat{M}(\alpha,0;\mu) = 0$ and $\frac{\partial}{\partial\nu}\hat{M}(\alpha,0;\mu) \neq 0$. Thus condition (ii) in Theorem 5.5 is not satisfied. However, we shall show that there exists the transversal intersection of the stable and unstable manifolds of system (11.25). First recall, by using the notation in (5.16) and (5.17), that we have the following situation.

$$
\begin{array}{c|cc|c}
\mathcal{R}(I-Q(\alpha)) & & \alpha,\nu & \mathcal{R}(P(\alpha)) \\
& \mu_2^u & m_1^s(\alpha,\nu) & \\
\mathcal{R}(Q(\alpha)) & m_2^u(\alpha,\nu,\mu_2^u) & m_2^s(\alpha,\nu) & \mathcal{R}(I-P(\alpha))
\end{array}
$$

(11.31) $\qquad DF^u = \begin{pmatrix} 1 & 0 & 0 \\ 0 & 1 & 0 \\ 0 & 0 & 1 \\ \frac{\partial}{\partial\alpha}m_2^u & \frac{\partial}{\partial\nu}m_2^u & \frac{\partial}{\partial\mu_2^u}m_2^u \end{pmatrix},$

(11.32) $\qquad DF^s = \begin{pmatrix} 1 & 0 \\ 0 & 1 \\ \frac{\partial}{\partial\alpha}m_1^s & \frac{\partial}{\partial\nu}m_2^s \end{pmatrix}.$

Therefore if $\frac{\partial}{\partial\alpha}m_1^s \neq 0$, we have the transversal intersection even though $\frac{\partial}{\partial\alpha}\hat{M}(\alpha,0 : \mu) = 0$ which means that $\frac{\partial}{\partial\alpha}m_2^u = \frac{\partial}{\partial\alpha}m_2^s$. Note that the linearized system of the unperturbed system of (11.25) has an unbounded solution $(0, 0, 0, 0, ce^t)$ on $[\alpha, \infty)$ where c is a nonzero constant. Let

(11.33) $$\phi_3^*(t) = (0, 0, 0, 0, ce^{-t}).$$

Then

(11.34)
$$\begin{aligned}
m_1^s(\alpha) &= \phi_3^*(\alpha) m^s(\alpha) \\
&= \int_\infty^\alpha ce^{-t} \cos \omega(t - \alpha) dt \\
&= \frac{c}{1 + \omega^2} e^{-\alpha}.
\end{aligned}$$

So we have

(11.35) $$\frac{d}{d\alpha} m_1^s(\alpha) = \frac{c}{1 + \omega^2} e^{-\alpha} \neq 0 \text{ for any } \alpha.$$

Thus, in this example, there exists the transversal intersection but the condition by the Melnikov vector cannot be used to show it. In other words, the Melnikov vector cannot give complete information about the transversal intersection. This limitation of the Melnikov vector in higher dimensional cases comes from the fact that the Melnikov vector is the projection of the real distance between the stable and unstable manifolds to the space of the completely unbounded solutions, i.e., the complement subspace of $\mathcal{R}(P(\alpha)) + \mathcal{R}(I - Q(\alpha))$. Therefore the Melnikov vector drops the information about the projection of the real distance to other subspaces. See the decomposition in Section 5 of $T_{\gamma(\alpha)}^\perp \mathbb{R}^n$.

Finally, we examine an example for which the Melnikov function cannot be applied to detect the intersection of the stable and unstable manifolds. Before doing this, we recall system (11.1) in Example 1.

The first approximation of the Melnikov function of this system was

$$\hat{M}(\alpha) = -c \sin \alpha$$

where $c \neq 0$ is a constant. The reason that $\hat{M}(\alpha)$ can be used to detect the intersection of the stable and unstable manifolds of this system is that the distance $\tilde{d}$ between the stable and unstable manifolds is expressed as

$$(11.36) \qquad \tilde{d} = \epsilon(\hat{M}(\alpha) + 0(|\epsilon|)).$$

That is, $\hat{M}(\alpha)$ constitutes the leading term.

Now we consider the following rapidly forced system.

$$(11.37) \qquad \ddot{x} - x + \frac{3}{2}x^2 = \epsilon \cos\left(\frac{t}{\epsilon_1}\right)$$

where $\epsilon << 1$ and $\epsilon_1 << 1$. In this case the Melnikov function takes the form

$$(11.38) \qquad \hat{M}(\alpha, \epsilon) = -\frac{\pi}{\epsilon} \operatorname{cosech}\left(\frac{\pi}{\epsilon}\right) \sin\left(\frac{\alpha}{\epsilon}\right).$$

Hence $\hat{M}(\alpha, \epsilon)$ cannot be the leading term of the expansion of $\tilde{d}$ in terms of ϵ. See also Holmes, Marsden and Scheule [11]. This is a serious limitation of the perturbation method used in the theory of the Melnikov function we developed before and in fact it relates to one of the fundamental problems in dynamics since the time of Poincaré. Resolution of this difficulty has to wait for future study.

REFERENCES

[1] Arnold, V.I., *Mathematical Methods of Classical Mechanics*, Springer-Verlag, Berlin, New York, 1978.

[2] Arnold, V.I., *Instability of dynamical systems with several degrees of freedom*, Sov. Math. Dokl., **5**(1964), 581-585.

[3] Birkhoff, G.D., *Nouvelles recherches sur les systemes dynamiques*, Mem. Pont. Acad. Sci. Novi Lyncaei, **1**(1935), 85-216.

[4] Chow, S.N., Hale, J.K., and Mallet-Paret, J., *An example of bifurcation to homoclinic orbits*, J. Diff. Eqns., **37**(1980), 351-373.

[5] Coppel, W.A., *Dichotomies in Stability Theory*, Lecture Notes in Mathematics, **629**, Springer-Verlag, Berlin, New York, 1978,

[6] Gruendler, J., *The existence of homoclinic orbits and the method of Melnikov for systems in* $\mathbb{R}^n$, SIAM J. Math. Anal., **16** (5)(1985), 907-931.

[7] Hale, J.K. *Ordinary Differential Equations*, Krieger, 1980.

[8] Hale, J.K., and Lin, X.-B., *Heteroclinic orbits for retarded functional differential equations*, J. Diff. Eqns. **65**(1986), 175-202.

[9] Holmes, P.J., *Averaging and chaotic motions in forced oscillations*, SIAM J. Appl. Math. **38**(1980), 65-80.

[10] Holmes, P.J., and Marsden, J.E., *Melnikov method and Arnold diffusion for perturbations of integrable Hamiltonian systems*, J. Math. Physics, 4(1982), 669-675.

[11] Holmes, P.J., Marsden, J.E., and Scheurle, J., *Exponentially small splitting of separatrices*, Preprint, 1987.

[12] Melnikov, V.K., *On the stability of the center for time periodic perturbations*, Trans. Moscow Math. Soc. **12**(1963), 1-57.

[13] Meyer, K.R., and Sell, G.R., *Melnikov transforms, Bernoulli bundles, and almost periodic perturbations*, Oscillation, Bifurcation and Chaos, CMS Conference Proceedings, **8**, 1987.

[14] Moser, J., *Stable and Random Motions in Dynamical Systems*, Princeton University Press, 1973.

[15] Newhouse, S.E., *Diffeomorphisms with infinitely many sinks*, Topology **13**(1974), 9-18.

[16] Palmer, K.J., *Exponential dichotomies and transversal homoclinic points*, J. Diff. Eqns. **55**(1984), 225-256.

[17] Poincaré, H., *Les Methodes Nouvelles de la Mecanique celeste*, t.III, Gauthier-Villars, Paris, 1899.

[18] Sacker, R.J., *The splitting index for linear differential systems*, J. Diff. Eqns., **33**(1979), 368-405.

[19] Smale, S., *Diffeomeophisms with many periodic points*, Diff. and Comb. Topology (S.S. Chern, ed.), Princeton University Press (1963), 63-80.

[20] Wiggings, S., *A generalization of the method of Melnikov for detecting chaotic invariant sets*, Preprint.

[21] Wiggings, S., and Holmes, P., *Homoclinic orbits in slowly varying oscillators*, Preprint.

Nonlinear Waves

Andrea Donato
Department of Mathematics
University of Messina, Italy

Abstract

A quasilinear hyperbolic system which is invariant under scaling is transformed into a system whose stationary solutions in canonical variables are the similarity solutions. The propagation of weak discontinuities in this special non-constant state is considered. The transformed system may have a constant state solution that becomes a non-constant state solution in the original variables. The possibility for a similarity line to be a discontinuity line is also investigated.

1 Introduction

A precise definition of *wave*, which embraces all the physical phenomena commonly referred to by this name, is unattainable. From an intuitive point of view, any kind of disturbance which propagates in a medium with a finite speed as time increases may be considered as a wave. The signal associated with the perturbation may distort, change its magnitude and velocity but it is still clearly identifiable.

A diversity of wave propagation problems are described by partial differential equations of hyperbolic type. However, wave phenomena in the above mentioned general sense can also be described by nonlinear parabolic equations which arise out of asymptotic analysis of more complicated systems.

In many physical situations involving wave interactions, the appropriate describing partial differential equations are nonlinear. In the case of hyperbolic systems, the nonlinearity commonly leads to the solution developing discontinuities (shocks) even for smooth and small initial data.

In the case of parabolic equations, the effect of nonlinearity is balanced by dispersive or dissipative effects. The major problem in the study of nonlinear wave phenomena is related to the loss of the principle of superposition.

In the sequel we are mainly interested on the propagation of weak discontinuities compatible with first order quasilinear hyperbolic systems. These are piece-wise continuous solutions with discontinuities in the first order derivatives occurring across a smooth surface said the wave front. Such kind of discontinuities may occur only across characteristic surfaces which evolve in time. Ahead of the wave front we have the "undisturbed solution" characterized by some known solution, whilst on the other side we have the unknown perturbed solution.

There exists a large literature concerning the propagation of weak discontinuities in many physical contexts. In dealing with applications, the "unperturbed state" is very often considered a "constant state" as the most simple solution of the nonlinear governing equations. On the other hand it is well known the difficulty to characterize solutions of nonlinear systems.

A systematic approach in order to characterize exact solutions of partial differential equations is the group analysis that in the meantime allows us to put in evidence physical symmetries. The particular solutions so obtained usually referred to as similarity solutions are restricted in the sense that they must satisfy special initial and boundary conditions. Nevertheless they play an important role as a vehicle of information for the description of more general physical contexts.

Similarity solutions arise in a natural way in axi-symnmetric problems as well as in regions with inhomogeneities. Two orders of problems may be considered. First we ask the possibility of wave propagation compatible with the similarity assumptions and whether, in the one-dimensional case, the similarity lines may be considered as discontinuity lines across which the jump conditions involve similarity variables only. As a second problem we are interested to consider the propagation of weak discontinuities in a non-constant state characterized by a known similarity solution.

Studies of blast waves in fluids give rise to problems of this type and the occurrence of a secondary shock can be determined.

In both the above mentioned problems the canonical variables play a fundamental role. In terms of these variables the transformation group admitted by the governing system may be expressed as a traslation. When the group of invariance is the dilatation group, by using in a proper way the canonical variables, the constant solution of the transformed system may appear to be a special non-constant solution in the original variables.

Consequently, the procedure outlined allows not only to study wave propagation into a "non-constant state" as a propagation into a "constant state" in the new variables, but also to characterize special similarity solutions. The usual invariant solutions are characterized as stationary solutions of the transformed system.

2 Hyperbolicity

Wave phenomena of hyperbolic type are described by first order quasilinear systems of equations in matrix form, namely:

$$A^0(\mathbf{u},\mathbf{x},t)\frac{\partial \mathbf{u}}{\partial t} + A^i(\mathbf{u},\mathbf{x},t)\frac{\partial \mathbf{u}}{\partial x_i} = \mathbf{B}(\mathbf{u},\mathbf{x},t) \tag{1}$$

where $\mathbf{u}(\mathbf{x},t)$ is an unknown column vector with elements $(u_1(\mathbf{x},t),$ $u_2(\mathbf{x},t), \ldots, u_N(\mathbf{x},t))$ defined for $\mathbf{x} = (x_1, x_2, \ldots, x_n) \in R^n$, $t \in R^+$. The $N \times N$ matrices A^α ($\alpha = 0, 1, 2, \ldots, n$) and the column vector $\mathbf{B}$ of N elements are assumed given. On introduction of the matrix $A_n = \sum_{i=1}^{n} A^i n_i$, $\mathbf{n} \in R^n$ then we introduce the following definition:

Definition 1 *The system 1 is said to be hyperbolic in the t-direction if*

$$\det(A^0) \neq 0 \tag{2}$$

and the eigenvalue problem

$$A_n \mathbf{d}_K = \lambda A^0 \mathbf{d}_K \tag{3}$$

has, for every $\mathbf{n} \in R^n$, *only real eigenvalues* $\lambda(\mathbf{u},\mathbf{n})$ *and a corresponding complete family of right eigenvectors* $\mathbf{d}_K(\mathbf{u},\mathbf{n})$ *together with a complete family of left eigenvectors* l_J *satisfying*

$$l_J A_n = \lambda l_J A^0 \quad , \tag{4}$$

$(J, K = 1, 2, \ldots, p$ *if* λ *has multiplicity* p). *The system is said to be strictly hyperbolic if, moreover,*

$$\det(A_n - \lambda A^0) = 0 \tag{5}$$

has N *distinct real roots.*

The hyperbolicity so defined depends not only on the point $(\mathbf{x}, t) \in R^n \times R^+$ but also on the solution $\mathbf{u}(\mathbf{x}, t)$. Moreover, the assumed conditions constitute a necessary requirement for the Cauchy problem for the system 1 to be well posed [1].

Physical processes are commonly modeled by balance laws

$$\frac{\partial \mathbf{F}^0(\mathbf{u})}{\partial t} + \frac{\partial \mathbf{F}^i(\mathbf{u})}{\partial x_i} = \mathbf{f}(\mathbf{u}) \tag{6}$$

which involve the unknown field $\mathbf{u}(\mathbf{x}, t)$. These are particular cases of the system 1 when :

$$A^0 = \frac{\partial \mathbf{F}^0}{\partial \mathbf{u}} = \nabla_\mathbf{u} \mathbf{F}^0 \qquad A^i = \frac{\partial \mathbf{F}^i}{\partial \mathbf{u}} = \nabla_\mathbf{u} \mathbf{F}^i \tag{7}$$

Important examples of such conservation laws occur in gas dynamics, magnetofluid dynamics and nonlinear elasticity ([2], [3], [4], [5], [6]).

If the matrices A^α are symmetric and, moreover, A^0 is definite positive then, by elementary linear algebra, it may be seen that the system 6 that results is automatically hyperbolic. For such a system the resulting Cauchy problem is locally well-posed under suitable smooth initial data [7].

We now assume that the system 6 is compatible with a supplementary scalar conservation law:

$$\frac{\partial h(\mathbf{u})}{\partial t} + \frac{\partial h^i(\mathbf{u})}{\partial x_i} = g(\mathbf{u}) \tag{8}$$

where $h(\mathbf{u})$ is a strictly convex function of the field $\mathbf{u}$ in some open domain.

This requirement is often fulfilled in dealing with physical models. Thus, a conservation law such as 8 arises as the *Entropy Principle* in Continuum Mechanics when $g \leq 0$ or as the *energy equation* in the Maxwell system of the Electromagnetic theory. Scalar *conservation laws* derived from variational principles also adopt such a form [8].

The equation 8 may be considered as a differential constraint to be satisfied by the solutions of the system 6. Restrictions on the structure of the system are thereby implied. Such problems have been studied in [9], [10], [11], [12], [13]. As A_0 is non-singular we may choose $\mathbf{v} = \mathbf{F}^0(\mathbf{u})$ as field variable and the compatibility conditions of 6 and 8 are:

$$\nabla_\mathbf{v} h \nabla_\mathbf{v} \mathbf{F}^i = \nabla_\mathbf{v} h^i \qquad\qquad \nabla_\mathbf{v} h \cdot f = g \qquad\qquad (9)$$

On introduction of the transformation of variables :

$$\begin{aligned} h' &= h - \mathbf{u}'^T \mathbf{F}^0 \\ h'^i &= h^i - \mathbf{u}'^T \mathbf{F}^i \\ \mathbf{u}'^T &= \nabla_\mathbf{v} h \end{aligned} \qquad\qquad (10)$$

we get

$$\mathbf{F}^{0T} = \frac{\partial h'}{\partial \mathbf{u}'} \qquad\qquad \mathbf{F}^{iT} = \frac{\partial h'^i}{\partial \mathbf{u}'} \qquad\qquad (11)$$

and the system 6 takes the form

$$\frac{\partial^2 h'}{\partial \mathbf{u}' \partial \mathbf{u}'} \frac{\partial \mathbf{u}'}{\partial t} + \frac{\partial^2 h'^i}{\partial \mathbf{u}' \partial \mathbf{u}'} \frac{\partial \mathbf{u}'}{\partial x_i} = \mathbf{f} \qquad\qquad (12)$$

This is a symmetric and hyperbolic system.

The Euler equations for ideal Gas Dynamics neglecting external forces take the form 6 with

$$\mathbf{F}^0(\mathbf{u}) = \begin{bmatrix} \rho \\ \rho\mathbf{q} \\ \epsilon \end{bmatrix} \qquad \mathbf{F}^i(\mathbf{u}) = \begin{bmatrix} \rho q_i \\ \rho q_i \mathbf{q} + p\mathbf{c}_i \\ (\epsilon + p)\rho q_i \end{bmatrix} \qquad \mathbf{f} = 0 \quad (13)$$

where $\epsilon = \rho q^2/2 + \rho e$, $i = e + p/\rho$, $di = Tds + dp/\rho$, $\mathbf{q}$ is the particle velocity, i the enthalpy, $p(\rho,e)$ the pressure depending on the density ρ and the internal energy e of the fluid; T will be the absolute temperature and $S(\rho,e)$ the entropy.

The supplementary conservation law to be associated to the system in this special case is of the form 8 with:

$$h = -\rho S \qquad\qquad h^i = -\rho S q_i \qquad\qquad g = 0 \qquad\qquad (14)$$

expressing the law of conservation of entropy.

Taking into account that

$$TdS = de - \frac{p}{\rho^2}d\rho \quad , \qquad\qquad (15)$$

the following relation is valid:

$$Tdh = \mathbf{q}d(\rho\mathbf{q}) + (G - \frac{q^2}{2})d\rho - d\epsilon \qquad\qquad (16)$$

where $G = i - TS$ is the free enthalpy.

Consequently, by 11 we obtain:

$$\mathbf{u}'^T = \left[\frac{G - q^2/2}{T}, \frac{\mathbf{q}}{T}, -\frac{1}{T}\right] \qquad\qquad (17)$$

The function h may be shown [14] to be a convex function of the field variables under the usual restrictions for a gas in the thermodynamic equilibrium. Again, by 11 it follows:

$$h' = \frac{p}{T} \qquad\qquad h'^i = \frac{pq_i}{T} \quad . \qquad\qquad (18)$$

3   Discontinuity waves

The condition $\det(A^0) \neq 0$ valid for hyperbolic systems allows us, by multiplying 1 by the inverse matrix $(A^0)^{-1}$, to write the system as:

$$\frac{\partial \mathbf{u}}{\partial t} + A^i(\mathbf{u})\frac{\partial \mathbf{u}}{\partial x_i} = \mathbf{B}(\mathbf{u}, \mathbf{x}, t) \qquad\qquad (19)$$

where now $A^i(\mathbf{u}, \mathbf{x}, t)$ and $\mathbf{B}(\mathbf{u}, \mathbf{x}, t)$ are the result of the original quantities multiplied by $(A^0)^{-1}$.

For simplicity we assume that the matrices A^i do not depend on $\mathbf{x}$ and t. Consider a smooth moving surface $\phi(\mathbf{x}, t) = 0$ (wave front) separating $R^n \times R^+$ into two subspaces in each of which $\mathbf{u}(\mathbf{x}, t)$ is a C^1 function of its arguments. Ahead of the wave front we have a known solution $\mathbf{u}_0(\mathbf{x}, t)$ and behind the wave front the unknown perturbed field $\mathbf{u}(\mathbf{x}, t)$. Across the wave front $\mathbf{u} = \mathbf{u}_0$ but there are discontinuities in the first order derivatives along the normal to the wave front. If the jump is denoted by

$$[\cdot] = (\cdot)_{\phi=0+} - (\cdot)_{\phi=0-} \tag{20}$$

then

$$\mathbf{u} - \mathbf{u}_0 = [\mathbf{u}] = 0 \qquad\qquad [\frac{\partial \mathbf{u}}{\partial \phi}] = \Pi \neq 0 \tag{21}$$

The normal speed of propagation and the unit normal vector of the wave front are:

$$\frac{d\mathbf{x}}{dt}\mathbf{n} = \lambda = -\frac{\phi_t}{|grad\phi|}$$

$$\phi_t = \frac{\partial\phi}{\partial t} \tag{22}$$

$$\mathbf{n} = \frac{grad\phi}{|grad\phi|}$$

The well known Hadamard relations now yield:

$$[\frac{\partial \mathbf{u}}{\partial t}] = \phi_t \Pi$$

$$\tag{23}$$

$$[\frac{\partial \mathbf{u}}{\partial \mathbf{x}}] = grad\phi \Pi$$

On use of 23 and 24 in 19 we readily obtain the following compatibility conditions to be satisfied by the discontinuity vector Π:

$$(A_n(\mathbf{u}_0) - \lambda I)\Pi = 0 \tag{24}$$

The normal velocities λ are found as eigenvalues of the matrix $A_{0n} = A_n(\mathbf{u}_0)$ evaluated in $\mathbf{u}_0(\mathbf{x}, t)$, the known unperturbed field, while the jump vector $\mathbf{\Pi}$ is a linear combination of the right eigenvectors (corresponding to λ) evaluated in $\mathbf{u}_0$:

$$\mathbf{\Pi} = \sum_{K=1}^{p} \Pi^K \mathbf{d}_{0K} \tag{25}$$

On the wave front $\lambda = \lambda(\mathbf{u}_0, \mathbf{n}) = \lambda_0$, whereas, by 23

$$\Psi_0(\mathbf{u}_0, \phi_t, grad\phi) = \phi_t + \lambda_0|grad\phi| = 0 \tag{26}$$

The latter partial differential equation can be solved by introducing the characteristic rays defined by

$$t = \sigma$$
$$\Lambda = \frac{d\mathbf{x}}{d\sigma} = \frac{\partial \Psi_0}{\partial(grad\phi)}$$

$$\tag{27}$$

$$\frac{d\phi_t}{d\sigma} = -\frac{\partial \Psi_0}{\partial t}$$
$$\frac{d(grad\phi)}{d\sigma} = \frac{\partial \Psi_0}{\partial \mathbf{x}}$$

where σ denotes the time along the rays and Λ the ray velocity in terms of λ expressed by

$$\Lambda = \lambda\mathbf{n} - (\frac{\partial \lambda}{\partial \mathbf{n}}\mathbf{n})\mathbf{n} + \frac{\partial \lambda}{\partial \mathbf{n}} \tag{28}$$

Given the initial surface $\phi_0(\xi) = 0$ on the n-dimensional sub-manifold in parametric form $\mathbf{x} = \mathbf{x}(\xi)$, it is possible, at least in principle, to solve the Cauchy problem associated with the system 28 in the form:

$$t = \sigma$$
$$\mathbf{x} = \mathbf{x}(\xi, \sigma) \tag{29}$$

Moreover, provided $\det \|\partial x_i/\partial \xi_j\| \neq 0$ it is possible to solve 29 with respect to ξ obtaining the wave surface $\phi = \phi_0(\mathbf{x}, t) = 0$. The discontinuity vector $\mathbf{\Pi}$ propagates along the characteristic rays (the so called bi-characteristics of the system 19) with velocity Λ [15].

Details concerning the evolution equation for $\mathbf{\Pi}$ which obeys, in the general n-dimensional case, a differential system of Bernoulli type, may be found in [15].

In the one-dimensional case, that will be considered in the sequel, the evolution law for the components of Π in 25 is [16]:

$$\Delta_{0IK}\frac{d\Pi_K}{d\sigma} + \Delta_{0IK}\phi_x a_J \Pi_J \Pi_K + h_{IK}\Pi_K = 0 \qquad (30)$$

where

$$\Delta_{0IK} = \mathbf{l}_{0I}\mathbf{d}_{0K}$$
$$a_I = (\tfrac{\partial\lambda}{\partial\mathbf{u}}\mathbf{d}_I)_0$$
$$h_{IK} = ((\mathbf{l}_I\tfrac{\partial\mathbf{u}}{\partial x})(\tfrac{\partial\lambda}{\partial\mathbf{u}}\mathbf{d}_K) + \mathbf{d}_K((\tfrac{\partial\mathbf{l}_I}{\partial\mathbf{u}})^T - \tfrac{\partial\mathbf{l}_I}{\partial\mathbf{u}})\tfrac{d\mathbf{u}}{d\sigma} \qquad (31)$$
$$+ \tfrac{d\Delta_{IK}}{d\sigma} - \tfrac{\partial}{\partial\mathbf{u}}(\mathbf{l}_I\mathbf{B})\mathbf{d}_K + \mathbf{l}_I\tfrac{\partial\mathbf{d}_K}{\partial\mathbf{u}}\tfrac{d\mathbf{u}}{d\sigma})_0$$
$$\tfrac{d}{d\sigma} = \tfrac{\partial}{\partial t} + \lambda\tfrac{\partial}{\partial x}$$

Since the system 19 is hyperbolic, the matrix $\Delta_{0IK} = \mathbf{l}_{0I}\mathbf{d}_{0K}$ must be non-singular and the left and the right eigenvectors may be chosen in order that it becomes the unitary matrix. The system of equations may be formally integrated to give

$$\Pi_K = \frac{\eta_K}{\varphi} \qquad (32)$$

where η_K and φ are solutions of the following ordinary differential equations:

$$\tfrac{d\eta_K}{d\sigma} + h_{KI}(\sigma)\eta_I = 0$$
$$\tfrac{d\varphi}{d\sigma} + \phi_x a_I \eta_I = 0 \qquad (33)$$
$$\eta_K(0) = \Pi_K(0) \qquad \phi(0) = 1$$

where ϕ_x , by 28, satisfies

$$\frac{d\phi_x}{d\sigma} + (\frac{\partial\lambda}{\partial\mathbf{u}}\mathbf{u}_x)_0\phi_x = 0 \qquad \phi_x(0) = \phi_{0x} \qquad (34)$$

For strictly hyperbolic systems or for systems with a characteristic velocity of multiplicity 1 we have simply (choosing $\Delta = \mathbf{l}_0\mathbf{d}_0 = 1$):

$$\Pi = \frac{\Pi(0)\exp(-\int_0^\sigma h(\sigma)d\sigma}{1 + \Pi(0)\int_0^\sigma a(\sigma)\phi_x \exp(-\int_0^\sigma h(s)ds)d\sigma} \qquad (35)$$

It can be easily shown that if $a_I \neq 0$ there may exist a time t_c such that the denominator of 32 vanishes and the discontinuities become unbounded.

This corresponds, usually, to the formation of shocks resulting from the non-linearity. A discontinuity wave propagating with velocity λ such that $a_I = \frac{\partial \lambda}{\partial \mathbf{u}} \mathbf{d}_I = 0$ is said an *exceptional wave* since the discontinuity behaves as in the linear case. It is worthwhile to note that, for conservative systems, every discontinuity wave having a multiplicity more than one is exceptional [17], [18]. This is illustrated by material waves in three-dimensional fluid dynamics.

We remark that all the above considerations are of practical interest if a solution $\mathbf{u}_0(\mathbf{x}, t)$ in the unperturbed state is known. The most obvious solutions to seek are constant solutions satisfying the condition $\mathbf{B}(\mathbf{u}_0) = 0$. These are common in the literature. Here, however, we would like to consider the case when the propagation is into a non-constant state characterized by an exact solution of the system 19 invariant with respect to a one-parameter group of transformations corresponding to the stretching group. Problems such as these arise in the analysis of wave propagation in media with cylindrical or spherical symmetry as well as in regions with inhomogeneities [4], [19], [20]

4   Canonical variables

Here we limit ourselves to one-dimensional problems described by first order systems in the form

$$\frac{\partial \mathbf{u}}{\partial t} + A(\mathbf{u}, x, t)\frac{\partial \mathbf{u}}{\partial x} = \mathbf{B}(\mathbf{u}, x, t) \qquad\qquad A = \frac{\partial \mathbf{F}}{\partial \mathbf{u}} \qquad (36)$$

These we assume to be invariant with respect to a one-parameter infinitesimal group of transformations characterized by the generators:

$$\begin{aligned} T &= T(x, t, \mathbf{u}) \\ X &= X(x, t, \mathbf{u}) \\ \eta &= \eta(x, t, \mathbf{u}) \end{aligned} \qquad (37)$$

This allows us to find the finite invariance group G:

$$
\begin{aligned}
t^* &= t^*(x,t,\mathbf{u};\epsilon)\\
x^* &= x^*(x,t,\mathbf{u};\epsilon)\\
\mathbf{u}^* &= \mathbf{u}^*(x,t,\mathbf{u};\epsilon)
\end{aligned}
\tag{38}
$$

by solving a Cauchy problem as is stated by Lie's theorem [21], [22], [23].

In principle it is always possible to consider a canonical transformation:

$$
\begin{aligned}
\tau &= \tau(x,t,\mathbf{u})\\
\xi &= \xi(x,t,\mathbf{u})\\
\mathbf{w} &= \mathbf{w}(x,t,\mathbf{u})
\end{aligned}
\tag{39}
$$

in terms of which the group G reduces simply to a translation in the variable τ:

$$
\begin{aligned}
\tau^* &= \tau+\epsilon\\
\xi^* &= \xi\\
\mathbf{w}^* &= \mathbf{w}
\end{aligned}
\tag{40}
$$

and the associated generators are :

$$
\begin{aligned}
\Theta &= 1\\
\chi &= 0\\
\Gamma &= 0
\end{aligned}
\tag{41}
$$

The infinitesimal operator of the group G

$$
\zeta\cdot\partial(\cdot) = X(x,t,\mathbf{u})\frac{\partial(\cdot)}{\partial x} + T(x,t,\mathbf{u})\frac{\partial(\cdot)}{\partial t} + \eta_A(x,t,\mathbf{u})\frac{\partial(\cdot)}{\partial u_A}
\tag{42}
$$

in the new variables becomes

$$
\begin{aligned}
\zeta'\cdot\partial(\cdot) &= \Theta\frac{\partial(\cdot)}{\partial\tau} + \chi\frac{\partial(\cdot)}{\partial\xi} + \Gamma_A\frac{\partial(\cdot)}{\partial w_A} =\\
&= (\zeta\cdot\partial)\tau\frac{\partial(\cdot)}{\partial\tau} + (\zeta\cdot\partial)\xi\frac{\partial(\cdot)}{\partial\xi} + (\zeta\cdot\partial)\Gamma_A\frac{\partial(\cdot)}{\partial w_A}
\end{aligned}
\tag{43}
$$

where, in order to have a translation, we require

$$
\begin{aligned}
(\zeta \cdot \partial)\tau &= 1 \\
(\zeta \cdot \partial)\xi &= 0 \\
(\zeta \cdot \partial)\Gamma_A &= 0 \quad .
\end{aligned}
\tag{44}
$$

These represent a system of differential equations to determine the canonical variables ξ, τ and w_A.

The method of characteristics produces the solution in the form

$$
\begin{aligned}
\tau &= \int \frac{dt}{T[x(t,\sigma,w_1,\ldots,w_N),t,u_A(t,\sigma,w_1,\ldots,w_N)]} \\
\xi &= \xi(\sigma,W_1,\ldots,W_N) \\
w_A &= w_A(\sigma,W_1,\ldots,W_N)
\end{aligned}
\tag{45}
$$

where the parameters σ, W_1, ..., W_N are obtained by integration of the characteristic equations:

$$
\frac{dx}{X(x,t,\mathbf{u})} = \frac{dt}{T(x,t,\mathbf{u})} = \frac{du_1}{\eta_1(x,t,\mathbf{u})} = \cdots = \frac{du_N}{\eta_N(x,t,\mathbf{u})} = \frac{d\tau}{1}
\tag{46}
$$

A possible choice of canonical variables rendering the group G a τ-translation is:

$$
\begin{aligned}
\tau &= \int \frac{dt}{T} \\
\xi &= \sigma \\
w_A &= W_A
\end{aligned}
\tag{47}
$$

For instance, if the group G is the dilatation group having generators

$$
\begin{aligned}
T &= \gamma t \\
X &= x \\
\eta_A &= \alpha_A u_A
\end{aligned}
\tag{48}
$$

with α_A and $\gamma \neq 1$ suitable constants, then the canonical variables are:

$$
\tau = \frac{1}{\gamma}\ln t
$$

$$\xi \;=\; \frac{x}{t^{1/\gamma}} \tag{49}$$

$$w_A \;=\; t^{-\alpha_A/\gamma} u_A$$

We remark that, by expressing the system 36 in terms of canonical variables the solutions depending only on ξ are, infact, the invariant solutions with respect to the group G since the invariance condition is

$$X\frac{\partial u_A}{\partial x} + T\frac{\partial u_A}{\partial t} = \eta_A \tag{50}$$

Let us consider now the case when the system 36 is invariant with respect to the dilatation group. In such a case $\mathbf{F(u)}$ and $\mathbf{B(u,}x,t)$ must satisfy restrictions in their functional forms. These have been determined in [24]. Thus, if we assume that at least one of the α_A is different from zero, say $\alpha_1 = \alpha$, with the corresponding component of $\mathbf{u}$ denoted by $u_1 = u$, the invariance requires the following expressions for $\mathbf{F}$ and $\mathbf{B}$:

$$F_A \;=\; u^{\rho_A/\alpha} f_A(U_1, U_2, \ldots, 1, \ldots, U_N)$$

$$\tag{51}$$

$$B_A \;=\; u^{(\rho_A-1)/\alpha} b_A(U_1, U_2, \ldots, 1, \ldots, U_N, x/u^{1/\alpha}, t/u^{\gamma/\alpha})$$

where $U_A = u_A/u^{\alpha_A/\alpha}$, $\rho_A = \alpha_A - \gamma + 1$, $A = 1, 2, \ldots, N$.

To these relations the transformations of variables 50 is applied such that

$$U_A \;=\; w_A/w^{\alpha_A/\alpha}$$

$$w_1 \;=\; w \tag{52}$$

$$u \;=\; t^{\alpha_A/\gamma} w$$

whereupon 52 becomes

$$F_A \;=\; t^{\rho_A/\gamma} w^{\rho_A/\alpha} f_A(U_1, \ldots, 1, \ldots, U_N) = t^{\rho_A/\gamma} \tilde{F}_A$$

$$B_A \;=\; t^{(\rho_A-1)/\gamma} w^{(\rho_A-1)/\alpha} b_A(U_1, \ldots, 1, \ldots, U_N, \frac{\xi}{w^{1/\alpha}}, \frac{1}{w^{\gamma/\alpha}})$$

$$\;=\; t^{(\rho_A-1)/\gamma} \tilde{B}_A \tag{53}$$

In the new variables 50, the governing system 36 in component form, becomes

$$\frac{\partial w_A}{\partial \tau} - \xi \frac{\partial w_A}{\partial \xi} + \gamma \frac{\partial \tilde{F}_A}{\partial \xi} = \gamma \tilde{B}_A - \alpha_A w_A \qquad (54)$$

It is now assumed that $\tilde{B}_A$ may be written as

$$\tilde{B}_A = \hat{B}_A/\xi = \frac{w^{\rho_A/\alpha}}{\xi} \tilde{b}_A(U_1, \ldots, 1, \ldots, U_N, \frac{w^{(1-\gamma)/\alpha}}{\xi}) \qquad (55)$$

In the new variables

$$\begin{aligned} w_A &= \xi^{\alpha_A/(1-\gamma)} v_A \\ \eta &= \ln \xi \end{aligned} \qquad (56)$$

the system 54 becomes :

$$\frac{\partial v_A}{\partial \tau} - \frac{\partial v_A}{\partial \eta} + \gamma \frac{\partial \hat{F}_A}{\partial \eta} = \gamma \hat{B}_A + \frac{\gamma}{\gamma - 1}(\alpha_A v_A - \rho_A \hat{F}_A) \qquad (57)$$

where $\hat{F}_A = \tilde{F}_A(v_A)$, $\hat{B}_A = \tilde{B}_A(v_A)$.

Consequently, under the assumed conditions, the transformations 53 and 56 change 36 into 57, with independent variables that do not appear explicitly in the coefficients. The condition 56 means that the system 54 is invariant with respect to the dilatation group

$$\begin{aligned} \tau^* &= \tau \\ \xi^* &= \mu\xi \qquad\qquad \mu \in R - \{0\} \\ v_A^* &= \mu^{\alpha_A/(1-\gamma)} v_A \end{aligned} \qquad (58)$$

Since, in a certain sense, this result generalizes to systems of equations the ideas of L. Dresner [20], we call this group the *associated group* of the principal dilatation invariance group for the system 36:

$$\begin{aligned} t^* &= \epsilon^\gamma t \\ x^* &= \epsilon x \\ u_A^* &= \epsilon^{\alpha_A} u_A \qquad (A = 1, \ldots, N) \end{aligned} \qquad (59)$$

In terms of the original variables the existence of the *associated group* , as can be easily verified, is assured by requiring that B_A in 52 is expressed by:

$$B_A = \frac{u^{\rho_A/\alpha}}{x} \tilde{b}_A(U_1, \ldots, 1, \ldots, U_N), u^{(1-\gamma)/\alpha}\frac{t}{x}) \tag{60}$$

In other words the functions $\tilde{b}_A$ must be homogeneous of degree -1 in the last two variables .

In the more usual case when B_A does not depend on t, the form (4.16) remains the same with the last variable, a combination of u, x and t, dropped.

From the relations, 50, 53 and 56 it follows that the system 36 admits a solution of the form

$$u_A = (\frac{x}{t})^{\alpha_A/(1-\gamma)} v_A \tag{61}$$

provided that the *associated group* is admitted. The similarity solutions $v_{0A}(\eta)$ are obtained by solving the autonomous system of ordinary differential equations which results from 57 when the dependence on τ is omitted, namely:

$$-\frac{1}{\gamma}\frac{dv_{0A}}{d\eta} + \frac{d\hat{F}_{0A}}{d\eta} = \gamma \hat{B}_{0A} + \frac{1}{\gamma - 1}(\alpha_A v_{0A} - \rho_A \hat{F}_{0A}) = B_{0A} \tag{62}$$

There "$_0$" means that a quantity is evaluated for $v_A = v_{0A}$. If the matrix

$$\hat{A}_{0MC} = \frac{\partial \hat{F}_{0M}}{\partial v_{0C}} \tag{63}$$

has N real and distinct eigenvalues $\hat{\lambda}_0$ with the corresponding left and right eigenvectors linearly independent, that is the governing system is strictly hyperbolic, the system 62 may be solved with respect to $dv_{0A}/d\eta$ to obtain, by the Cramer rule:

$$\frac{dv_{0A}}{d\eta} = \frac{\Delta_A}{(\frac{1}{\gamma} - \hat{\lambda}_{01})(\frac{1}{\gamma} - \hat{\lambda}_{02})\ldots(\frac{1}{\gamma} - \hat{\lambda}_{0N})} \tag{64}$$

$$\Delta_A = B_{0M}C_{MA}$$

where C_{MA} is the transpose of the matrix formed with the cofactors of $A_{0MC} - \delta_{MC}/\gamma$ and $\hat{\lambda}_{01}$, $\hat{\lambda}_{02}$, ... , $\hat{\lambda}_{0N}$ the eigenvalues of $\hat{A}_{0MC}$. There is a possible singularity on a curve $\eta = constant$ if one of the $\hat{\lambda}_0$ takes the value $1/\gamma$. It is well known that singularities can occur only through characteristics of the original system expressed by :

$$\frac{dx}{dt} = \lambda = \frac{x}{t}\hat{\lambda}_0 \tag{65}$$

where λ is an eigenvalue of the matrix A in 36. Since on a line $\eta = constant$ we have simply

$$\frac{dx}{dt} = \frac{x}{\gamma t} \tag{66}$$

it follows $\hat{\lambda}_0 = 1/\gamma$ in order for this line to be a characteristic. To avoid this discontinuity we have to require that also $\Delta_A = 0$ when $det(\hat{A}_0 - \frac{1}{\gamma}I) = 0$.

From the system 62 it follows that

$$(\hat{\lambda}_0 - \frac{1}{\gamma})\hat{l}_{0A}\frac{dv_{0A}}{d\eta} = \hat{l}_{0A}B_{0A} \tag{67}$$

with $\hat{l}_0$ the left eigenvector of the matrix $\hat{A}_0$ corresponding to $\hat{\lambda}_0$ that can be chosen in one of the alternative forms

$$\hat{l}_{0A} \propto C_{A1}^\lambda \propto C_{A2}^\lambda \propto C_{A3}^\lambda \propto ... \propto C_{AN}^\lambda \tag{68}$$

where C_{AM}^λ is the transpose of the matrix formed with the cofactors of $\hat{A}_0 - \lambda_0 I$.

It can be easily seen that if one of the $\hat{\lambda}_0 = \frac{1}{\gamma}$ then the corresponding left hand-side of 67 is zero and, consequently, there is one of the Δ_A in 65 that is automatically zero since

$$(C_{AM}^\lambda)_{\hat{\lambda}_0=1/\gamma} = C_{AM} \quad . \tag{69}$$

The *limiting characteristic* where the singularity appears is characterized by a value of η, say η_c, such that $\exp(\eta_c) = x/t^\gamma$ corresponding to the $\hat{\lambda}_0(v_{0L}) = 1/\gamma$ that makes the denominator of 65 zero. In order that the solution to be continued for value of η close to η_c , since one of the Δ_A is zero, it is enough to require that the others $N - 1$ Δ_A are zero. In order to determine η_c we assume to have at least one boundary condition associated with the system 62 (see [20], [25]) of the type $v_{01}(0) = v_0$. Now we guess a value $\tilde{\eta}_c$ and calculate $\tilde{v}_{0L}(\tilde{\eta}_c)$ by using $det(\hat{A}_0 - \hat{\lambda}_0 I) = 0$ and $\Delta_A = 0$. The $\tilde{v}_{0L}$ so determined may be used as boundary conditions to integrate the system 62 inwards to the origin $\eta = 0$ obtaining $\tilde{v}_{0L}(0)$. By the invariance conditions we may determine μ as $\tilde{v}_{01} = \mu^{\alpha_1/(1-\gamma)} v_0$ and by using $\xi^* = \mu\xi$ we get:

$$\tilde{\eta}_c = \eta_c + \ln|\mu| \quad . \tag{70}$$

5   Wave propagation in a "non-constant state"

Here we would like to study the propagation of weak discontinuities, in the field variables v_L compatible with the system 57 across a curve $\Phi(\tau, \eta) = 0$ in a non-constant state characterized by a similarity solution $v_{0L}(\eta)$ satisfying 62 [24].

This problem requires that the non-constant state, in which the discontinuity propagates, be dependent only on one independent variable called the *space variable*.

For simplicity the system 57 studied hereafter is supposed to be strictly hyperbolic with the characteristic velocities solutions of

$$det(\gamma\hat{A} - (\Lambda + 1)I) = 0$$
$$\Lambda = -\frac{\Phi_\tau}{\Phi_\eta} = \frac{d\eta}{d\tau} \tag{71}$$
$$\hat{A} = \nabla_v \hat{\mathbf{F}}$$

consequently, the link between the characteristic velocities Λ and the eigenvalues $\hat{\lambda}$ of the matrix $\hat{A}$ are given by

$$\Lambda = \gamma\hat{\lambda} - 1 \tag{72}$$

Under these assumptions and with $\hat{\mathbf{d}}_{0L} = \hat{\mathbf{d}}_L(v_{0L})$ after 25 it follows

$$\delta v_A = \pi \hat{d}_{0A}$$

$$\delta(\cdot) \;=\; (\frac{\partial(\cdot)}{\partial\Phi})_{\Phi=0^+} - (\frac{\partial(\cdot)}{\partial\Phi})_{\Phi=0^-} \quad . \tag{73}$$

In the original variables u_A is given by 61 and its first derivatives are discontinuous across $\Phi^*(x,t) = \Phi(\tau(t),\eta(x,t))$ with

$$\delta^*(\cdot) = (\frac{\partial(\cdot)}{\partial\Phi^*})_{\Phi^*=0^+} - (\frac{\partial(\cdot)}{\partial\Phi^*})_{\Phi^*=0^-} \tag{74}$$

there follows

$$\delta^* u_A = (\frac{x}{t})^{\alpha_A/(1-\gamma)} \delta v_A \tag{75}$$

Moreover as

$$\frac{\partial}{\partial t} = \frac{1}{\gamma t}(\frac{\partial}{\partial\tau} - \frac{\partial}{\partial\eta}) \qquad \frac{\partial}{\partial x} = \frac{1}{x}\frac{\partial}{\partial\eta} \tag{76}$$

there follows that

$$\frac{dx}{dt} = \lambda = \frac{x}{\gamma t}(\Lambda + 1) = \frac{x}{t}\hat{\lambda} \quad . \tag{77}$$

The characteristic rays associated with 57 are

$$\tau = \sigma \quad \tau - \tau_0 = \int_{\eta_0}^{\eta} \frac{1}{\Lambda_0} d\eta \tag{78}$$

so instead of using σ, the time along the rays, we may use η in order to compute the discontinuity in our case given by

$$\Pi = \frac{\Pi(0)\exp(-\int_{\eta_0}^{\eta} h(s)\frac{1}{\Lambda_0}ds}{1 + \Pi(0)\int_{\eta_0}^{\eta} \frac{1}{\Lambda_0} a(\eta)\phi_\eta \exp(-\int_{\eta_0}^{\eta} h(s)\frac{ds}{\Lambda_0})d\eta} \tag{79}$$

where

$$
\begin{aligned}
a(\eta) &= \gamma(\nabla_{\mathbf{v}}\hat{\lambda}\hat{\mathbf{d}})_0 \\
h &= \frac{\gamma}{(\hat{\mathbf{l}}\hat{\mathbf{d}})_0}\{(\hat{\mathbf{l}}\frac{d\mathbf{v}}{d\eta})(\frac{\partial\hat{\lambda}}{\partial\mathbf{v}}\hat{\mathbf{d}}) + \frac{\Lambda}{\gamma}\hat{\mathbf{d}}((\frac{\partial\hat{\mathbf{l}}}{\partial\mathbf{v}})^T - \frac{\partial\hat{\mathbf{l}}}{\partial\mathbf{v}})\frac{d\mathbf{v}}{d\eta} \\
&\quad + \frac{\Lambda}{\gamma}\frac{d\hat{\mathbf{l}}\hat{\mathbf{d}}}{d\eta} - \frac{\partial\hat{\mathbf{l}}B}{\partial\mathbf{v}}\hat{\mathbf{d}}\}_0
\end{aligned}
\tag{80}
$$

For what concerns ϕ_η, given the initial discontinuity line $\phi_0(\eta)$, after 78 , we get $\phi = \phi_0(\tau - \int_{\eta_0}^0 \frac{1}{\Lambda_0} d\eta)$ so it follows $\phi_\eta = -\frac{1}{\Lambda_0}\phi_0'$. Moreover, we may use 65 in 81 in order to eliminate $\frac{d\mathbf{v_0}}{d\eta}$. In the case when $h = 0$ we have the following relation to determine the discontinuity

$$\Pi = \frac{\Pi(0)}{1 + \Pi(0) \int_{\eta_0}^\eta \frac{1}{\Lambda_0} a(\eta)\phi_\eta d\eta} \tag{81}$$

6 Propagation in a "constant state": physical applications

Let $v_A = v_{0A} = constant$ be a solution of

$$\mathcal{B}(v_{0A}) = \hat{B}_{0A} + \frac{1}{\gamma - 1}(\alpha_A v_{0A} - \rho_A \hat{F}_{0A}) = 0 \quad . \tag{82}$$

Because of 61 this corresponds to a non-constant particular solution of the original system 36, in fact a self-similar solution. As will be seen, the propagation of weak discontinuities in the original variables in such "non-constant" states may be studied as a propagation in a "constant state" in the new variables. In the present special case the characteristic rays are

$$\begin{aligned} \tau &= \sigma \\ \eta &= \eta_0 + \Lambda_0(\tau - \tau_0) \end{aligned} \tag{83}$$

with $\Lambda_0 = constant$. As $\tau = \frac{1}{\gamma}lnt$ and $\eta = \ln\xi = \ln\frac{x}{t^{1/\gamma}}$ we have

$$\frac{x}{x_0} = (\frac{t}{t_0})^{(\Lambda_0+1)/\gamma} \quad . \tag{84}$$

Consequently from 79, 81 there results:

$$\pi = \frac{\pi_0(\frac{t}{t_0})^{-\frac{h_0}{\gamma}}}{1 - \pi_0\frac{a_0}{b_0}\left((\frac{t}{t_0})^{-\frac{h_0}{\gamma}} - 1\right)} \tag{85}$$

where, without loss of generality, we choose $\phi_\eta = 1$ as we are considering a propagation in a constant state.

From 36 the time t_c when the discontinuity wave evolves into a shock wave is given by:

$$t_c = t_0 \left(\frac{h_0 + \pi_0 a_0}{\pi_0 a_0} \right)^{-\frac{\gamma}{h_0}} \tag{86}$$

This occurs when the initial discontinuity is such that:

i) $\pi_0 a_0 > 0$, $h_0 > -\pi_0 a_0$

ii) $\pi_0 a_0 < 0$, $h_0 < |\pi_0 a_0|$.

In the case when $h_0 = 0$ the equation 85 must be substituted by

$$\pi = \frac{\pi_0}{1 + \pi_0 \ln(t/t_0)^{a_0}}) \tag{87}$$

As a first physical example [27], [28] a spherical symmetric motion of a polytropic gas, neglecting any dissipation mechanism, is studied. The equations of motion take the form 36 with:

$$\mathbf{u} = \begin{bmatrix} \rho \\ u \\ p \end{bmatrix} \qquad A = \begin{bmatrix} u & \rho & 0 \\ 0 & u & \frac{1}{\rho} \\ 0 & \Gamma p & u \end{bmatrix} \qquad \mathbf{B} = \begin{bmatrix} -\frac{2\rho u}{r} \\ 0 \\ -\frac{2\Gamma p u}{r} \end{bmatrix} \tag{88}$$

where ρ is the density, u is the particle velocity, p the gas pressure, r the distance from the center of symmetry and Γ the gas index.

Upon identifying $u_1 = \rho$, $u_2 = u$, $u_3 = p$, there is invariance with respect to the dilatation group provided that

$$\begin{aligned} \alpha_2 &= 1 - \gamma \\ \alpha_3 &= \alpha_1 + 2(1 - \gamma) \end{aligned} \quad . \tag{89}$$

If the initial density of mass in the medium at rest is:

$$\rho_0(r) = \bar{\rho}_0 r^{-\Omega} \qquad \Omega = \frac{7 - \Gamma}{1 + \Gamma} \tag{90}$$

then $\alpha_1 = \alpha = -\Omega$. Moreover, if a strong explosion takes place at the origin, then, considering the motion to be self-similar, it follows from energy conservation

$$E = \int_0^{(t)} 4\pi \left(\frac{p}{\Gamma - 1} + \frac{1}{2} \rho u^2 \right) r^2 dr = E_0 \tag{91}$$

that γ is given by:

$$\gamma = \frac{3\Gamma - 1}{\Gamma + 1} \quad . \tag{92}$$

Under the transformations

$$
\begin{aligned}
p(r,t) &= \bar{p}_0 t^{\alpha/\gamma} \dot{R}^2(t) P(\xi,\tau) \\
\rho(r,t) &= \bar{\rho}_0 t^{\alpha/\gamma} g(\xi,\tau) \\
u(r,t) &= \dot{R}(t) v(\xi,\tau)
\end{aligned}
\tag{93}
$$

where

$$\xi = \frac{r}{R(t)} \qquad R(t) = t^{1/\gamma} \qquad \tau = \frac{1}{\gamma} \ln t \tag{94}$$

the system becomes

$$\frac{\partial \mathbf{w}}{\partial t} + \tilde{A}\frac{\partial \mathbf{w}}{\partial \xi} = \tilde{\mathbf{B}} \tag{95}$$

where

$$
\mathbf{w} = \begin{bmatrix} g \\ v \\ z \end{bmatrix}
$$

$$
\tilde{A} = \begin{bmatrix} v - \xi & g & 0 \\ \frac{z}{g\Gamma} & v - \xi & \frac{1}{\Gamma} \\ 0 & (\Gamma - 1)z & v - \xi \end{bmatrix} \tag{96}
$$

$$
\tilde{\mathbf{B}} = \begin{bmatrix} -g(\frac{2v}{\xi} + \frac{\Gamma - 7}{\Gamma + 1}) \\ 2v(\frac{\Gamma - 1}{\Gamma + 1}) \\ 2(\Gamma - 1)z(\frac{2}{\Gamma + 1} - \frac{v}{\xi}) \end{bmatrix}
$$

In fact this is the case studied in [28], [29]. The system 52 is invariant with respect to the "associated" group

$$
\begin{aligned}
g^* &= \omega^A g \\
v^* &= \omega v \\
z^* &= \omega^2 z \\
\tau^* &= \tau \\
\xi^* &= \omega \xi
\end{aligned}
\tag{97}
$$

The new variable transformation

$$g = \xi G$$
$$v = \xi V$$
$$z = \xi^2 Z \qquad (98)$$
$$\eta = \ln \xi$$

changes 95 into the form

$$\frac{\partial \mathbf{v}}{\partial \tau} + \hat{A}\frac{\partial \mathbf{v}}{\partial \eta} = \hat{\mathbf{B}} \qquad (99)$$

where

$$\mathbf{v} = \begin{bmatrix} G \\ V \\ Z \end{bmatrix}$$

$$\hat{A} = \begin{bmatrix} V-1 & G & 0 \\ \frac{Z}{G\Gamma} & V-1 & \frac{1}{\Gamma} \\ 0 & (\Gamma-1)Z & V-1 \end{bmatrix} \qquad (100)$$

$$\hat{\mathbf{B}} = \begin{bmatrix} 4G(\frac{2}{\Gamma+1} - V) \\ 2V\frac{\Gamma-1}{\Gamma+1} - V(V-1) - \frac{3Z}{\Gamma} \\ Z(3\Gamma-1)(\frac{2}{\Gamma+1} - V) \end{bmatrix}$$

The system 99 has a constant solution given by:

$$G = G_0 = \frac{\Gamma+1}{\Gamma-1}$$
$$V = V_0 = \frac{2}{\Gamma+1} \qquad (101)$$
$$Z = Z_0 = \frac{2\Gamma(\Gamma-1)}{(\Gamma+1)^2}$$

This solution, when substituted into 98, gives rise to the so-called "Sedov solution" [28], [29]. It is worth of note that 102 correspond to the Rankine-Hugoniot relations for sharp shocks in terms of the transformed variables [2].

Routine calculations generate the eigenvalues of $\hat{A}$:

$$\Lambda_1 = V-1$$
$$\Lambda_2 = V-1+\sqrt{Z} \qquad (102)$$
$$\Lambda_3 = V-1-\sqrt{Z}$$

The values a_0 and h_0 occurring in 85 corresponding to Λ_2 and Λ_3 are, respectively

$$a_0 = \pm\frac{(1+\Gamma)\sqrt{2\Gamma(\Gamma-1)}}{2(3\Gamma-1)}$$

$$h_0 = -\frac{\mp 6\sqrt{2\Gamma(\Gamma-1)}+3\Gamma-5}{2(3\Gamma-1)}$$

$$(103)$$

that allow us to obtain, because of 86 the critical time when the secondary shock would appear.

For the special value of $\Gamma = 1.0463$ corresponding to the characteristic propagating with the velocity Λ_2, we get $a_0 = 0.148479$. Consequently, the evolution of the discontinuity is completely determined by (5.3) in this special case.

Other physical examples where the above mentioned procedure to study the propagation of weak discontinuities in a non-constant state characterized by a similarity solution may be found in [30], [31], [32], [33].

Bibliography

[1] A. Jeffrey, *Quasi-linear hyperbolic systems and waves,* Research Notes in Mathematics **5**, Pitman Publishing, London 1976.

[2] G. B. Whitham, *Linear and nonlinear waves,* John Wiley and Sons, New York, 1974.

[3] R. Courant and K. O. Friedrichs, *Supersonic flow and shock waves,* Interscience Publishers Inc., New York, 1967.

[4] B. L. Rozdestvenskii and N. N. Janenko, *Systems of Quasilinear Equations and Their Applications to Gas Dynamics,* Translations of Mathematical Monographs, **55**, A.M.S., 1983.

[5] A. Majda, *Compressible Fluid Flow ans Systems of Conservation Laws in Several Space Variables,* Applied Mathematical Sciences, **53**, Springer-Verlag, New York, 1984.

[6] A. Jeffrey and T. Taniuti, *Nonlinear wave propagation*, Accademic Press, New York, 1964.

[7] A. Fischer and D. P. Marsden, *The Einstein evolution equations as a first order quasilinear simmetric hyperbolic system*, Comm. Math. Phys., **28** p. 1–38, 1972.

[8] S. K. Godunov, *An interesting class of quasilinear systems*, Sov. Math., **2**, p. 947–949, 1961.

[9] K. O. Friedrichs and P. D. Lax, *Systems of conservations equations with a convex exstension*, Proc. Nat. Acad. Sc. USA, **68**, p. 1686–1688, 1971.

[10] G. Boillat, *Sur l'éxistence et la recherche d'équations de conservation supplementaires pour les systèmes hyperboliques*, C. R. Acad. Sci. Paris, **278A**, p. 909–912, 1974.

[11] G. Boillat and T. Ruggeri, *Simmetric form of nonlinear mechanics equations and entropy growth across a shock*, Acta Mechanica, **35**, p. 271–274, 1980.

[12] T. Ruggeri, *Symmetric hyperbolic system of conservative equations for a viscous heat conducting fluid*, Acta Mechanica, **47**, p. 167–183, 1983.

[13] A. Donato, *On a supplementary conservation law for the balance laws in thermoelastic bodies*, Meccanica, **10**, p. 229–232, 1983.

[14] D. Fusco, *Alcune considerazioni sulle onde d'urto in fluidodinamica*, Atti Sem. Mat. Fis. Univ. Modena, **28**, p. 223–236, 1979.

[15] G. Boillat, *La propagation des ondes*, Gauthier-Villars, Paris, 1965.

[16] G. Boillat and T. Ruggeri, *On the evolution law of weak discontinuites for hyperbolic quasilinear systems*, Wave Motion, **1**, p. 149–152, 1979.

[17] G. Boillat, *Chocs charactéristiques,* C. R. Acad. Sc. Paris, **274A**, p. 1018–1021, 1972.

[18] G. Boillat, *Discontinuités de contact,* C. R. Acad. Sc. Paris, **275A**, p. 1255–1258, 1972.

[19] G. I. Barenblatt, *Similarity, self-similarity and intermediate asymptotics,* Consultant Bureau, New York, 1979.

[20] L. Dresner, *Similarity solutions of nonlinear partial differential equations,* Research Notes in Math., **88**, Pitman, 1983.

[21] W. F. Ames, *Nonlinear partial diffrential equations in engineering,* Vol. II, Academic Press, New York, 1972.

[22] L. V. Ovsiannikov, *Group analysis of differential equations,* (English Edition, edited by W. F. Ames), Academic Press, Boston, 1982.

[23] G. W. Bluman and J. D. Cole, *Similarity methods for differential equations,* Applied. Math. Sciences, Springer-Verlag, **13**, New York, 1974.

[24] W. F. Ames and A. Donato, *On the evolution of weak discontinuites in a state characterized by invariant solutions,* Int. J. Non-linear Mechanics, **23**, p. 167–174, 1988.

[25] A. Donato, *Similarity analysis and nonlinear wave propagation,* Int. J. Non-linear Mech., **22**, p. 307–314, 1987.

[26] A. Donato, *Invariant solutions and nonlinear wave propagation,* Physics of Earth and Planetary Interiors, **50**, p. 52–55, Amsterdam, 1987.

[27] L. I. Sedov, *Similarity and dimensional methods in mechanics,* Academic Press, New York, 1959.

[28] S. G. Tagare, *Evolution of discontinuity in self-similar flows,* Il Nuovo Cimento **11B**, 73–82, 1972.

[29] N. Virgopia and A. Ferraioli, *Evolution of weak discontinuities waves in self-similar flows and formation of secondary shocks. The point explosion model,* J. Appl. Math. Phys. (ZAMP), **33**, p. 63–80, 1982.

[30] A. Donato and M. C. Nucci, *Similarity solutions and spherical discontinuity waves in hyperelastic materials subjected to a non-constant deformation,,* Meccanica, **23**, p. 156–159, 1988.

[31] W. F. Ames, A. Donato and M. C. Nucci, *Analysis of the thread-line equations,* Nonlinear wave motion, A. Jeffrey ed., Pitman mon., **43**, Longman Scientific and Technical, p. 1–10, 1989.

[32] N. Manganaro and F. Oliveri, *Group analysis approach in magnetohydrodynamics: weak discontinuity propagation in a non-constant state,* Meccanica, **24**, p. 71–78, 1989.

[33] M. Torrisi, *Similarity solutions and growth of weak discontinuities in a non-constant state for a reactive polytropic gas,* Int. J. Non-linear Mech., **24**, p. 441–449, 1989.

Integrable Nonlinear Equations

A.S. Fokas
*Department of Mathematics and Computer Science
and The Institute for Nonlinear Studies
Clarkson University
Potsdam, New York 13699-5815, U.S.A.*

1 Introduction

In the last twenty years or so, many physically important nonlinear models have been solved exactly. Such models arise in many branches of classical physics, in classical and quantum field theories, in particle physics, in relativity, in statistical physics, etc. They take the form of ordinary differential equations, partial differential equations, singular integrodifferential equations, operator equations, differential-difference equations, spin systems, etc. In this article I will limit my discussion to methods of solution of *nonlinear evolution equations in* $1+1$ *(i.e. one temporal and one spatial dimensions) and in* $2+1$ *(i.e. one temporal and two spatial dimensions)*.

Emphasis is placed on the basic ideas and on motivation. Because of economy of presentation, although the relevant formalism is presented in some detail, the associated rigorous aspects are only briefly summarized in §7.

Regarding methods of solution there exist two different approaches. The first, involves finding ways of generating large classes of solutions of integrable equations, without caring about what initial or boundary conditions these solutions satisfy. Among such methods the most well known are:

(a) The use of Lie-point symmetries [1].

(b) The use of Bäcklund transformations [2].

(c) The bilinear approach of Hirota [3].

(d) The more general τ-function approach of the Japanese School [4].

(e) The Dressing Method of Zakharov-Shabat [5].

(f) The Riemann-Hilbert direct method of Zakharov-Shabat [6], and the $\bar{\partial}$ direct method of Zakharov and Manakov [7].

(g) The Direct Linearizing method of the author and Ablowitz [8], [11]-[13].

The second approach involves developing ways of solving initial value problems. The most well known initial value problems solved are regarding initial data which:

(a) Are decaying at infinity.

(b) Are periodic in the spatial variable [9].

(c) Are self similar [10].

The integrable nonlinear equations discussed in this article are occasionally referred to as *soliton* equations, or as equations solvable by the *Inverse Scattering Method*. It should be pointed out that there exist certain nonlinear equations in $1 + 1$ which can be linearized via an explicit change of variables. The most well known such equation is the Burgers equation,

$$u_t = u_{xx} + 2uu_x. \tag{1.1}$$

Using the Cole-Hopf transformation,

$$u = \frac{\varphi_x}{\varphi}, \tag{1.2}$$

equation (1.1) becomes the heat equation:

$$\left(\frac{\varphi_x}{\varphi}\right)_t = \left(\frac{\varphi_t}{\varphi}\right)_x = \left[\frac{\varphi_{xx}}{\varphi} - \left(\frac{\varphi_x}{\varphi}\right)^2 + \left(\frac{\varphi_x}{\varphi}\right)^2\right]_x,$$

or

$$\varphi_t = \varphi_{xx}. \qquad (1.3)$$

Although equations like the Burgers equation share some common features with equations solvable by the inverse scattering, the method of solving the latter equations is far more complicated. The Inverse Scattering Method is not based on an explicit transformation like equation (1.2), but on *the association of the nonlinear equation under consideration with a pair of linear equations known as Lax pair.* Actually one of the two equations is time independent, and can be thought of as a *linear eigenvalue problem.* It is precisely this equation which is at the heart of integrability.

In this article I will use appropriate linear eigenvalue problems to solve several physically significant initial (and initial-boundary) value problems. The main mathematical tools used are the so called *Riemann-Hilbert (RH) problem* [15]-[17] for equations in $1 + 1$, the *non-local Riemann-Hilbert problem* for some equations in $2 + 1$ and the $\bar{\partial}$ *problem* for some equations in $2 + 1$. In §2 I use the KPI equation to illustrate the dressing method. In §3,§4 and §5 I use the KdV, the N-wave interaction equations in $2 + 1$, and the KPII, to illustrate the *inverse scattering transform* (IST) method for solving Cauchy problems with decaying initial data for equations in $1 + 1$ (KdV), $2 + 1$ of the nonlocal RH type (N-wave interactions), and in $2 + 1$ of the $\bar{\partial}$ type (KPII).

It is well known that arbitrary decaying initial data for equations in $1+1$ will decompose in general into a number of solitons for large t. Hence solitons for equations in $1 + 1$ are generic. This, together with the fact that solitons have interesting collision properties, are the reason why solitons have been very useful in the physical applications of one-dimensional solvable systems. However, in multidimensions, dispersion dominates over nonlinearity. Thus solitons "leak away" and an arbitrary initial disturbance disperses away for large t. This situation can change provided that there exists a mechanism to add energy into the system. For DSI the boundary conditions provide a source of energy, and hence DSI can support localized, exponentially decaying solutions. These solutions have several novel features not found in the usual solitons and have been named *dromions* by Santini and the author. The solution of an initial boundary value problem of DSI and the dromions are discussed in §6.

A summary of rigorous considerations and open questions regarding the formal results of §3-6 is given in §7.

As it was mentioned earlier, periodic and self similar problems have also been considered for integrable equations. The periodic problems yield to consideration of Riemann Theta functions, while the self similar solutions yield to Painlevé equations. A brief introduction of these results in given in §8.

2 The Dressing Method

The dressing method introduced by Zakharov and Shabat has been a powerful tool for obtaining new integrable nonlinear equations as well as characterizing large classes of solutions of these equations. This method is applicable to both equations in $1 + 1$ and in $2 + 1$. There exist several formulations of the dressing method; for a recent discussion of the relationship between these formulations see [14].

In its simplest form the dressing method starts with a RH or $\bar{\partial}$ problem and then algorithmically implies: (a) A Lax pair, (b) A nonlinear integrable equation, (c) Large classes of solutions of the nonlinear equation. We illustrate this method starting with a nonlocal RH problem, although the same philosophy can be used for both local RH problems and $\bar{\partial}$ problems.

Consider the scalar non-local RH problem

$$\mu^+(x,y,t,k) = \int_R dl\mu^-(x,y,t,l)F(x,y,t,k,l), \mu \sim I + \frac{\mu^{(1)}}{k} + O(\frac{1}{k^2}), k \to \infty$$
$$(2.1)$$

In equation (2.1), $+(-)$ denotes holomorphicity in the upper (lower) half k-complex plane, and $\int_R$ denotes integration over the real axis. We assume that (2.1) has a unique solution. Our aim is to find F, the scattering data, in such a way that (2.1) can be used to generate solutions of some nonlinear evolution equation. Let D_x, D_y, D_t be defined by

$$D_x = \partial_x + ik, \quad D_y = \partial_y + ik^2, \quad D_t = \partial_y + ik^3. \quad (2.2)$$

We determine F by the requirement that D_x, D_y, D_t commute with

$$P_F \mu(k) = \int_R d\mu(l) F(k,l).$$

This implies

$$F_x = i(l-k)F, \quad F_y = i(l^2 - k^2)F, \quad F_t = i(l^3 - k^3)F. \quad (2.3)$$

Note that $iD_y \mu - D_x^2 \mu \sim O(1)$ as $k \to \infty$. Hence $iD_y \mu - D_x^2 \mu - q\mu$, $q \doteq -2i\mu_x^{(1)}$ is of $O(\frac{1}{k})$ and the unique solvability of (2.1) implies that it must be zero. Thus

$$i\mu_y = \mu_{xx} + 2ik\mu_x + q\mu, \quad q = -2i\mu_x^{(1)}. \quad (2.4)$$

Similarly

$$D_t \mu = -D_x^3 \mu + vD_x\mu + w\mu,$$

or

$$\mu_t = -(\mu_{xxx} + 3ik\mu_{xx} - 3k^2\mu_x) + v(\mu_x + ik\mu) + w\mu \quad (2.5)$$

or, using (2.4) to eliminate $3k^2\mu_x$,

$$\mu_t = -\mu_{xxx} + v\mu_x + w\mu + ik \left[v\mu + \frac{3}{2}\mu - \frac{3}{2}i\mu_y - \frac{3}{2}\mu_{xx} \right].$$

The $O(k)$ and $O(1)$ terms of this equation imply

$$v = 3i\mu_x^{(1)} = -\frac{3}{2}q, \quad (2.6)$$

$$w = \frac{3}{2}i\mu_{xx}^{(1)} - \frac{3}{2}\mu_y^{(1)} = -\frac{3}{4}q_x - \frac{3}{4}i\partial_x^{-1}q_y. \quad (2.7)$$

Using equation (2.4) to eliminate k we find

$$\mu_t = -\frac{1}{4}\mu_{xxx} + v\mu_x + w\mu + \frac{3}{4}(q\mu)_x + \frac{3}{4}\partial_x^{-1}\mu_{yy} + \frac{3i}{4}\partial_x^{-1}(q\mu)_y.$$

Expressing v, w, q in terms of $\mu^{(1)}$ (equations (2.4b), (2.6a), (2.7a)) and considering the $O(\frac{1}{k})$ of the above equation we find

$$\mu_t^{(1)} = -\frac{1}{4}\mu_{xxx}^{(1)} + \frac{3}{2}i\mu_x^{(i)2} + \frac{3}{4}\partial_x^{-1}\mu_{yy}^{(1)}. \tag{2.8}$$

Equation (2.8) reduces to the KPI for q,

$$q_t + \frac{1}{4}q_{xxx} + \frac{3}{2}qq_x - \frac{3}{4}\partial_x^{-1}q_{yy} = 0. \tag{2.9}$$

In Summary. The KP I equation (2.9) is associated with the Lax pair (2.4a) and (2.5). Let F be defined by (2.3) and let $\mu = 1 + \frac{\mu^{(1)}}{k} + O(\frac{1}{k^2})$ be the unique solution of (2.1). Then $q = -2i\mu_x^{(1)}$ solves KPI.

The usual dressing method for equations in $1+1$ yields solutions which can be thought of as perturbations of the zero solution. The starting local RH problem is, in general, inadequate for capturing solutions which are perturbations of an arbitrary exact solution. It is interesting that in $2+1$, one is able to use the same analytic structure associated with decaying solutions (namely the non-local RH problem (2.1)) to capture bounded but non-decaying solutions. It seems that this is a consequence of the decaying in one of the two dimensions. An extended dressing method capturing such solutions is given in [18]. In particular this method is capable of capturing solutions which are perturbations of line-solitons.

3   A Cauchy Problem for the KdV

The KdV equation

$$u_t + u_{xxx} + 6uu_x,$$

is associated with the Lax Pair

$$w_{xx} + (u + k^2)w = 0 \tag{3.1}$$

$$u_t = (u_x + \nu)w + (4k^2 - 2u)w_x, \tag{3.2}$$

where ν is an arbitrary constant. Indeed, using $(w_{xx})_t = (w_t)_{xx}$ and assuming $k_t = 0$, it follows that u satisfies the KdV. Because of $k_t = 0$, equation (3.1) is called an isospectral eigenvalue problem. In order to solve a Cauchy problem for KdV one needs to analyze equation (3.1).

(a) Analytic Eigenfunctions

The first step of the Inverse Scattering Method is to find appropriate solutions of equation (3.1) which are analytic in the variable k. It turns out that equation (3.1) does not have solutions analytic in the entire complex k-plane. However, it does have solutions analytic in the upper and lower half-planes. We will denote such solutions with superscripts $+$ and $-$ respectively.

As $|x| \to \infty$, $u \to 0$, thus $w \sim e^{\pm ikx}$. Hence there exist solutions of (3.1) which are characterized by the following boundary conditions:

$$x \to -\infty: \quad \phi \to e^{-ikx} \quad x \to +\infty: \quad \Psi \to e^{ikx}$$

$$\hat{\phi} \to e^{ikx} \qquad\qquad \hat{\Psi} \to e^{-ikx} \tag{3.3}$$

Since we are looking for analytic solutions it is convenient to get rid of the exponentials by using

$$\Phi = \phi e^{ikx} \qquad \Psi = \psi e^{-ikx}$$

$$\hat{\Phi} = \hat{\phi} e^{-ikx} \quad \hat{\Phi} = \hat{\psi} e^{ikx}. \tag{3.4}$$

It turns out that Φ, Ψ are $+$ functions, i.e. are analytic in the upper half complex k-plane, while $\hat{\Phi}$, $\hat{\Psi}$ are $-$ functions. To establish analyticity we replace (3.1) by appropriate integral equations. Consider for example Φ: Since ϕ solves (3.1), Φ solves

$$\Phi_{xx} - 2ik\Phi_x = -u\Phi. \qquad (3.5)$$

Treating $u\Phi$ as a forcing, equation (3.5) is associated with the Green's function $G(x, \xi, k, t)$, where

$$G_{xx} - 2ikG_x = -\delta(x - \xi), \quad G \to 0 \text{ as } x \to -\infty. \qquad (3.6)$$

Equation (3.6) yields $G = \frac{i}{2k}\left(-1 + e^{2ik(x-\xi)}\right)\theta(x - \xi)$, where θ denotes the Heavyside distribution. Hence, Φ solves

$$\Phi(k, x, t) = 1 + \frac{i}{2k}\int_{-\infty}^{x} d\xi \left(-1 + e^{2ik(x-\xi)}\right) u(\xi, t)\Phi(k, \xi, t). \qquad (3.7)$$

Equation (3.7), for $u\epsilon L_1$ is a Volterra integral equation. Furthermore, its kernel is a $+$ function with respect to k; this follows from the fact that $x - \xi$ is always non negative. Hence its solution Φ is also a $+$ function. Similarly Ψ solves

$$\Psi(k, x, t) = 1 + \frac{i}{2k}\int_{x}^{\infty} d\xi \left(-1 + e^{-2ik(x-\xi)}\right) u(\xi, t)\Psi(k, \xi, t). \qquad (3.8)$$

which implies that Ψ is also a $+$ function.

Under the transformation $k \to -k$, equation (3.1) remains invariant, while the boundary condition of Ψ goes to that of $\hat{\Psi}$. Hence $\Psi(k, x, t) = \hat{\Psi}(-k, x, t)$. Similarly, $\phi(k, x, t) = \hat{\phi}(-k, x, t)$. Thus

$$\hat{\Phi}(k, x, t) = \Phi(-k, x, t), \quad \hat{\Psi}(k, x, t) = \Psi(-k, x, t). \qquad (3.9)$$

(b) A Riemann-Hilbert Problem

The second step of inverse scattering involves formulating a Riemann-Hilbert Problem [15-17], i.e. finding a relationship between $+$ and $-$ functions. Using the linear independence property of the solutions of the second order ODE (3.1), we obtain $\phi = a\hat{\psi} + b\psi$. This equation together with (3.4), (3.9) implies

$$\frac{\Phi(k,x,t)}{a(k,t)} = \Psi(-k,x,t) + \rho(k,t)e^{2ikx}\Psi(k,x,t), \quad \rho(k,t) \doteq \frac{b(k,t)}{a(k,t)}.$$
(3.10)

The functions a, b are constant with respect to x, and can be interpreted as the transmission and reflection coefficients. The functions Φ and Ψ and + functions, hence $\Psi(-k,x,t)$ is a − function. What about a,b? It turns out that b has no analytic properties in k but $1/a$ is a + function. This follows from the following expressions for a, b:

$$a = 1 - \frac{i}{2k}\int_R d\xi u(\xi,t)\Phi(k,\xi,t), \quad b = \frac{i}{2k}\int_R d\xi u(\xi,t)\Phi(k,\xi,t)e^{-2ik\xi}.$$
(3.11)

To derive these expressions let $\Delta = \Phi(k,x,t) - a(k,t)\Psi(-k,x,t)$. Then equations (3.8), (3.9) imply that Δ satisfies

$$\Delta = 1 - a + \frac{i}{2k}\int_R d\xi \left(-1 + e^{2ik(x-\xi)}\right) u\Phi - \frac{i}{2k}\int_x^\infty d\xi \left(-1 + e^{2ik(x-\xi)}\right) u\Delta.$$

On the other hand, using (3.10), $\Delta = be^{2ikx}\Psi$ and hence it satisfies (using (3.9)),

$$\Delta = be^{2ikx} - \frac{i}{2k}\int_x^\infty d\xi \left(-1 + e^{2ik(x-\xi)}\right) u\Delta.$$

Comparing the two equations for Δ, equations (3.11) follow.

It can be shown [19] that $a(k,t)$ may have simple zeros $k_1, \cdots, k_N$ in the upper half k-complex plane. Hence in general Φ/a will be meromorphic in the upper half k-plane. So let

$$\frac{\Phi}{a} = \Phi^+(k,x,t) + \sum_{j=1}^N \frac{A_j(x,t)}{k - k_j},$$
(3.12)

where Φ^+ is a + function. Equations (3.12), (3.10) suggest that $A_j = c_j(t)e^{2ik_jx}\Psi(k_j, x, t)$. This can be verified by using an argument similar to the one used to derive (3.11). Thus equation (3.10) becomes

$$\Phi^+(k, x, t) = \Psi(-k, x, t) + \sum_{j=1}^{N} \frac{c_j(t)e^{2ik_jx}\Psi(k_j, x, t)}{k - k_j} + \rho(k, t)e^{2ikx}\Psi(k, x, t).$$

$$(3.13)$$

Equation (3.13) defines a Riemann-Hilbert problem: Given c_j, k_j, ρ, find Φ^+, Ψ. Applying a $-$ projection of (3.13), and letting $k \to -k$ we find

$$\Psi(k, x, t) - \frac{1}{2\pi i}\int_R \frac{dl\rho(l, t)e^{2ilx}\Psi(l, x, t)}{l + k + i0} = 1 - \sum_{l=1}^{N} \frac{c_l(t)e^{2ik_lx}}{k + k_l}\Psi(k_l, x, t).$$

$$(3.14)$$

Equation (3.14) and the equations obtained by evaluating (3.14) at $k = k_j$, define a system of linear integral equations for Ψ.

Equation (3.14) expresses Ψ in terms of the *scattering data* c_j, k_j, ρ. Since u can be found in terms of Ψ from equation (3.1), u also can be expressed in terms of the scattering data. Rather than using (3.1), the following expression is more convenient,

$$u = \partial_x\left[-\frac{1}{\pi}\int_R dk\rho(k, t)e^{2ikx}\Psi(k, x, t) + 2i\sum_{j=1}^{N} c_j(t)e^{2ik_jx}\Psi(k_j, x, t)\right].$$

$$(3.15)$$

Let us summarize: Given the scattering data $\{c_j(t), k_j\}_{j=1}^{N}$, $\rho(k, t)$, equation (3.14) yields Ψ and then equation (3.15) implies u. This scheme will be useful for solving a Cauchy problem for KdV provided that one can determine the scattering data from knowledge of $u(x, 0)$. This is indeed the case, and is a consequence of the fact that c_j, ρ have a simple time evolution (recall that $k_{j_t} = 0$):

$$c_j(t) = c_j(0)e^{8ik_j^2t}, \quad \rho(k, t) = \rho(k, 0)e^{8ik^3t}. \qquad (3.16)$$

To obtain the evolution of ρ note that since $\phi \sim e^{-ikx}$ as $x \to -\infty$ equation (3.2) yields $\nu = 4ik^3$. Also $\phi \sim ae^{-ikx} + be^{ikx}$ as $x \to \infty$; substituting this expression into (3.2) with $\nu = 4ik^3$, it follows that

$$a_t = 0, \quad b_t = 8ik^3 b.$$

The evolution of $c_j(t)$ is obtained in a similar way.

In summary: The initial data $u(x,0)$ implies $k_j, c_j(0), \rho(k,0)$. Then equations (3.16) yield the scattering data for arbitrary t, which in turn (through Ψ) imply $u(x,t)$.

Remarks

1. Pure soliton solutions correspond to $\rho = 0$. Hence they are characterized through the linear system of algebraic equations

$$\Psi_j = 1 - \sum_{l=1}^{N} \frac{c_l(0)e^{2ik_l x + 8ik_l^3 t}}{k_j + k_l} \Psi_l, \quad u = -2i\partial_x \sum_{j=1}^{N} c_j(0)e^{2ik_j x + 8ik_j^3 t} \Psi_j.$$

$$(3.17)$$

2. To establish the correction with the direct linearization note that (3.14) can be written as

$$\Psi(k,x,t) - \frac{1}{2\pi i} \int_L d\zeta(l) \frac{e^{2ilx + 8il^3 t}\Psi(l,x,t)}{l + k + i0} = 1, \qquad (3.18)$$

where

$$d\zeta(l) = \begin{bmatrix} \rho(l,0)dl & \text{for } l \text{ real} \\ \\ -2\pi i c_j(0)\delta(l - k_j)dl & \text{for } l \text{ imaginary.} \end{bmatrix}$$

3. In the original formalism of inverse scattering the basic linear integral equation was not equation (3.14) but the Gel'fand-Levitan-Marchenko equation. To derive this equation assume that

$$\Psi(k,x,t) = 1 + \int_x^\infty dsK(x,s,t)e^{ik(s-x)}. \qquad (3.19)$$

Substituting (3.19) in (3.18) and operating with $\frac{1}{2\pi}\int_R dke^{ik(x-y)}$ one obtains

$$K(x,y,t) + F(x+y,t) + \int_x^\infty dsK(x,s,t)F(s+y,t) = 0, \quad y > x,$$
$$(3.20)$$

where

$$F(x,t) \doteq \frac{1}{2\pi}\int_L d\zeta(k)e^{ikx+4ik^3t}.$$

Also equation (3.15) now becomes

$$u(x,t) = 2\partial_x K(x,x,t). \qquad (3.22)$$

4. Perhaps the most important consequence of the inverse scattering method is that it allows one to evaluate the long time behavior of $u(x,t)$. As $t \to \infty$, stationary phase analysis implies that only the part of (3.14) associated with the discrete spectrum survives (the continuous spectrum gives a contribution which decays like $\frac{1}{\sqrt{t}}$). Hence equations (3.14), (3.15) for large t reduce to equations (3.17). Thus any initial data will decompose into a number of solitons as $t \to \infty$. This is in my opinion why solitons are so important: They describe the long time behavior of the nonlinear equation under consideration.

4   A Cauchy Problem for N-Wave Interactions in $2+1$.

The N-wave interaction equation in $2+1$,

$$Q_{ij_t} = a_{ij}Q_{ij} + (C_i - J_ia_{ij})Q_{ij_y} + \sum_{k \neq j, k=1}^{N} (a_{ik} - a_{kj})Q_{ik}Q_{kj}, \quad (4.1)$$

where $a_{ij} \doteq \frac{C_i - C_j}{J_i - J_j}, i \neq j$ and C_i, J_i are arbitrary real constants is associated with the following linear eigenvalue problem

$$\frac{\partial \mu}{\partial x} - J\frac{\partial \mu}{\partial y} = ik\hat{J}\mu + Q\mu, \quad \hat{J}A = [J, A]. \quad (4.2)$$

In equation (4.2): (i) μ is an $N \times N$ matrix-valued function on $\mathbb{R}^2(N \geq 2)$, (ii) J is a constant, real diagonal $N \times N$ matrix with entries $J_1 > J_2 > \cdots > J_N$ and Q is off-diagonal with rapidly decaying component functions.

Actually, equation (4.2) is also the time-independent part of the Lax pair of the Davey-Stewartson I (DS) equation and modified KPI equation. Equation (4.1) has been considered in [20]. Here we follow [21].

Writing equation (4.2) in component form we find

$$(\mu_{ab})_x - J_a(\mu_{ab})_y - ik(J_a - J_b)\mu_{ab} = (Q\mu)_{ab}. \quad (4.3)$$

Let $\tilde{f}_{ab}(\xi, \eta_a) := f_{ab}(\xi, \eta_a - J_a\xi)$, then equation (4.3) becomes

$$(\tilde{\mu}_{ab})_\xi - ik(J_a - J_b)\tilde{\mu}_{ab} = (\widetilde{Q\mu})_{ab},$$

which can be converted into the integral equation

$$\tilde{\mu}_{ab}(\xi, \eta_a, k) = \zeta_{ab}(\eta_a, k) + \int_{\alpha_{ab}}^{\xi} d\xi' (\widetilde{Q\mu})_{ab}(\xi', \eta_a, k)e^{ik(J_a - J_b)(\xi - \xi')},$$

where α_{ab} are arbitrary constants and ζ_{ab} are arbitrary functions of (η_a, k). Writing this equation in the original coordinates we obtain

$$\mu_{ab}(x, y, k) = \zeta_{ab}(y + xJ_a, k)e^{ik(J_a - J_b)x}$$

$$+ \int_{\alpha_{ab}}^{x} dx'(Q\mu)_{ab}\left(x', y + (x - x')J_a, k\right) e^{ik(J_a - J_b)(x - x')}. \qquad (4.4)$$

By a proper choice of ζ_{ab} and α_{ab} we can define eigenfunctions $\mu^{\pm}$ which are holomorphic in $\mathbb{C}^{\pm}$:

$$\mu^{\pm}(x, y, k) = I + \left[(\Pi_0 + \Pi_{\pm})\int_{-\infty}^{x} dx' + \Pi_{\mp}\int_{+\infty}^{x} dx'\right]$$

$$e^{ik(x - x')J}(Q\mu^{\pm})(x', y + (x - x')J, k) \qquad (4.5)^{\pm}$$

for $k \in \overline{\mathbb{C}^{\pm}}$, where $f_{ab}(x', y + (x - x')J] := f_{ab}(x', y + (x - x')J_a)$. We will denote equation $(4.5)^{\pm}$ by

$$\mu^{\pm} = I + \mathsf{N}_{Q,k}^{\pm}\mu^{\pm}.$$

Note that $(4.5)^{\pm}$ are equivalent to (4.2) with the asymptotic conditions

$$\begin{cases} \lim_{x\to-\infty} \Pi_0\mu^{\pm} = I, \\[2mm] \lim_{x\to-\infty} \Pi_{\pm}\mu^{\pm} = 0, \\[2mm] \lim_{x\to+\infty} \Pi_{\mp}\mu^{\pm} = 0. \end{cases} \qquad (4.6)^{\pm}$$

Let the scattering data $T_{\pm}$ be defined by

$$T_{\pm}(l, k) = \frac{1}{2\pi}\Pi_{\pm}\int_{\mathbb{R}^2} d\xi d\zeta e^{-il(\xi J + \zeta I)}Q(\xi, \zeta)\mu^{\mp}(\xi, \zeta, k)e^{ik(\xi J + \zeta I)}$$

$$(4.7)^{\pm}$$

Hereafter we shall use $\Pi_{+}A, \Pi_{-}A$ and $\Pi_0 A$ to denote the strictly upper triangular, the strictly lower triangular and the diagonal part of the matrix A, respectively. We note that

$$\int_{\mathbb{R}} dx' f\left(x', y + (x - x')J\right) = \frac{1}{2\pi}\int_{\mathbb{R}^3} dx' dy' dl' e^{il'(x - x')J + il'(y - y')I} f(x', y'),$$

thus equation $(4.5)^\pm$ becomes

$$\mu^\pm(x,y,k) = I + \frac{1}{2\pi} \left(\int_{-\infty}^{x} dx' - \Pi_\mp \int_{\mathbb{R}} dx' \right)$$

$$\int_{\mathbb{R}^2} dy' dl' e^{i(l'+k)(x-x')J+il'(y-y')I}(Q\mu^\mp)(x',y',k)e^{-ik(x-x')J}.$$

Letting $l' + k = l$ in these equations and taking $x \to -\infty$ we obtain

$$\lim_{x\to-\infty} \left(\mu^\pm(x,y,k) - I \right) e^{ik(xJ+yI)} = -\int_{\mathbb{R}} dl e^{il\eta} T_\mp(l,k). \qquad (4.8)^\pm$$

Throughout this paper the large x limit of the ab component is taken along the characteristic line $y + xJ_a = \eta_a$, where $\eta_1, \cdots, \eta_N$ are the compartments of the diagonal matrix η.

The eigenfunction μ^0 is defined by

$$\mu^0(x,y,k) = I + \int_{-\infty}^{x} dx' e^{ik(x-x')\hat{J}}(Q\mu^0)(x', y+(x-x')J, k), \, k \in \mathbb{R}. \qquad (4.9)$$

We will denote equation (4.9) by

$$\mu^0 = I + \mathsf{N}_{Q,k}^0 \mu^0.$$

Note that (4.9) is equivalent to (4.2) with the asymptotic condition

$$\lim_{x\to-\infty} \mu^0 = I. \qquad (4.10)$$

Using the asymptotics of $\mu^\pm$ and μ^0 as $x \to -\infty$ it is straightforward to relate these eigenfunctions:

$$\mu^\pm(x,y,k) = \mu^0(x,y,k) - \int_{\mathbb{R}} dl \mu^0(x,y,l) e^{il(xJ+yI)} T_\mp(l,k) e^{-ik(xJ+yI)}. \qquad (4.11)^\pm$$

Adding $\mathbb{P}_\mp (4.11)^\pm$, we find

$$\mu^0(x, y, \cdot) = I + \mathsf{T}_{(x,y)}\mu^0(x, y, \cdot); \; \mathsf{T}_{(x,y)} := \mathsf{P}_+\mathsf{T}^+_{(x,y)} + \mathsf{P}_-\mathsf{T}^-_{(x,y)},$$
$$(4.12)$$

where $\mathsf{P}_\pm$ are the usual projection operators in k and

$$(\mathsf{T}^\pm_{(x,y)}f)(k) := \int_{\mathbb{R}} dl f(l) e^{il(xJ+yI)} T_\pm(l, k) e^{-ik(xJ+yI)}. \qquad (4.13)^\pm$$

Conversely, given inverse data $T_\pm$, equation (4.12) yields μ^0. Then the potential Q can be reconstructed from

$$Q(x, y) = -i \lim_{|k| \to \infty} k \hat{J} \mu^0(x, y, k)$$

$$= \frac{\hat{J}}{2\pi} \int_{\mathbb{R}^2} dk dl \mu^0(x, y, l) e^{ik(xJ+yI)} \left[T_+(l, k) - T_-(l, k)\right] e^{-ik(xJ+yI)}.$$
$$(4.14)$$

The first equation in (4.14) follows from the large k asymptotics of (4.9), while the second equation follows from the large k asymptotics of (4.12).

The above analysis shows that Q can be reconstructed from knowledge of T. If Q evolves in time according to the N-wave equations (4.1), then it is easy to show that the time evolution of T is given by

$$\frac{\partial T}{\partial t} = ilCT - ikTC, \quad T(l, k, 0) = T_0(l, k). \qquad (4.15)$$

5   A Cauchy Problem for the KPII Equation

The KP equation is given by

$$(q_t + 6qq_x + q_{xxx})_x + 3\sigma^2 q_{yy} = 0 \qquad (5.1)$$

where $\sigma^2 = \pm 1$. The choice $\sigma^2 = -1$ is frequently refered to as KPI, $\sigma^2 = 1$ as KPII. The KP equation plays the role in 2+1 that the KdV plays in $1 + 1$. It arises in the study of plasma waves, surface waves etc. Both KPI and KPII admit line soliton solutions. For $\sigma^2 = -1$ these solutions are unstable (with respect to slow perturbations in the y-direction), where for $\sigma^2 = 1$ they are stable. On the other hand KPI admits lump type solutions, while such solutions do not exist for KPII [22].

KPII is associated with the following linear eigenvalue problem

$$\sigma \mu_y + \mu_{xx} + 2ik\mu_x + q(x,y)\mu = 0, \sigma_R \neq 0. \qquad (5.2)$$

It can be shown that, for q decaying, there exists a unique solution μ of (5.2) which is bounded for all complex k and tending to 1 at $r = \sqrt{x^2 + y^2} \to \infty$. This μ is given by

$$\mu(x,y,k) = 1 + \int_{\mathbb{R}^2} dx'dy' G(x-x', y-y', k)q(x',y')\mu(x',y',k), \qquad (5.3a)$$

where

$$G(x,y,k_R,k_I) = \frac{1}{(2\pi)^2} \int_{\mathbb{R}^2} d\xi d\eta \frac{e^{i(\xi x + \eta y)}}{\xi^2 + 2\xi k - i\sigma\eta}. \qquad (5.3b)$$

To derive this result we note that the Green's function satisfies

$$\sigma G_y + G_{xx} + 2ikG_x = -\delta(x)\delta(y).$$

Seeking a representation of G in the form

$$G = \frac{1}{(2\pi)^2} \int_{\mathbb{R}^2} d\xi d\eta \hat{G}(\xi,\eta,k)e^{i(\xi x + \eta y)},$$

we find (5.3b).

The singularities of $(\xi^2 + 2\xi k - i\sigma\eta)^{-1}$ are integrable in the ξ,η plane for all σ, with $\sigma_R \neq 0$. This implies that G depends on k_R, k_I as opposed to $k_R + ik_I$, i.e. G is in general non-analytic. In other words sectionally holomorphic eigenfunctions do not exist for (5.2). Despite this, an inverse problem can still be formulated. The key idea is to compute the "departure from analyticity of μ".

We will show that if μ solves (5.3) then

$$\frac{\partial \mu}{\partial \bar{k}} = \frac{sgnk_0}{2\pi|\sigma_R|}e^{i\beta(x,y,k_R,k_I)}T(k_R,k_I)\mu(x,y,\hat{k}_0,k_I), \qquad (5.4)$$

where

$$k_0 \doteq k_R + \frac{\sigma_I}{\sigma_R}k_I, \hat{k}_0 \doteq -\left(k_R + 2\frac{\sigma_I}{\sigma_R}k_I\right), \beta(x,y,k_R,k_I) \doteq -2\left(x + 2\frac{k_I}{\sigma_R}y\right),$$
$$(5.5)$$

and the *inverse data* T is defined by

$$T(k_R,k_I) \doteq \int_{\mathbb{R}^2} dxdy e^{-i\beta(x,y,k_R,k_I)}q(x,y)\mu(x,y,k_R,k_I). \qquad (5.6)$$

To derive (5.4), note that

$$\frac{\partial}{\partial \bar{k}}\frac{1}{k - k_0} = \pi\delta(k - k_0)$$

implies

$$\frac{\partial G}{\partial \bar{k}} = -\frac{sgn(k_0)}{2\pi|\sigma_R|}e^{i\beta(x,y,k_R,k_I)}. \qquad (5.7)$$

Differentiating (5.3) w.r.t. $\bar{k}$ and using (5.7) we find

$$\frac{\partial \mu}{\partial \bar{k}} = \frac{sgnk_0}{2\pi|\sigma_R|}T(k_R,k_I)N(x,y,k_R,k_I), \qquad (5.8)$$

where N solves

$$N(x,y,k_R,k_I) = e^{i\beta(x,y,k_R,k_I)} + \int_{\mathbb{R}^2} dx'dy'GqN.$$

Multiplying the above equation by $e^{-i\beta}$ and using the symmetry condition

$$e^{-i\beta(x,y,k_R,k_I)}G(x,y,k_R,k_I) = G(x,y,\hat{k}_0,k_I)$$

we find

$$N(x, y, k_R, k_I) = e^{i\beta}\mu(x, y, \hat{k}_0, k_I).$$

Equation (5.4) implies the integral equation

$$\mu(x, y, k_R, k_I) = 1 + \frac{1}{\pi}\int_{\mathbb{R}^2}dk_R'dk_I'\frac{\frac{\partial\mu}{\partial\bar{k}'}(x, y, k_R', k_I')}{k - k'}. \qquad (5.9)$$

Hence, given inverse data T, equations (5.4) and (5.9) yield μ. Then q can be found from

$$q(x, y) = -\frac{2i}{\pi}\frac{\partial}{\partial x}\int_{\mathbb{R}^2}dk_Rdk_I\frac{\partial\mu}{\partial\bar{k}}(x, y, k_R, k_I). \qquad (5.10)$$

Equation (5.9) follows from the generalized Cauchy formula by using $\mu \sim 1$ as $|k| \to \infty$. Also using $\mu \sim 1 + \frac{\mu^{(1)}}{k} + O(\frac{1}{k^2})$ in (5.2) we find $q = -2i\mu_x^{(1)}$. But equation (5.9) implies

$$\mu^{(1)} = \frac{1}{\pi}\int_{\mathbb{R}^2}dk_Rdk_I\frac{\partial\mu}{\partial\bar{k}},$$

and hence equation (5.10) follows.

If $q(x, y, t)$ evolves according to equation (5.1), then T satisfies

$$\frac{\partial T}{\partial t} = 8ik_0(6kk_0 - 4k_0^2 - 3k^2)T. \qquad (5.11)$$

In the special case of KPII, i.e. $\sigma = -1$,

$$\frac{\partial T}{\partial t} = 8ik_R(3k_I^2 - k_R^2)T.$$

6 DAVEY-STEWARTSON I

In this section we summarize the results of [24] see also [23], [47]. We consider the Davey-Stewartson I equation

$$iq_t + \frac{1}{2}(q_{xx} + q_{yy}) - \left(\varphi_x - \varepsilon|q|^2\right)q = 0, \quad \varphi_{xx} - \varphi_{yy} - 2\varepsilon|q|_x^2 = 0, \quad \varepsilon = \pm 1. \qquad (6.1)$$

This equation provides a two-dimensional generalization of the celebrated nonlinear Schrödinger equation and can be derived from rather general asymptotic considerations. Physical applications include plasma physics, nonlinear optics and water waves. In the above $q(x, y, t)$ is the amplitude of a surface wave packet and φ is the velocity potential of the mean flow interacting with the surface wave. One assumes small amplitude, nearly monochromatic, nearly one-dimensional waves. In the context of water waves equation (6.1) is the shallow water limit of the Benney-Roskes equation, where one assumes dominant surface tension; $\varepsilon = -1$ and $\varepsilon = 1$ indicate focusing and defocusing regimes respectively.

Introducing characteristic coordinates, $\xi = x + y, \eta = x - y$, equations (6.1) yield

$$
iq_t + q_{\xi\xi} + q_{\eta\eta} + (U_1 + U_2)q_1 = 0, \quad \varphi_\eta = -U_1 + \frac{\varepsilon}{2}|q|^2, \varphi_\xi = -U_2 + \frac{\varepsilon}{2}|q|^2,
\tag{6.2}
$$

where

$$
U_1 = -\frac{\varepsilon}{2} \int_{-\infty}^{\xi} d\xi'|q|_\eta^2 + u_1(\eta, t), U_2 = -\frac{\varepsilon}{2} \int_{-\infty}^{\eta} d\eta'|q|_\xi^2 + u_2(\xi, t). \tag{6.3}
$$

We shall solve an initial-boundary value problem for the above system, i.e. given initial data $q(\xi, \eta, 0)$ and boundary data $u_1(\eta, t), u_2(\xi, t)$, we shall determine $q(\xi, \eta, t)$. In particular we are interested in finding the effect on the surface wave of imposing an external mean flow $(u_1(\eta, t), u_2(\xi, t))$ far upstream in the (ξ, η) plane. We assume that $q(\xi, \eta, 0), u_1(\eta, t), u_2(\xi, t)$ are decaying for large values of ξ, η and bounded for all t.

Equation (6.2a) is associated with the Lax equation

$$
(\partial_x + J\partial_y)\Psi + Q\Psi = 0, \quad J \doteq \begin{pmatrix} 1 & 0 \\ 0 & -1 \end{pmatrix}, \quad Q \doteq \begin{pmatrix} 0 & q \\ \varepsilon q^* & 0 \end{pmatrix}. \tag{6.4}
$$

We let $\Psi = M exp[ik(Jx - y)]$, we use characteristic coordinates and we define the analytic eigenfunction M^+ as the solution of the Volterra integral equations

$$M_{11}^+ = 1 - \tfrac{1}{2} \int_{-\infty}^{\xi} d\xi' q M_{21}^+ \qquad M_{12}^+ = -\tfrac{1}{2} \int_{-\infty}^{\xi} d\xi' q M_{22}^+ e^{ik(\xi - \xi')}$$

$$M_{21}^+ = \tfrac{\varepsilon}{2} \int_{\eta}^{\infty} d\eta' q^* M_{11}^+ e^{ik(\eta' - \eta)} \qquad M_{22}^+ = 1 - \tfrac{\varepsilon}{2} \int_{-\infty}^{\eta} d\eta' q^* M_{12}^+.$$
$$(6.5)$$

M^+ satisfies (6.4a) and it is also analytic in the upper half k-complex plane. Similarly if M^- satisfies equations similar to those of (6.5), with the integrals in M_{21}^+, M_{12}^+ replaced by $- \int_{-\infty}^{\eta}$ and $- \int_{\xi}^{\infty}$ respectively, it follows that M^- is analytic in the lower half k-complex plane. The eigenfunctions M^+, M^- are related via the scattering equations

$$M_{11}^+(k) - M_{11}^-(k) = -\varepsilon \int_R d\ell S^*(\ell, k) e^{-i\ell\xi - ik\eta} M_{12}^+(\ell)$$
$$(6.6)$$
$$M_{12}^+(k) - M_{12}^-(k) = - \int_R d\ell S(k, \ell) e^{i\ell\eta + ik\xi} M_{11}^-(\ell),$$

where the scattering data $S(k, \ell)$ is defined by

$$S(k, \ell) = \frac{1}{4\pi} \int_{R^2} d\xi d\eta q(\xi, \eta) e^{-i\ell\eta - ik\xi} M_{22}^-(\xi, \eta, k). \qquad (6.7)$$

Given S, the nonlocal RH problem (6.6) yields M^-, then q can be found from

$$q(\xi, \eta) = \frac{1}{\pi} \int_{R^2} dk d\ell S(k, \ell) e^{i\ell\eta + ik\xi} M_{11}^-(\xi, \eta, \ell). \qquad (6.8)$$

We note that equations (6.7), (6.8) can be thought of as a nonlinear Fourier transform. These equations are basic for solving both an initial value problem and an initial-boundary value problem for DSI. In the case of an initial value problem the evolution of $S(k, \ell, t)$ satisfies $iS_t + (k^2 + \ell^2)S = 0$. However, if the boundary conditions u_1, u_2

are nonzero, S satisfies a more complicated equation: Let $\hat{S}(\xi, \eta, t)$ be the Fourier transform of $S(k, \ell, t)$, then $\hat{S}$ satisfies

$$i\hat{S}_t + \hat{S}_{\xi\xi} + \hat{S}_{\eta\eta} + (u_2(\xi, t) + u_1(\eta, t))\,\hat{S} = 0,$$

$$\hat{S}(\xi, \eta, t) \doteq \frac{1}{2\pi} \int_{R^2} dk\,d\ell\, e^{ik\xi + i\ell\eta} S(k, \ell, t). \qquad (6.9)$$

Given $q(\xi, \eta, 0)$ one first computes $\hat{S}(\xi, \eta, 0)$; then, since $u_1(\eta, t)$, $u_2(\xi, t)$ are given, equation (6.9) yields $\hat{S}(\xi, \eta, t)$. Then the RH problem (6.6) yields M^-, and (6.8) yields $q(\xi, \eta, t)$. In order to investigate the structure of the solution $q(\xi, \eta, t)$, and in particular the behavior of $q(\xi, \eta, t)$ as $t \to \infty$, one needs to analyze equation (6.9). We have achieved this, using the spectral theory of the time-dependent Schrödinger equation. Indeed, separation of variables suggests that the analysis of (6.9) is intimately related to the analysis of

$$i\Psi_t + \Psi_{xx} + u(x, t)\Psi = 0, \quad u\epsilon\mathbb{R}. \qquad (6.10)$$

In particular the long time behavior of $\hat{S}$ depends only on the discrete spectrum of (6.10). Suppose that u_1, u_2 contain L, M discrete eigenfunctions respectively, corresponding to eigenvalues λ_j, μ_j, i.e.,

$$u_1(\eta, t) : Y_j(\eta, t), \quad \lambda_j, \quad j = 1, \cdots, L$$

$$u_2(\xi, t) : X_j(\xi, t), \quad \mu_j, \quad j = 1, \cdots, M. \qquad (6.11)$$

Then

$$\hat{S}(\xi, \eta, t) \sim \sum_{j=1, r=1}^{M, L} \rho_{rj} X_j(\xi, t) Y_r(\eta, t), \quad t \to \infty. \qquad (6.12)$$

It is quite interesting that for any initial-boundary conditions, the scattering data become degenerate as $t \to \infty$. Then the RH problem (6.6) degenerates to a system of linear algebraic equations. The asymptotic value of q can be calculated in closed form provided that X_j, Y_j can be found in closed form. For example, suppose u_1, u_2

are reflectionless, then X_j, Y_j also satisfy linear algebraic equations:
In $u_1(\eta, t)$ is given by

$$u_1(\eta, t) = -2\partial_\eta \sum_{j=1}^{L} \ell_j^* e^{-\lambda_j^*(\eta + i\lambda_j^* t)} Y_j(\eta, t), \qquad (6.13)$$

then the $Y_j's$ solve the linear algebraic system

$$Y_r + \sum_{j=1}^{L} (C^\eta)_{rj} Y_j = \ell_r e^{-\lambda_r(\eta - i\lambda_r^2 t)}, \quad (C^\eta)_{rj} \doteq \frac{\ell_r \ell_j^*}{\lambda_r + \lambda_j^*} e^{-(\lambda_r + \lambda_j^*)\eta + i(\lambda_r^2 - \lambda_j^{*2})t}.$$

$$(6.14)$$

Similarly if $u_2(\xi, t)$ is given by

$$u_2(\xi, t) = -2\partial_\xi \sum_{j=1}^{M} m_j^* e^{-\mu_j^*(\xi + i\mu_j^* t)} X_j(\xi, t), \qquad (6.15)$$

then

$$X_r + \sum_{j=1}^{M} (C^\xi)_{rj} X_j = m_r e^{-\mu_r(\xi - i\mu_r^2 t)}, (C^\xi)_{rj} = \frac{m_r m_j^*}{\mu_r + \mu_j^*} e^{-(\mu_r + \mu_j^*)\xi + i(\mu_r^2 - \mu_j^{*2})t}.$$

$$(6.16)$$

In the above equations $\ell_i, m_i, \lambda_i, \mu_i \epsilon C$ and $\lambda_{iR}, \mu_{iR} \epsilon R^+$. Using (6.12), where X_j, Y_j are given by (6.14), (6.16) and solving the inverse problem in closed form we find that q is given by,

$$q = 2 \sum_{i=1, j=1}^{M, L} X_i(\xi, t) Y_j(\eta, t) Z_{ij}(\xi, \eta, t). \qquad (6.17)$$

Z_{ij} solves

$$Z_{ij} - \varepsilon \sum_{r=1}^{M} A_{ir} Z_{rj} = \rho_{ij}, \qquad (6.18)$$

and the matrix A is defined by

$$A \doteq \rho(I + C^{\eta})^{-1} \left[(I + C^{\xi})^{-1} \rho^* \right]^T, \qquad (6.19)$$

where superscript T denotes the transpose of a matrix, and the matrix ρ can be found from initial data

$$\rho_{ij} = \int_{R^2} d\xi d\eta \hat{S}(\xi, \eta, 0) X_i^*(\xi, 0) Y_j^*(\eta, 0). \qquad (6.20)$$

We call the above solution an (M, L) dromion. In the special case that u_1, u_2 are time independent, then $\lambda_{jI} = \mu_{jI} = 0$ and the above solution degenerates to an (M, L) breather. The solutions obtained by Boiti et al correspond to certain choices of ρ.

The following elegant formulae are valid,

$$|q|^2 = 4\partial_\xi \partial_\eta \ln det(I - \varepsilon A), \qquad (6.21)$$

where the matrix A is defined by (6.19). Furthermore

$$E \doteq \int_{R^2} d\xi d\eta |q|^2 = 4 \ln det(I - \varepsilon \rho \rho^+), \qquad (6.22)$$

where $(\rho^+)_{ij} = (\rho^*)_{ji}$.

The $(1,1)$ dromion is given by

$$q = \frac{4\rho\sqrt{\lambda_R \mu_R} e^{-[\lambda_R(\hat{\eta}-\bar{\eta})+\mu_R(\hat{\xi}-\bar{\xi})]+i[-(\lambda_I \hat{\eta}+\mu_I \hat{\xi})+(|\mu|^2+|\lambda|^2)t+arg(\ell m)]}}{\left(1 + e^{-2\lambda_R(\hat{\eta}-\bar{\eta})}\right) \left(1 + e^{-2\mu_R(\hat{\xi}-\bar{\xi})}\right) + |\rho|^2}, \qquad (6.23)$$

where

$$\hat{\eta} \doteq \eta + 2\lambda_I t, \hat{\xi} \doteq \xi + 2\mu_I t, \bar{\eta} = \frac{1}{\lambda_R} \ln \frac{|\ell|}{\sqrt{2\lambda_R}}, \bar{\xi} = \frac{1}{\mu_R} \ln \frac{|\mu|}{\sqrt{2\mu_R}}.$$

Thus if $u_1(\eta, t) = 2\lambda_R^2 / cosh^2 \lambda_R(\hat{\eta}-\bar{\eta})$ and $u_2(\xi, t) = 2\mu_R^2 / cosh^2 \mu_R(\hat{\xi}-\bar{\xi})$ then $q(\xi, \eta, t)$ is given by (6.23) with ρ uniquely determined from $q(\xi, \eta, 0)$. The velocity of q in the (ξ, η) directions is given by $(-2\mu_I, -2\lambda_I)$

therefore it is completely determined by the velocities of u_2, u_1 respectively. The initial data, i.e. ρ, only affect the motion of the maximum of q. The $(2,2)$ dromion corresponds to $L = M = 2$. As $t \to \pm\infty, u_1(\eta, t)$ consists of two one-dimensional solitons traveling with speeds $-2\lambda_{1I}, -2\lambda_{2I}$. Similarly $u_2(\xi, t)$ consists of two one-dimensional solitons traveling with speeds. $-2\mu_{1I}, -2\mu_{2I}$. By analyzing the solution $q(\xi, \eta, t)$ as $t \to \pm\infty$ we find that q consists of 4 localized entities $q_{ij}^{\pm}, i, j = 1, 2$ each travelling with velocity $(-2\mu_{iI}, -2\lambda_{jI})$ (see Fig. 1.1-1.3). However, in contrast to the one dimensional case, $q_{ij}^{+} \neq q_{ij}^{-}$, i.e. the two-dimensional dromions do not retain their form upon interaction (unless $\rho_{12} = \rho_{21} = 0$). Thus it appears that in addition to the exchange of energy between the mean flow and the surface waves, the localized lumps on the surface can also exchange energy among themselves. The complete investigation of the asymptotic behavior of the (L, M) dromion is given in [47]. Several other exact solutions of the DS equation are also analyzed in [47].

We have also analyzed the ability of the dromions to be driven by the boundaries. In particular we have considered the case where the motion of each boundary is given by a single soliton, which changes velocity at $t = t_0$. Let w_i, w_f and v_i, v_f be the initial and final speeds associated with the motion of $u_1(\eta, t)$ and $u_2(\xi, t)$ respectively. We have first shown that as $t \to \infty$ the dromion follows the motion of the boundaries. Furthermore if E denotes the energy of the corresponding dromion generated by the this motion of the boundaries then

$$E_f - E_i = 4\ln\left(\frac{1 + |\rho_f|^2}{1 + |\rho_i|^2}\right), \quad \left|\frac{\rho_f}{\rho_i}\right| = \frac{V}{\sinh V}\frac{W}{\sinh W}, \qquad (6.24)$$

where $V = \left(\frac{v_i - v_f}{4\mu_R}\right)\pi$, $W = \left(\frac{w_i - w_f}{4\lambda_R}\right)\pi$, and $2\lambda_R^2, 2\mu_R^2$ are the maximum amplitutes of the solitons describing the motion of u_1, u_2 respectively. Thus if the motion of the boundaries is not uniform the dromions radiate energy. Similar results can be obtained when the boundaries move in a more complicated way.

7   Rigorous Considerations

The eigenvalue problem (4.2) is considered rigorously by Sung and
the author in [21] where it is assumed that Q belongs to the Schwartz
class and then it is shown that: (a) The direct problem is always
solvable. Namely the integral equations for $\mu^{\pm}$ and μ^0 are always
solvable without a small norm assumption. (b) If

$$\varepsilon_T \doteq \max_{1 \leq b \leq N} \sum_{a=1}^{N} |||\check{T}_{ab}||| < 1,$$

where

$$\check{T}(l,k) = T(l+k,k), |||g||| \doteq \frac{1}{2\pi} \int_{\mathbb{R}^2} dl d\zeta | \int_{\mathbb{R}^2} dk e^{-ik\zeta} g(l,k)|,$$

then the inverse problem is also solvable. (c) The norm of T can be
estimated from a norm of Q as follows:

$$\max_{1 \leq a,b \leq N} |||\check{T}_{ab}||| \leq \frac{\max_{1 \leq a,b \leq N} |||Q_{ab}|||}{1 - \delta_Q}, \delta_Q \doteq \max_{1 \leq a \leq N} \left(\sum_{i=1}^{N} |||Q_{ab}||| \right).$$

Thus in general the inverse problem is solvable only under a small
norm assumption. However, if Q have some symmetry, then the
small norm assumption can be relaxed. In particular if Q is skew
Hermitian, i.e. $Q^* = -Q$, then it can be shown that the inverse
problem is always solvable.

The above results can be used to prove that the skew-Hermitian
N wave interactions ($Q^* = -Q$) and the "focusing" DSI equation
($\varepsilon = -1$ in equation (6.2)) have global solutions for any Schwartz
initial data.

With respect to KPI: Rigorous results for the direct problem are
given by Segur [25]. Solvability is established only if a certain norm
is small. Zhou [26] has recently shown that the inverse problem is
solvable without a small norm assumption.

With respect to KPII: Wickerhauser [27] has established solvability of the direct problem under a certain small norm assumption.

For the inverse problem one does not need a small norm assumption. Indeed, for $\sigma = -1$, equation (5.4) yields

$$\frac{\partial \mu}{\partial \bar{k}} = \frac{sgnk_R}{2\pi} e^{-2i(x-2k_Iy)k_R} T(k_R, k_I)\mu(x, y, -\bar{k}). \qquad (7.1)$$

However, if q is real then $\mu(x, y, -\bar{k}) = \bar{\mu}(x, y, k)$ and equation (7.1) shows that μ is a generalized analytic function [28].

With respect to DSII: For the defocusing case ($Q^* = Q$), Beals and Coifman [29] have shown solvability without a small norm assumption. This is a consequence of the following facts: (a) Both the direct and inverse problems satisfy similar kinds of 2×2 matrix $\bar{\partial}$ problems. (b) These $\bar{\partial}$ problems are equivalent to scalar $\bar{\partial}$ problems for generalized analytic functions. For the focusing case ($Q^* = -Q$) of DSII rigorous results assuming that a certain norm is small, are given in [30]. Also this is a special case of a general $N \times N$ problem in multidimensions rigorously studied by Sung and the author [31].

8   Self Similar and Periodic Solutions

8.1   The Connection with Painlevé Equations

It is well known that similarity reductions of integrable PDE's in $1+1$ lead to ODE's of Painlevé type. For example [32] the transformation

$$q(x, T) = (3T)^{-2/3}u(t), \quad t = (3T)^{-1/3}x, \qquad (8.1)$$

maps the KdV equation

$$q_T + 6qq_x + q_{xxx} = 0, \qquad (8.2)$$

to an ODE for $u(t)$ of third order. This ODE in turn can be mapped to the second Painlevé equations (PII),

$$u'' = u^3 + tu + \theta, \quad \theta \text{ constant.} \qquad (8.3)$$

Similarly, the Lie point group

$$x' = x + 6T\lambda, \quad T' = T + \lambda/\alpha, \quad q' = q + \lambda, \qquad (8.4)$$

leaves equation (8.2) invariant and gives rise to the transformation

$$q(x, T) = \alpha T + u(t), \quad t = x - 3\alpha T^2. \tag{8.5}$$

The transformations (8.5) map (8.2) to

$$u''' + 6uu' + \alpha = 0,$$

i.e. to Painlevé I equation,

$$u'' = -3u^2 + \alpha t. \tag{8.6}$$

Flaschka and Newell [33] and Jimbo, Miwa and Ueno [34] have introduced a powerful approach for studying the initial value problem of equations of the Painlevé type. Their approach can be thought of as an extension of the inverse scattering transform method for solving ODE's: They have shown that solving such an initial value problem is essentially equivalent to solving an inverse problem for an associated isomonodromic linear equation. This inverse problem can be formulated in terms of monodromy data which can be calculated from initial data. The author and Ablowitz [35] have shown that the inverse problem can be formulated as a matrix, singular, discontinuous Riemann-Hilbert (RH) problem defined on a complicated contour. Hence techniques from RH theory can be employed to study the solvability of certain nonlinear ODE's. The above method is called inverse monodromic transform (IMT), and can be thought of as a nonlinear analogue of the Laplace's method for solving linear ODE's. A rigorous methodology for studying the RH problems associated with isomonodromic equations has been recently introduced by the author and Zhou [36]. Using this methodology it can be shown that Painlevé type equations admit in general global, meromorphic in t solutions. Furthermore, for special relations among the monodromy data and for t on certain rays, these solutions are bounded for finite t.

8.2 Periodic Solutions

The initial value problem for periodic initial data for several equations in $1 + 1$ have been solved using techniques from algebraic geometry (see for example [37]-[41]). For equations in $2 + 1$ there exist only weaker results (except for KPII): Krichever and Novikov [42]-[44] have found an effective way of generating periodic solutions for solvable equations in $2 + 1$. Their method can be thought of as an extension of the dressing method from the complex plane to an arbitrary Riemann surface. In what follows we summarize the main construction. The interested reader is refered to [45].

Let Γ be a compact Riemann surface of genus $g \geq 1$. It is well known [46] that on Γ there exist g holomorphic differentials $w_1, \cdots, w_g$ normalized by the conditions

$$\oint_{a_k} w_j = 2\pi i \delta_{jk}, \quad j, k = 1, \cdots, g, \qquad (8.7)$$

where $a_1, \cdots, a_g$ and $b_1, \cdots, b_g$ denote a basis of canonical cycles. The matrix of b-periods, the so called Riemann matrix is defined by

$$B_{jk} \doteq \oint_{b_k} w_j, \quad j, k = 1, \cdots, g. \qquad (8.8)$$

It can be shown that B is symmetric and has negative definite real parts. Let $J(\Gamma)$ be the Jacobi variety associated with Γ. The Abel map

$$A : \Gamma \to J(\Gamma), \quad A(P) = (A_1(P), \cdots, A_g(P)),$$

is defined by

$$A_k(P) = \int_{P_0}^{P} w_k, \quad k = 1, \cdots, g. \qquad (8.9)$$

An important role in the theory of periodic solutions is played by the Riemann theta function which is defined by:

$$\theta(z) = \sum_{N \epsilon \mathbb{Z}^g} e^{\frac{1}{2}\langle BN, N \rangle + \langle z, N \rangle}, \qquad (8.10)$$

where $z = (z_1, \cdots, z_g) \epsilon \mathbb{C}^g$ is a complex vector, B is a Riemann matrix and $\langle N, z \rangle = \sum_{i=1}^{g} N_i z_i$. The summation in (8.10) is taken over the lattice of integer vectors $N = (N_1, \cdots, N_g)$.

$\theta(z)$ is analytic in $\mathbb{C}^g$ furthermore it possesses the following important properties:

(a)

$$\theta(z + 2\pi i N + BM) = e^{-\frac{1}{2}\langle BM,M \rangle - \langle z,M \rangle}, \forall M, N \epsilon \mathbb{Z}^g. \qquad (8.11)$$

(b)

$$\theta(A(P) - A(D) - K) \quad \text{has exactly} \quad g \quad \text{zeros} \quad P_1, \cdots, P_g, \qquad (8.12)$$

where the vector K has components

$$K_j \doteq \frac{2\pi i + B_{jj}}{2} - \frac{1}{2\pi i} \sum_{l \neq j} \int_{a_l} \left(w_l(P) \int_{P_0}^{P} w_j \right), j = 1, \cdots g$$

and the vector $A(D)$ has components

$$A(D)_j = \sum_{l=1}^{g} \int_{P_0}^{P_l} w_j.$$

Using the above notion it is straightforward to construct periodic solutions of integrable equations. The construction is based on the *Baker-Akhiezer* function $\Psi(P)$: Let $P_1, \cdots, P_g$ be g generic points on Γ (more precisely assume that the divisor $D = P_1 + \cdots P_g$ is non spatial). Let Q be a point on Γ corresponding to ∞, i.e. $z = z(P)$ with $z(Q) = 0$ in the neighborhood of Q, and $k = \frac{1}{z}$. Let $q(k)$ be an arbitrary given polynomial. Then there exist a unique function $\Psi(P)$ with the following properties: (a) $\Psi(P)$ is meromorphic everywhere on Γ, except at $P = Q$. (b) Its only poles on Γ/Q are at the points $P_1, \cdots, P_g$. (c) $\Psi(P)$ has an essential singularity of the form $\Psi(P) \sim c e^{q(k)}$ near $P = Q$ (c is a constant).

The uniqueness is a simple consequence of the fact that the devisor is non-special. It is simple to show that

$$\Psi(P) = e^{\int_{P_0}^{P} \Omega} \frac{\theta(A(P) - A(D) + U - K)}{\theta(A(P) - A(D) - K)}, \qquad (8.13)$$

where the vector U has components

$$U_j = \oint_{b_j} \Omega, j = 1, \cdots, g \qquad (8.14)$$

and Ω is a normalized differential of the second kind on Γ with principal part at Q of the form $dq(k)$, i.e.

$$\oint_{a_j} \Omega = 0, j = 1, \cdots, g \quad \text{and} \quad \int_{P_0}^{P} \Omega \sim q(k) \quad \text{as} \quad P \to Q. \qquad (8.15)$$

Indeed, using (8.12) it follows that Ψ has poles at $P_1, \cdots, P_g$. Also it is obvious that Ψ has the correct behavior at $P = Q$. To show that Ψ is meromorphic we note that if we choose another path of integration then $\int_{P_0}^{P} \to \int_{P_0}^{P} + \oint_\gamma$ where γ is an N multiple of the a cycles and an M multiple of the b cycles. Thus

$$\int_{P_0}^{P} \Omega \to \int_{P_0}^{P} \Omega + \langle N, M \rangle, A(P) \to A(P) + 2\pi i N + BM.$$

Hence, the additional terms in (8.13) cancel (using (8.11)) and the meromorphicity of $\Psi(P)$ is established.

Taking $q(k) = kx + k^2 y + k^3 t$ and using the dressing method on Ψ it follows that

$$\Psi = e^{kx + k^2 y + k^3 t} \left(1 + \frac{\Psi^{(1)}}{k} + O(\frac{1}{k^2}) \right)$$

satisfies the Lax pair associated with the KP. In particular

$$-(\partial_y + \partial_x^2 + q)\Psi = 0,$$

where $q \doteq -2\Psi_x^{(1)}$ solves

$$\left[q_t - \frac{1}{4}(q_{xxx} + 6qq_x) \right]_x - \frac{3}{4}q_{yy} = 0.$$

Acknowledgements

This work was partially supported by the Office of Naval Research under Grant Number N00014-88K-0447, National Science Foundation under Grant Number DMS-8803471, and Air Force Office of Scientific Research under Grant Number 87-0310.

Bibliography

[1] *Symmetries and Nonlinear Phenomena*, World Scientific, eds. D. Levi and P. Winternitz (1988).

[2] *Bäcklund Transformations*, Lecture Notes in Mathematics 515, ed. R. Miura, Springer Verlag (1976).

[3] R. Hirota, *Direct Method of Finding Exact Solutions of Nonlinear Evolution Equations*, Lecture Notes in Mathematics, **515**, 40 (1976); *Direct Methods in Soliton Theory, in Solitons*, Topics in Current Physics, **17**, eds. R. Bullough, P. Caudrey, **157**, (1980).

[4] M. Kashiwara and T. Miwa, Proc. Japan Acad. **57A**, 342 (1981); E. Date, M. Kashiwara, and T. Miwa, Proc. Japan Acad. **57A**, 387 (1981); M. Jimbo, E. Date, M. Kashiwara, and T. Miwa, J. Phys. Soc. of Japan, **50**, 3806 (1981); and RIMS preprints 359 and 360.

[5] V.E. Zakharov and P.B. Shabat, Funct. Anal. Appl. **8**, 226-235 (1974).

[6] V.E. Zakharov and P.B. Shabat, Funct. Anal. Appl. **13**, 166 (1979).

[7] V.E. Zakharov and S.V. Manakov, Funct. Anal. Appl. **19**, No. 2, 11 (1985).

[8] A.S. Fokas and M.J. Ablowitz, Phys. Rev. Lett. **47**, 1096 (1981).

[9] V.E. Zakharov, S.V. Manakov, S.P. Novikov, and L.P. Pitayevskil, *Theory of Solitons. The Method of the Inverse Scattering Problem*, Nauka, Moscow (in Russian), 1980.

[10] A.R. Its and V.Y. Novokshenov, *The Isomonodromic Deformation Method in the Theory of Painlevé Equation*, Springer-Verlag, 1191, (1986).

[11] R. Rosales, Stud. Appl. Math. **59**, 117-151 (1978).

[12] P. Santini, M.J. Ablowitz, and A.S. Fokas, J. Math. Phys. **25**, 2614 (1984).

[13] F.W. Nijhoff, G.R.W. Quispel, J. Van der Linden, H.W. Capel, Physica A, **119**, 101 (1983); F.W. Nijhoff, H.W. Capel, G.L. Wiersma, G.R.W. Quispel, Phys. Lett. A, **103**, 293 (1984).

[14] V.E. Zakharov, On the Dressing Method, to appear (1990).

[15] F.D. Gakhov, *Boundary Value Problems*, Pergamon, 1966.

[16] N.I. Muskhelishvili, *Singular Integral Equations*, Noordhoff, Groningen, 1953.

[17] N.P. Vekua, *Systems of Singular Integral Equations*, Gordon and Breach, 1967.

[18] A.S. Fokas and V.E. Zakharov, The Dressing Method and Nonlocal Riemann-Hilbert Problems, preprint, Clarkson University, INS #137, (1989).

[19] P. Deift and E. Trubowitz, Comm. Pure Appl. Math., **32**, 121-125 (1979).

[20] D.J. Kaup, Physica **3D**, 45 (1980).

[21] L.-Y. Sung and A.S. Fokas, Inverse Problems of $N \times N$ Hyperbolic Systems on the Plane and the N Wave Interactions, to appear in Comm. Pure and Appl. Math. (1990).

[22] A.S. Fokas and M.J. Ablowitz, Stud. Appl. Math., **69**, 211-228 (1983); M.J. Ablowitz, D. Bar Yaacov, and A.S. Fokas, Stud. Appl. Math., **69**, 135-143 (1983).

[23] M. Boiti, J. Leon, L. Martina, F. Pempinelli, Phys. Lett. A, **132**, 432 (1988).

[24] A.S. Fokas and P.M. Santini, Phys. Rev. Lett. **63**, 1329 (1989); Dromions and a Boundary Value Problem for the Davey-Stewartson I Equation, to appear in Physica D.

[25] H. Segur, AIP Conference Proceedings, **88**, 211 (1982).

[26] X. Zhou, Inverse Scattering Transform for the Time Dependent Schrödinger Equation and KPI, to appear in Comm. in Math. Phys.

[27] M.V. Wickerhauser, Comm. Math. Phys., **108**, 67 (1987).

[28] Y.L. Rodin, Generalized Analytic Functions on Riemann Surfaces, Lecture Notes in Mathematics, (1988), Springer Verlag, (1987).

[29] R. Beals and R.R. Coifman, "The Spectral Problem for the Davey-Stewartson and Ishimori Hierarchies", Proc. Conf. on Nonlinear Evolution Equations: Integrability and Spectral Methods, Como, University of Manchester, 1988.

[30] R. Beals and R.R. Coifman, "Multidimensional Inverse Scattering and Nonlinear Partial Differential Equations", Proc. Symp. Pure Math. **43**, Amer. Math. Soc. Providence, pp. 45-70 (1985).

[31] L.-Y. Sung and A.S. Fokas, Inverse Problems in Multidimensions, INS #98, preprint, Clarkson University, (1990).

[32] A.S. Fokas and M.J. Ablowitz, J. Math. Phys. **23**, 2033 (1982).

[33] H. Flaschka and A.C. Newell, Commun. Math. Phys. **76**, 67 (1980).

[34] K. Ueno, Proc. Jpn. Acad. **56A**, 97 (1980); M. Jimbo, T. Miwa and K. Ueno, Physica **2D**, 306 (1981); M. Jimbo and T. Miwa, Physica **2D**, 407 (1981); **4D**, 47 (1981); M. Jimbo, Prog. Theor. Phys. **61**, 359 (1979).

[35] A.S. Fokas and M.J. Ablowitz, Comm. Math. Phys. **91**, 381 (1983).

[36] A.S. Fokas and X. Zhou, On the Solvability of Painlevé II and IV, INS #148, preprint, Clarkson University (1990).

[37] P.D. Lax, Comm. Pure Appl. Math. **28**, 141 (1975).

[38] H.P. McKean and P. van Moerbeke, Invent. Math. **30**, 217 (1975).

[39] H.P. McKean and E. Trubowitz, Comm. Pure Appl. Math. **29**, 143 (1977).

[40] A.P. Its and V.B. Matveev, Teoret. Mat. Fiz. 2̲3, 51 (1975).

[41] B.A. Dubrovin, V.B. Matveev and S.P. Novikov, Uspekhi Mat. Nauk **31**, 1, 55 (1976). Russian Math. Surveys **31**, 1, 59 (1976).

[42] I.M. Krichever and S.P. Novikov, Uspekhi Mat. Nauk **35**, 6, 47 (1980). Russian Math. Surveys, **35**, 6, 53 (1980).

[43] I.M. Krichever and S.P. Novikov, Uspekhi Mat. Nauk **32**, 6, 183 (1977). Russian Math. Surveys, **32**, 6, 185 (1977).

[44] I.M. Krichever and S.P. Novikov, Funktsional. Anal. i Prilozhen, **11**, 1, 15 (1977). Functional Anal. Appl. **11**, 1, 12 (1977).

[45] B.A. Dubrovin, Uspekhi Mat. Nauk, **36**, 2, 11 (1981). Russian Math. Surveys, **36**, 2, 11 (1981).

[46] G. Springer, Introduction to Riemann Surfaces, Addison Wesley, MA (1957).

[47] P.M. Santini, Physica D, (1990).

210   A. S. Fokas

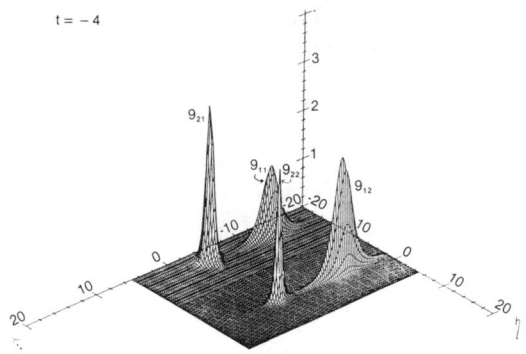

Figure 1.1

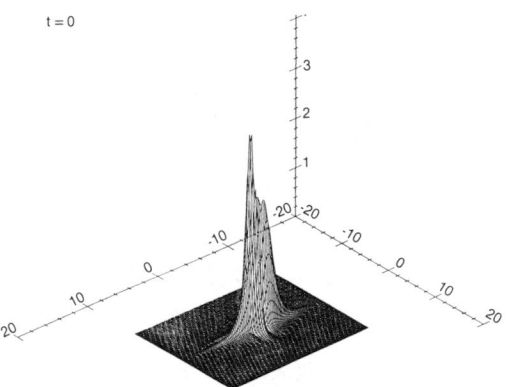

Figure 1.2

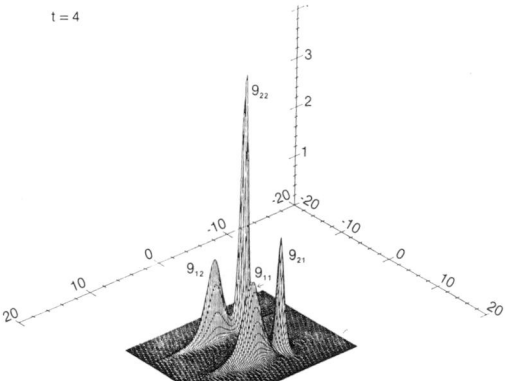

Figure 1.3

Hamiltonian Structure and Integrability

Benno Fuchssteiner
University of Paderborn
D 4790 Paderborn
Germany

1 Introduction

Whenever a quantity, or a set of quantities, evolves with time then we call this a **dynamical system**. The evolution of the universe certainly is a dynamical system, however a complicated one. The laws of evolution which govern such a system are called the **dynamical laws**.

To describe dynamical systems we usually make suitable approximations in the hope of finding valid descriptions of their characteristic quantities. But even after such approximations we mostly cannot write down explicitly how these quantities depend on time, usually such a dependence is much to complicated to be computed explicitly. Therefore we commonly write down dynamical systems in their **infinitesimal form**.

Considering a dynamical system in its infinitesimal form has many advantages. The principal one is that such an infinitesimal description is possible even in those cases where a **global** description is not feasible at all. Technically speaking, an infinitesimal description leads to a differential equation, which in many cases has nonlinear terms due to the interaction between different quantities. To find such a differential equation we only have to know a suitable set of dynamical laws. However, solving such a nonlinear differential equation for arbitrary starting points (initial conditions) is often a hopeless endeavor.

Fortunately, the infinitesimal description sometimes gives an insight into the essential structures for the dynamics of the system, or at least into those parts of the dynamics which can be described locally.

Speaking from an abstract viewpoint the main objects of our interest are equations of the form

$$u_t = K(u) \tag{1.1}$$

where $K(u)$ is a **vector field** on some manifold M and where u denotes the general point on this manifold. Since we do not restrict the size of the dimension of the manifold M this equation still comprises an abundance of possible dynamical systems. For example u, could be the collection of all relevant data of an economy, then equation (1.1) describes the evolution of that economy. With regard to size of the manifold, this would be a rather simple dynamical system since the manifold certainly has finite dimension whereas most systems we consider later on will describe systems on **infinite dimensional manifolds**. Most notions which we use in the study of equation (1.1) do have a very intuitive meaning. For example, we call equation (1.1) a **flow** on the underlying manifold. Thus we imagine that a point is flowing along its path on the manifold. Such a path is called an **orbit** of the system. Since $K(u)$ describes the change in the position of u for infinitesimal times,

$K(u)$ must be tangential to the orbit of u. A simplified picture for the situation under consideration is:

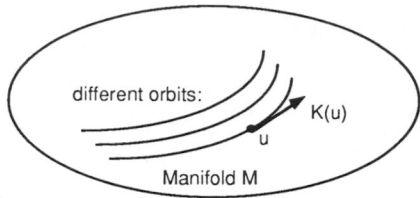

different orbits:

$K(u)$

u

Manifold M

Fig. 1: Flow on a manifold M

Systems of particular importance are those describing the dynamics of particles in classical mechanics. For these systems the dynamical laws are determined by the total energy of the system. As an example, we consider the case when the energy $H = V + T$ is the sum of potential energy V and kinetic energy T, where $V = V(\vec{x}_1, ..., \vec{x}_n)$ only depends on the positions of the different particles and where the kinetic energy is

$$T = \frac{1}{2} \sum m_i \dot{\vec{x}}_i^2, \quad (m_i = mass\ of\ particle\ i) .$$

In order to eliminate the masses m_i (which are irrelevant for the structure of the system) we introduce new coordinates in the space $\mathbb{R}^{2n}$ (phase space)

$$q_i = \vec{x}_i, \ p_i = m_i \dot{\vec{x}}_i, \ i = 1, ..., n. \tag{1.2}$$

Using Newton's law we find

$$\dot{q}_i = \frac{\partial H}{\partial p_i}, \ \dot{p}_i = -\frac{\partial H}{\partial q_i} \tag{1.3}$$

or, if we introduce the formal field variable $\vec{u}$ to be the transposed of the position-momentum vector $\vec{u} = (q_1, ..., q_n, p_1, ..., p_n)^T$, then the dynamics has the form

$$\vec{u}_t = \begin{pmatrix} 0 & I \\ -I & 0 \end{pmatrix} \nabla H \tag{1.4}$$

where I is the identity in n-space.

Notable is that in this equation only ∇H the *gradient* of the energy (taken in the phase space) enters. These equations, either in the form (1.3) or (1.4), are prototypes of *Hamiltonian systems.* [1] Systems of this form have remarkable properties. Not only do they seem to be solely determined by their energy, but also there is the surprising property that whenever the energy functional H is invariant under either translation or rotation then we have conservation of momentum or angular momentum, respectively. So there must

---

[1] Sir William Rowan Hamilton (August 4, 1805 - September 2, 1865) studied at Trinity College in Dublin where he obtained the Chair of Astronomy in the age of 22. Apart from his work in mechanics (principle of least action) he gave fundamental contributions to optics and mathematics. For example, he introduced complex numbers and the quaternions.

be some kind of relation between the conservation laws of the system and its symmetry structure. Indeed such a relation was revealed for classical Hamiltonian systems by Emmy Noether[2] in her habilitation [18] [3] and this relation is nowadays generalized to symmetries of so-called Lie-Bäcklund type for systems on infinite dimensional manifolds. The special form of (1.3) or (1.4) is due to the special coordinate system which was chosen, whereas the fundamental relation between symmetries and conserved quantities certainly must go beyond a structure which is the consequence of a special coordinate system. So, in studying Hamiltonian equations they must be analyzed in a differential geometric invariant setup such that their structure becomes independent of special charts which were chosen to parametrize the underlying manifold. We shall do that in the following sections 4 and 5.

The best known examples of Hamiltonian systems probably are the Harmonic Oscillator and the nonlinear pendulum, described below:

Example 1.1: Harmonic Oscillator
The evolution equations

$$x_t = y \ , y_t = -x \ ,$$

where $x(t), y(t) \in \mathbb{R}$, describe the time dependence of the harmonic oscillator. In matrix form this can be written as

$$\begin{pmatrix} x \\ y \end{pmatrix}_t = \begin{pmatrix} 0 & 1 \\ -1 & 0 \end{pmatrix} \begin{pmatrix} x \\ y \end{pmatrix} . \tag{1.5}$$

which certainly has the form (1.4) since

$$\begin{pmatrix} x \\ y \end{pmatrix} = \nabla H, \quad H = \frac{1}{2}(x^2 + y^2) \ .$$

The manifold under consideration is $M = \mathbb{R}^2$. Introducing the abbreviations

$$K(u) = Au, \ u = \begin{pmatrix} x \\ y \end{pmatrix}, \ A = \begin{pmatrix} 0 & 1 \\ -1 & 0 \end{pmatrix}$$

the evolution equation (1.5) is clearly an example for (1.1). Looking at this system we detect many characteristic features which carry over to some nonlinear systems. The evolution of this flow is of the form

$$\exp(tA) : \ u(0) \longrightarrow u(t) \tag{1.6}$$

which shows that the map from the initial condition $u(0)$ to $u(t)$ defines a set of diffeomorphisms on the manifold $M = \mathbb{R}^2$. Because of the exponential function these diffeomorphisms

---

[2] Emmy Noether (March 3, 1882 - April 4, 1935) was the daughter of the renowned mathematician Max Noether. With her contributions to the theory of ideals and the field of non-commutative algebra she influenced the development of modern algebra to a great extent. After she lost in 1933 her venia legendi (right to teach) in Nazi Germany she emigrated to Princeton.

[3] A work for which Emmy Noether herself had no high opinion (see [1]). Later on she refused to take any more notice of this work of fundamental importance and she even claimed that it had been lost ("verschollen"). Hermann Weyl had a completely different opinion and acknowledged this in his memorial address delivered in Bryn Mawr College on April, 26, 1935: "For two of the most significant sides of general relativity theory she gave at that time the genuine and universal mathematical formulation: First, the reduction of the problem of differential invariants to a purely algebraic one by use of "normal coordinates"; second, the identities between the left sides of Euler's equations of a problem of variation which occur when the (multiple) integral is invariant with respect to a group of transformations..."

form a representation of the additive group $(\mathbb{R}, +)$. Furthermore, we observe the advantage of introducing polar coordinates $r = \sqrt{x^2 + y^2}$ and $\varphi = \arctan(y/x)$. Then in this new coordinate space the system becomes a flow with constant velocity along the coordinate lines $r = constant$. Thus in this case we are able to split up the coordinates into two sets, one set (*action variables*) which remains constant under the flow, and another set (*angle variables*) which grows on the orbits linear with time. If such special coordinates having these properties can be introduced then we call such a system *completely integrable*. Looking back at our example we see that this notion of complete integrability must be related to the existence of one-parameter symmetry groups. This is because changing one of the coordinates and leaving the others unchanged moves orbits into orbits. So this movement along coordinate lines constitutes a *symmetry group*. □

Example 1.2: Pendulum
The time development in this case is

$$\varphi_{tt} + \sin(\varphi) = 0. \tag{1.7}$$

Introducing

$$q = \varphi, \ p = \varphi_t \ and \ u = \begin{pmatrix} q \\ p \end{pmatrix}$$

we see that (1.7) is of the form (1.1)

$$u_t = \begin{pmatrix} q \\ p \end{pmatrix}_t = \begin{pmatrix} p \\ -\sin(q) \end{pmatrix} = \begin{pmatrix} 0 & 1 \\ -1 & 0 \end{pmatrix} \begin{pmatrix} \sin(q) \\ p \end{pmatrix}. \tag{1.8}$$

The manifold under consideration again is $M = \mathbb{R}^2$. In contrast to (1.5) this equation constitutes a nonlinear flow. Again the dynamics has the form (1.4) since

$$\begin{pmatrix} \sin(q) \\ p \end{pmatrix} = \nabla H, \ \ H = 1 - \cos(q) + \frac{1}{2}p^2 \ .$$

Although this is a nonlinear flow it can be linearized locally by introducing a suitable coordinate system. But this coordinate system is no longer given by polar coordinates. Obviously the part of the coordinate lines which we called action variables should now be given by the lines $H = constant$ and the remaining part should be chosen in a suitable way. How to do this will be described later on. □

The scope of this article is to rephrase these simple observations which we made for the harmonic oscillator in a general framework so that they can be carried over to other more complicated systems. Furthermore, we want to formulate the corresponding notions and relations in such a way that they are independent of the coordinate systems which we choose.

We organize the article in the following way: In the next section, we introduce some basic notions which lead in Section 3 to a description of the connection between symmetries and conserved quantities. At that point we shall not yet choose the most abstract setup for the description. Instead of formulating everything in a differential geometric invariant way we still will work with coordinate systems. This we do in order to keep the level of abstraction at the beginning as low as possible. Results in these sections are mostly presented without proofs because later on proofs will be given in a short and concise way by using a higher level of abstraction. In Section 4 we introduce Lie algebra modules, Lie derivatives and tensors in order to have a notation which allows one to see which notions are geometrically invariant. In Section 5 we introduce the notion of bi-hamiltonian fields on a general level. Then, in the following section, we introduce compatibility, especially for hamiltonian pairs, and illustrate the power of this notion by a set of suitable examples (Section 7). In the final part, Section 8, we discuss complete integrability in the finite dimensional case and we show how that notion is connected to the situation considered before. In addition the action/angle structure of the multisoliton manifolds is given

2   Basic Notions in Chart Representation

I hope that most readers are acquainted with notions like manifolds, vector fields, tangent space, differentiability and so on. However, I do not believe that a knowledge of the theoretical background in manifold analysis is really necessary for understanding the concepts described in this article. For the most part a more intuitive grasp of infinitesimal calculus and a heuristic idea of manifolds as being something like smooth surfaces seems sufficient.

For the sake of completeness however, we include some remarks on this subject since notation will differ somewhat from the conventional notation, insofar as we avoid the calculus of exterior forms.

For infinite dimensional manifolds we will use the notion of Hadamard differentiability [26,] [27.] This is a fairly weak notion which nevertheless ensures the validity of the chain rule. A function $F : E_1 \rightarrow E_2$ between two linear spaces [4] is said to be *Hadamard-differentiable* at $u \in E_1$ if there is a continuous linear map $L : E_1 \rightarrow E_2$ such that

$$\lim_{\epsilon \to 0} \frac{1}{\epsilon}\{F(u + \epsilon v) - F(u) - \epsilon L[v]\} = 0 \qquad (2.1)$$

uniformly in v on each compact subset of E_1. The linear operator L, and its application $L[v]$ to v are then denoted by $F'(u)$ and $F'(u)[v]$, respectively. Of course, $F'(u)[v]$ is most easily computed from the *directional derivative* of F

$$F'[v] = F'(u)[v] = \frac{\partial}{\partial \epsilon}\bigg|_{\epsilon=0} F(u + \epsilon v). \qquad (2.2)$$

If not otherwise mentioned functions are usually assumed to be C^∞-functions, i.e. infinitely often differentiable.

If the manifold is a vector space $M = E$, then vector fields are the continuous maps $K : E \rightarrow E$ assigning to each $u \in M$ some vector $K(u) \in E$. Again, we assume vector fields

---

[4] All linear spaces E are assumed to be locally convex Hausdorff topological vector spaces. Usually we do not describe explicitly the topology on E. We rather introduce a vector space E^* of linear functionals on E, which separate points, and we assume that E is endowed with the weakest locally convex topology such that the elements of E^* are continuous (i.e. the weak topology with respect to E^*). Spaces $L(E_1, E_2)$ of linear maps $E_1 \rightarrow E_2$ are then endowed with the weakest topology given by the dual pairs E_1, E_1^* and E_2, E_2^*, i.e. the weakest convex topology such that all linear functionals μ on $L(E_1, E_2)$ given by $L \rightarrow \mu(L) = \rho(L(u))$ with $\rho \in E_2^*$, $u \in E_1$ are continuous.

to be C^∞. Thus they constitute a Lie algebra with respect to the **commutator** defined by

$$[K, G](u) = \frac{\partial}{\partial \epsilon}\Big|_{\epsilon=0} \{G(u + \epsilon K(u)) - K(u + \epsilon G(u))\}$$
$$= G'(u)[K(u)] - K'(u)[G(u)].$$
(2.3)

This Lie algebra is referred to as the **vector field Lie algebra** .

Recall that the definition of a Lie-algebra implies that the map $(K, G) \rightarrow [K, G]$ is bilinear, antisymmetric ($[K, G] = -[G, K]$) and such that for all K, G, L the **Jacobi identity**

$$[[K, G], L] + [[L, K], G] + [[G, L], K] = 0$$
(2.4)

holds. In fact this identity is easily verified by using the chain rule of differentiation.

If the manifolds M which we consider are not linear spaces, then derivatives are defined in the usual way by parametrizing, or modeling, manifolds by linear spaces. Although in most of our examples the underlying manifold is a vector space we briefly illustrate that procedure for the sake of completeness. Those readers who do not care for technicalities should skip the following paragraphs up to the introduction of conserved quantities.

Let M be some Hausdorff topological space and E some linear space, then we call M a C^∞-manifold if there are given an open covering $\{U_\alpha | \alpha \in some\ index\ set\}$ of M and homeomorphisms

$$p_\alpha : U_\alpha \rightarrow V_\alpha,\ V_\alpha\ open\ susets\ of\ E$$

such that for all α and β the overlap map $p_\alpha \circ p_\beta^{-1}$ is a C^∞-map $V_\beta \cap p_\beta(U_\alpha \cap U_\beta) \rightarrow E$. These p_α can be considered to be local coordinates for the corresponding U_α. The collection of these (U_α, p_α) is defined to be an *atlas*. Such an atlas allows transfer of all aspects of the differential structure from E to M. For example, a map φ on M is defined to be C^∞ if all the $\varphi \circ p_\alpha^{-1}$ are C^∞. Consider $u \in M$, then a *chart* around u is a homeomorphism p from an open neighborhood U of u into the model space E such that for all α the map $p \circ p_\alpha^{-1}$ defined on $p_\alpha(U) \cap V_\alpha$ is C^∞. Now, the notion of tangent space is easily introduced. The *tangent space* $T_u M$ at the point u is represented by the model space together with a chart. The formal definition has to be such that it does not depend on the special chart which is chosen, hence it must be given by equivalence classes with respect to different charts. So, for fixed u, we consider pairs (p, v) consisting of a chart around u and a vector v in the model space. In these pairs we introduce an equivalence relation $(p_1, v_1) \equiv (p_2, v_2)$ defined by $(p_2 \circ p_1^{-1})'(p_1(u))[v_1] = v_2$. Then $T_u M$ is defined to be the set of equivalence classes endowed with the obvious topology inherited from the model space, and TM denotes the collection of all these tangent spaces and is called the *tangent bundle*. However, working with these equivalence classes is not always very practical, so locally around u, we choose common representatives by fixing some chart p around u and representing the tangent spaces of the points around u jointly by (p, E). Then a map $K : M \rightarrow TM$ which assigns to each $u \in M$ some element of $T_u M$ is said to be a C^∞-vector field if it is locally C^∞ with respect to such a common representation. Similarly, we define locally the *co-vector space* $T_u^* M$ by (p, E^*) instead, and *co-vector fields* to be suitable C^∞-maps from M into the *cotangent bundle* (collection of all $T_u^* M$.) Of course, strictly speaking, elements of $T_u^* M$ are again equivalence classes as before (only in the definition above, one has to replace the derivatives by suitable adjoints in order to leave the application of a co-vector to a tangent vector invariant under coordinate changes). It should be observed that the choice of a common representation around some point $u \in M$ is the same as choosing a particular chart in the manifold given by the tangent bundle.

If the manifold under consideration is a linear space, then we do not really need all these constructions because we then model the manifold by itself and for simplicity we choose the canonical chart given by the identity function on the model space. The validity of the requirements above then follows from the usual transformation formulas of differential calculus. In this case the tangent spaces $T_u M$ and cotangent spaces $T_u^* M$ can be identified with E and E^*, respectively, and we are back in the situation $M = E$ which we studied at the beginning.

We have chosen this formal approach to manifolds in order to indicate that differential calculus on abstract manifolds is indeed an easy task and that nevertheless for practical computations it mostly is sufficient to do analysis on linear spaces.

To proceed, we consider again

$$u_t = K(u) , \quad u \in M , \quad M \ some \ manifold . \tag{1.1}$$

C^∞-maps from the manifold M into the scalars (either $\mathbb{R}$ or $\mathbb{C}$) are called *scalar fields*. A scalar field $I(u)$ is said to be a **conserved quantity** for (1.1) if

$$I'(u)[K(u)] = 0 \tag{2.5}$$

for all $u \in M$. The reason why this name has been chosen is obvious: Take an orbit $u(t)$ of (1.1), then by the chain rule we find

$$\frac{d}{dt} I(u(t)) = I'(u(t))[K(u(t)] = 0. \tag{2.6}$$

Hence, (2.5) guarantees that I is constant along the orbits of (1.1).

Observe that, for every $u \in M$, the quantity $I'(u)$ is a continuous linear functional on the tangent space $T_u M$, i.e. $I'(u)$ must be a cotangent vector. Derivatives of scalar fields are called **gradients** and f. Therefore we use for scalar quantities I the notation $\nabla I(u)$ instead of $I'(u)$. If we write $<,>$ for the duality between tangent and cotangent vectors, then (2.5) is written as

$$< \nabla I, K > = 0. \tag{2.7}$$

Sometimes there is some advantage in looking at conserved quantities which depend explicitly on time. A family $F(u,t)$ of scalar fields depending in a C^∞-way on the parameter t is said to be a **time dependent conserved quantity** if

$$F_t(u,t) + < \nabla F(u,t), K(u) > = 0. \tag{2.8}$$

Here sub-t denotes partial derivative with respect to t and $\nabla F = F'$ is taken by ignoring the parameter t. The notion makes sense because it gives

$$\frac{d}{dt} F(u(t),t) = 0 \tag{2.9}$$

which implies that $F(u(t),t)$ is constant along the orbits of (1.1). From the physical point of view such a quantity does not seem to be very significant, since it is not invariant with respect to a translation of time. Nevertheless it turns out that it is a rather interesting quantity from the computational point of view.

Of special interest are those conserved quantities which are linear in t. Let

$$F(u,t) \;=\; f_0(u) \;+\; f_1(u)t \tag{2.10}$$

be such a quantity. Inserting F into (2.8) we then obtain by comparison of coefficients

$$f_1 \;+\; <\nabla f_0, K> \;=\; 0 \;. \tag{2.11}$$

Hence, $F(u,t)$ is uniquely determined by its absolute term $f_0(u)$. Furthermore, the term $f_1(u)$ must be a conserved quantity which is time independent.

Related to conserved quantities are one-parameter groups of C^∞-diffeomorphisms on the manifold M. Recall that these are defined to be one-to-one C^∞-maps such that the inverse is again differentiable. A **one-parameter group** of diffeomorphisms is a map $(u,\tau) \to R(\tau)(u)$ which is differentiable on the product $M \times \mathbb{R} \;=\; \{(u,\tau)|u \in M, \tau \in \mathbb{R}\}$ and assigns to every $\tau \in \mathbb{R}$ some diffeomorphism $R(\tau) : M \to M$ such that

$$R(\tau_1 + \tau_2) \;=\; R(\tau_1) \circ R(\tau_2) \; and \; R(0) \;=\; I \tag{2.12}$$

for all $\tau_1, \tau_2 \in \mathbb{R}$. This implies that all the $R(\tau)$ do commute and that $R(-\tau)$ must be the inverse of $R(\tau)$. With other words: $R(\tau)$ defines a group representation of the additive group $(\mathbb{R}, +)$. One-parameter groups are completely determined by their τ-derivative at $\tau \;=\; 0$. To see this let $R(\tau)$ be a one-parameter group then

$$G \;=\; \frac{d}{d\tau} R(\tau)|_{\tau \,=\, 0} \tag{2.13}$$

is said to be its *infinitesimal generator*. Equation (2.13) is an abbreviation for $G(u) \;=\; (d/d\tau)\{R(\tau)(u)\}|_{\tau=0}$. Since $R(\tau)$ assigns to each point of the manifold another point, G must assign to each manifold point u a tangent vector at u . Hence G is a vector field. Because of the functional equation (2.12) the τ-derivative of $R(\tau)$ at arbitrary τ is easily expressed by G:

$$\frac{d}{d\tau} R(\tau) \;=\; G \circ R(\tau). \tag{2.14}$$

Hence $R(\tau)$ is uniquely determined by the vector field $G(u)$.

If the $R(\tau)$ are linear then G again is linear. Then the solution of the linear differential equation (2.14) can formally be written as $R(\tau) \;=\; \exp(\tau G)$. In general however, diffeomorphism groups are far from being groups of linear transformations. Nevertheless their structure is more or less given by the exponential function since by use of pull-backs equation (2.14) can be transformed into a linear differential equation (on some abstract manifold with rather high dimension, however).

To see this, consider $\mathcal{F} \;=\; C^\infty(M, \mathbb{R})$ or $C^\infty(M, \mathbb{C})$, respectively, the vector space of scalar fields. Let $R : M \to M$ be a C^∞-map then

$$(R^* f)(u) \;:=\; f(R(u)) \;, \qquad f \in F \tag{2.15}$$

defines a map $R^* : \mathcal{F} \to \mathcal{F}$ which is linear on $\mathcal{F}$. R^* is said to be the **pull-back** given by R. Similarly, if K is a vector field we define a map $L_K : \mathcal{F} \to \mathcal{F}$

$$(L_K f)(u) \;=<\nabla f, K(u)>, \quad f \in \mathcal{F} \tag{2.16}$$

by assigning to each $f \in \mathcal{F}$ its derivative in the direction K. L_K is said to be the **Lie-derivative** given by K. Again, this is a linear map on $\mathcal{F}$.

The space of all Lie-derivatives is a vector space. The usual commutator of linear maps

$$[L_K, L_G] \;=\; L_K \circ L_G - L_G \circ L_K \tag{2.17}$$

endows this vector space in a natural way with a Lie-algebra structure.

Observation 2.1: *The map $K \to L_K$ is a Lie-algebra isomorphism from the Lie algebra of vector fields onto the Lie-derivatives, i.e. we have*

$$L_{[K,G]} \;=\; [L_K, L_G] \tag{2.18}$$

for all vector fields K and G.

The proof is simple, since by differentiation we see that the commutator bracket is a representation of the vector field bracket. Moreover, the required fact that $L_K \neq L_G$ whenever $K \neq G$ follows from the observation that for any two different tangent vectors in $T_u M$ we can find a scalar field (by application of a suitable co-vector) having different derivatives in the direction of these tangent vectors. By similar arguments we find $R_1^* \neq R_2^*$ whenever $R_1 \neq R_2$. Hence, it suffices to study the pull-backs and the Lie-derivatives instead of the original objects. For these new quantities equation (2.14) translates into

$$\frac{d}{d\tau} R^*(\tau) \;=\; L_K R^*(\tau). \tag{2.19}$$

which is clearly a linear differential equation since only linear operations are involved. Therefore we write

$$R^*(\tau) \;=\; \exp(\tau L_K) \tag{2.20}$$

thus obtaining a representation of the one-parameter group in terms of its infinitesimal generator. Of course, this is a highly artificial representation, since $R^*(\tau)$ and L_K act on infinite-dimensional vector spaces even when M is finite dimensional.

A consequence of these considerations is that one-parameter groups and evolution equations, whether linear or nonlinear, are more or less the same objects. To see this, let $\{R_G(\tau)|\tau \in \mathbb{R}\}$ be such a one-parameter group of diffeomorphisms on M with infinitesimal generator G. Since G is a vector field assigning to each $u \in M$ the tangent vector $G(u) \in T_u M$ we look at evolution equation

$$u_t \;=\; G(u). \tag{2.21}$$

In fact, for any initial condition $u(0)$ a solution is easily found, namely

$$u(t) \;=\; R_G(t)(u(0)) . \tag{2.22}$$

This solution for the initial value problem of (2.21) must be unique since R_G is a group: To see this take another solution $\bar{u}(t)$ fulfilling the same initial condition $\bar{u}(t=0) = u(0)$. Then by differentiation we obtain that $R_G(-t)(\bar{u}(t))$ must be independent of t and hence equal to $u(0)$. Recalling that $R_G(-t)$ is the inverse of $R_G(t)$ we find $\bar{u}(t) = u(t)$.

This viewpoint shows that $R_G(t)$ can be understood as the flow operator of (2.21) assigning to each initial condition $u(0)$ the solution $u(t)$ at time t. Of course, not all evolution equations of the form (2.21) necessarily yield one-parameter groups, only those where every initial condition $u(t=0) = u(0)$ have a unique solution *for all* t such that the flow operator is a C^∞-map.

This interpretation shows, that notions and methods coming from one-parameter groups must lead right-away to the crux of the algebraic aspects of evolution equations. Therein lies the problem of commutativity for nonlinear flows. It is easily seen that this important property can be expressed in terms of infinitesimal generators. Looking at the exponential form of the pull-backs for these groups one discovers the infinitesimal equivalent for commutativity:

Observation 2.2: *Let $R_K(\tau)$ and $R_G(t)$ be two one-parameter groups of diffeomorphisms with infinitesimal generators K and G. These two groups commute, i.e.*

$$R_K(\tau)^\circ R_G(t) = R_G(t)^\circ R_K(\tau) \text{ for all } t \text{ and } \tau \text{ in } \mathbb{R}$$

if and only if $[K, G] = 0$, *i.e. their infinitesimal generators commute in the vector field Lie-algebra.*

In general, it is very hard to verify whether or not a vector field is really the infinitesimal generator of a one-parameter group because usually it is difficult to see if equation (2.21) has a unique solution for every initial condition. But one of the reasons for the success of mathematical analysis is that global conditions (like existence and commutativity of groups $R_K(t)$ and $R_G(\tau)$) can be rephrased, by use of infinitesimal arguments, as local conditions. Therefore it seems natural to put the concept of symmetries onto a purely algebraic and infinitesimal basis by taking the commutativity of vector fields as definition (even in those cases where (2.21) is not the infinitesimal form of some globally defined group).

So, we define the vector field $G(u)$ to be a **symmetry** for the evolution equation (1.1) if and only if $[K, G] = 0$. Here the notion *symmetry* is used as abbreviation for what correctly should be termed as *infinitesimal symmetry-generator*.

Note that when G is a symmetry for (1.1) then K also is a symmetry for $u_t = G(u)$. Using the Jacobi identity we see that whenever the vector fields G and L are symmetries for (1.1) then $[G, L]$ is again a symmetry for this evolution equation. So, the symmetries of (1.1) are a subalgebra of the Lie algebra of vector fields.

It will turn out, that introduction of the concept of time dependent symmetries constitutes an efficient tool. A family of vector fields $G(u,t)$ depending in a C^∞-way on the parameter t is said to be a **time-dependent symmetry** of (1.1) if

$$G_t + [K, G] = 0. \qquad (2.23)$$

Here, again $[K, G]$ is taken by ignoring the parameter t .

If $G(u,t)$ and $L(u,t)$ are time-dependent symmetries then $[G,L]$ is again a time-dependent symmetry. This is easily seen from (2.23) and the Jacobi identity. Hence, the time-dependent symmetries are again a subalgebra of the Lie algebra of vector fields.

The algebraic structure of time-dependent symmetries is very similar to the corresponding structure for conservation laws. For example if

$$G(u,t) \quad = \quad G_0(u) \quad + \quad G_1(u)t \tag{2.24}$$

is a time-dependent symmetry linear in t, then insertion into (2.23) and comparison of coefficients yields $G_1 + [K, G_0] = 0$. Hence $G(u,t)$ is uniquely determined by its absolute term $G_0(u)$. Furthermore, $G_1(u)$ must be a symmetry.

3   Poisson Brackets and Hamiltonian systems

If one compares equations (2.8) and (2.23) for the dynamical variables given by conserved quantities and symmetries one discovers that these equations look very similar. They both are linear evolution equations on some infinite dimensional manifold

But there is one essential difference between these two equations. A difference which is easily discovered if one looks for means of constructing new solutions. 'A priori', equation (2.23) has more structure than equation (2.8) since there is a Lie algebra involved. This is of considerable advantage because we can take the commutator of any two solutions to find a new solution. So, in order to complete the analogies between conserved quantities and symmetries it seems intriguing to look for Lie algebra structure among solutions of (2.8). Another viewpoint arises by looking at the time derivative in both cases. The time derivative is a special case of what usually is said to be a *derivation*, where derivation means the validity of the product rule (of which the Jacobi identity is a representation). Equation (2.23) tells us that this special time-derivative can be replaced by some *inner derivation*, where an inner derivation is something given by commutation with an element taken out of the structure under consideration. And inner derivations are, from the mathematical viewpoint, much nicer than outer derivations. For example, apart from the discovery that dynamical variables are operators rather than scalars, one of the reasons for the success of quantum mechanics was the ansatz that the time evolution of these operators is given by inner derivations. It is hard to imagine that quantum mechanics would have been feasible at its beginning without this assumption.

Therefore it is natural to ask whether in case of (2.8) the time derivative can be replaced by some inner derivation.

Fortunately, all these questions lead to the same structure, namely *Hamiltonian systems*. If one analyzes the situation further it all boils down to:

Problem 3.1: *Take some operator valued function $\Theta(u)$ mapping each manifold element u to some linear operator $\Theta(u) : T_u M^* \to T_u M$. Define a bracket among scalar fields F_1, F_2 by*

$$\{F_1, F_2\}_\Theta \quad = \quad < \nabla F_2, \Theta \circ \nabla F_1 > . \tag{3.1}$$

When is this a Lie-algebra? In addition, when is $\Theta \circ \nabla$ a Lie algebra homomorphism into the vector field Lie algebra, i.e. when do we have

$$\Theta \circ \nabla \{F_1, F_2\}_\Theta = [\Theta \circ \nabla F_1, \Theta \circ \nabla F_2] \quad ? \tag{3.2}$$

We easily find the complete answer to that problem:

Theorem 3.2: *The following are equivalent:*

(1) *The bracket $\{\ ,\ \}_\Theta$ defines a Lie algebra*

(2) *The bracket $\{\ ,\ \}_\Theta$ defines a Lie algebra such that $\Theta \circ \nabla$ fulfills (3.2), i.e. $\Theta \circ \nabla$ is a Lie algebra homomorphism into the vector fields*

(3) Θ *has the following properties*

    (i) Θ *is skew-symmetric with respect to the duality between cotangent space and tangent space, i.e. $\Theta = -\Theta^+$ or $< v_1^*, \Theta v_2^* > = - < v_2^*, \Theta v_1^* >$ for all cotangent vectors $v_1{}^*, v_2{}^*$.*

    (ii) *for all cotangent vectors $v_1^*, v_2^*, v_3^* \in T_u M^*$ the following identity holds*
$$< v_1^*, \Theta(u)'[\Theta(u)v_2^*]v_3^* > + < v_2^*, \Theta(u)'[\Theta(u)v_3^*]v_1^* > + < v_3^*, \Theta(u)'[\Theta(u)v_1^*]v_2^* > = 0.$$

Proof:
First we show the equivalence between (1) and (3).

    The skew-symmetry is certainly necessary and sufficient in order to guarantee that $\{\ ,\ \}_\Theta$ is antisymmetric. Computation of the double-bracket yields

$$
\begin{aligned}
\{F_1, \{F_2, F_3\}_\Theta\}_\Theta &= \ < \nabla \{F_2, F_3\}_\Theta, \Theta \nabla F_1 > \\
&= \ < \nabla < \nabla F_3, \Theta \nabla F_2 >, \Theta \nabla F_1 > \\
&= \ F_3''[(\Theta \nabla F_2, \Theta \nabla F_1)] - F_2''[(\Theta \nabla F_3, \Theta \nabla F_1)] + < \nabla F_3, \Theta'[\Theta \nabla F_1] \nabla F_2 > .
\end{aligned}
$$

Since second derivatives are symmetric with respect to their entries all second derivatives F'' cancel if $\{F_1, \{F_2, F_3\}_\Theta\}_\Theta + its\ cyclic\ permutations$ are taken. Therefore condition (3,ii) is equivalent to the Jacobi identity for $\{\ ,\ \}_\Theta$ which finishes the proof of the equivalence between (1) and (3).

    Since (2) implies (1) it only remains to prove that (3,ii) implies equation (3.2). To see this take two scalar fields F_1, F_2 and some arbitrary co-vector v^* . Since the second derivatives of F_1, F_2 are symmetric they all cancel in the following computation:

$$
\begin{aligned}
< v^*, \ - \ \Theta \nabla \{F_1, F_2\}_\Theta &+ [\Theta \nabla F_1, \Theta \nabla F_2] > \\
&= \ < \nabla \{F_1, F_2\}_\Theta, \Theta v^* > \ + \ < v^*, [\Theta \nabla F_1, \Theta \nabla F_2] > \\
&= \ < \nabla F_2, \Theta'[\Theta v^*] \nabla F_1 > + < v^*, \Theta'[\Theta \nabla F_1] \nabla F_2 - \Theta'[\Theta \nabla F_2] \nabla F_1 > \\
&= \ < \nabla F_2, \Theta'[\Theta v^*] \nabla F_1 > + < v^*, \Theta'[\Theta \nabla F_1] \nabla F_2 > \ + \ < \nabla F_1, \Theta'[\Theta \nabla F_2] v^* > .
\end{aligned}
$$

Now recall that in our vector space situation every fixed co-vector v^* is a gradient, a fact which is extremely easy to see: take the gradient of $< v^*, u >$ to obtain v^*. Then application of condition (3,ii) yields that the right hand side of this last equation es equal to zero. So we see that $< v^*, -\Theta\nabla\{F_1, F_2\}_\Theta + [\Theta\nabla F_1, \Theta\nabla F_2] >= 0$. Moreover, because v^* was arbitrarily chosen we obtain that the vector on the right side in this bracket is equal to zero, i.e.

$$-\Theta\nabla\{F_1, F_2\}_\Theta + [\Theta\nabla F_1, \Theta\nabla F_2] = 0$$

which shows that $\Theta\nabla$ is a Lie algebra homomorphism. ∎

Operators Θ having one of the equivalent properties of the last theorem are called **implectic operators** or **Poissson operators**, they play a fundamental role for dynamical systems. The corresponding bracket introduced in (3.1) then is termed **Poisson bracket**. The flow

$$u_t = K(u) \tag{1.1}$$

is called a **hamiltonian flow**[5] if there is some scalar field $H(u)$ and some implectic operator $\Theta(u)$ such that

$$u_t = \Theta(u)°\nabla H(u). \tag{3.3}$$

The scalar field H then is the so called **Hamiltonian** of the system. The Poisson brackets give a suitable frame for describing the dynamics of scalar fields with respect to the evolution given by (1.1). Using $K = \Theta\nabla H$ we find that the total time derivative of some $F = F(u(t), t)$ can be written as

$$\frac{d}{dt}F = \{\nabla H, F\}_\Theta . \tag{3.4}$$

Hence a scalar field is a conserved quantity if and only if it commutes (in the Lie algebra given by the Poisson brackets) with the hamiltonian of the flow. As a particular consequence of that we have that the hamiltonian H itself always is a conserved quantity. This quantity usually is called **energy**[6]. However, the most important consequence of the above theorem is that now we have a precise relation between conserved quantities and symmetries:

Theorem 3.3:   *Whenever $I(u)$ is a conserved quantity for the hamiltonian flow $u_t = \Theta(u)°\nabla H(u)$ then $\Theta°\nabla I$ is a symmetry of that flow.*

Proof: By Theorem 3.2 we have $[K, \Theta\nabla I] = \Theta\nabla\{H, I\}_\theta$. This expression is equal to zero since I is a conserved quantity.∎

This result we call *Noether's theorem* since it is a generalization of the classical result obtained by Emmy Noether ([18]). A simple exercise shows that it carries over to time-dependent conserved quantities and time-dependent symmetries as well.

Example 3.4: Pendulum

If the manifold is a vector space and Θ an antisymmetric operator which does not depend on the manifold point u then we obviously have $\Theta(u)' = 0$, hence Θ fulfills condition 3 of

---

[5]In the classical finite dimensional situation for hamiltonian flows it is usually required that Θ is invertible. But here we are mainly interested in infinite dimensional manifolds where, for topological reasons, invertibility is a little bit problematic, therefore we have skipped this restrictive condition.

[6]One has to be a little bit careful with this interpretation since it can happen, as we will see soon, that a flow has more than one hamiltonian formulation.

Theorem 3.2 and therefore must be implectic. A particular example for such an operator is the antisymmetric matrix appearing in equations (1.5) and (1.7). So, these equations give hamiltonian formulations for these systems and, as stated above, their hamiltonians are given by energy conservation.

It should be mentioned that whenever the manifold is finite dimensional and the implectic operator Θ is invertible then locally there is a coordinate transformation on the manifold such that in the new coordinates the implectic operator is an off-diagonal matrix having -1's in the upper half and $+1$'s in the lower half of the off-diagonal (see ([16, page 30])). This means that equations (1.3) represent the prototype of hamiltonian equations in finite dimension.

The importance of conserved quantities is seen from the fact that even for a nonlinear system like the pendulum knowing the energy allows to integrate that equation completely. To see this we first remark that we already know the orbits in phase space, since they are given as $H = constant$

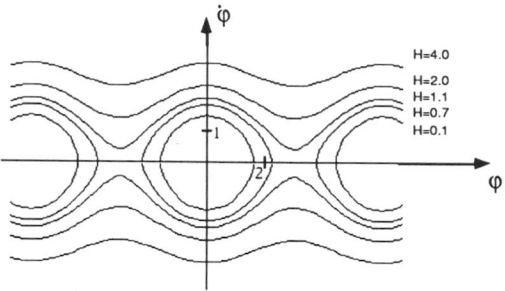

Fig. 2: Phase space orbits of the pendulum

To integrate the equation along these orbits we choose a fixed value E for this conserved quantity, then

$$H(\dot{\varphi},\varphi) \ = \ \frac{1}{2}\dot{\varphi}^2 \ - \ \cos(\varphi) = E \tag{3.5}$$

is a differential equation of first order and separation of variables yields that

$$t \ - \ \int^{\varphi} \frac{d\alpha}{\sqrt{2E \ + \ 2\cos(\alpha)}} = constant = F \tag{3.6}$$

must be a constant. This formula obviously gives the solution of (1.7) in implicit form. Expressing E again by $H(\dot{\varphi},\varphi)$ we obtain that

$$F \ = \ t \ - \ \int^{\varphi} \frac{d\alpha}{\sqrt{2H(\dot{\varphi},\varphi) \ + \ 2\cos(\alpha)}} \tag{3.7}$$

must be constant along any line on which $H(\dot{\varphi},\varphi)$ is constant. Hence F is constant along the orbits of (1.7). Rewriting, this in phase-space variables we have found a time-dependent conserved quantity for (1.8).

The pendulum provides the most simple which illustrates that knowing suitable and enough conserved quantities implies that an equation can be integrated. □

Example 3.5: Korteweg-de Vries equation
The Korteweg-de Vries equation [13(]KdV for short)

$$u_t = K(u) := 6uu_x + u_{xxx} \tag{3.8}$$

plays a central role in the history of completely integrable systems on infinite dimensional manifolds.

Usually, this equation is considered to be a flow on the space S of **tempered functions**. These are the C^∞-functions f in the real variable $x \in \mathbb{R}$ having the property that f and all its derivatives vanish at $\pm\infty$ faster than any rational function.

Define S^* to be the space of C^∞-functions in x such that all derivatives grow at most polynomially at $\pm\infty$. This space can be taken as a space of linear functionals on S by using an L^2 scalar product in the following way , namely

$$< U, u > = \int U(x)u(x)dx, \quad U \in S^*, u \in S. \tag{3.9}$$

We use the convention that if no boundaries are given then integrals always go over $\mathbb{R}$ or $\mathbb{R}^n$, respectively. As topology we take the weakest convex topology making all these functionals continuous. Then the scalar fields

$$I_0(u) = \int u(x)dx \tag{3.10}$$

$$I_1(u) = \int u(x)^2 dx \tag{3.11}$$

$$I_2(u) = \int (u^3 - \frac{1}{2}u_x{}^2)dx \tag{3.12}$$

are C^∞-functions $M \to \mathbb{R}$. The derivative of, for example, $I_2(u)$ is computed to be

$$I_3'[v] = \frac{\partial}{\partial\epsilon}\Big|_{\epsilon=0} \int \{(u + \epsilon v)^3 - \frac{1}{2}(u + \epsilon v)_x{}^2\}dx . \tag{3.13}$$

Integration by parts yields

$$I_3'[v] = \int (3u^2 + u_{xx})v dx. \tag{3.14}$$

Hence, provided the duality between tangent and cotangent space is represented by (3.9), then the gradient of I_2 can be identified with $3u^2 + u_{xx}$. The gradients of I_0, I_1 are given in the same way by the functions 1 and $2u$, respectively. We compute $I_1'[K(u)]$ for $K(u)$, the vector field given by the KdV. Integration by part (together with the fact that the function u is tempered) yields

$$I_2'(u)[K(u)] = \int 2u(u_{xxx} + 6uu_x)dx = 2\int (2u^3 + uu_{xx} - \frac{1}{2}u_x^2)_x dx. \tag{3.15}$$

The latter expression is equal to zero because the integrand is the derivative of a tempered function. So, I_1 is constant along the orbits of the KdV. The same holds true for the fields I_0 and I_2, whence all these quantities are conserved for the KdV. A time-dependent conserved quantity is for example

$$F(u,t) \ = \ \int \{xu \ - \ 3tu^2\}dx \ . \tag{3.16}$$

We can write the Korteweg-de Vries equation in the following way

$$u_t \ = \ \Theta^\circ \nabla H$$

where

$$\Theta \ = \ D \ \text{differential operator with respect to } x$$

and

$$H \ = \ \int (u^3 - \frac{1}{2}u_x{}^2)dx.$$

Since Θ is an antisymmetric operator which does not depend on the manifold points, Θ must be implectic and this is a hamiltonian formulation. So one is inclined to call H the energy of the system. However, another way to write the KdV is the following

$$u_t \ = \ \Theta^\circ \nabla H$$

where

$$H \ = \ \frac{1}{2} \int u^2 dx \ ,$$

and where

$$\Theta \ = \ D^3 \ + \ 2(Du \ + \ uD) \tag{3.17}$$

is again antisymmetric and is shown to fulfill condition (3. ii) of Theorem 3.2 (see Section 7). Hence this is a second hamiltonian formulation for the KdV.□

This last example shows that for some systems hamiltonian formulations are not unique. It will turn out, however, that this non-uniqueness is a highly desired property which will help to construct suitably many conserved quantities and thus will enable us to integrate the equation. The main idea for generating infinitely many conservation laws from two different hamiltonian formulations goes back to F. Magri ([14]) who proposed that one hamiltonian formulation should be used for going from a conserved quantity to a symmetry and then by the second one one should go back to another conserved quantity. Thus an infinite sequence of conservation laws would be generated. There is one apparent difficulty with this concept, namely, that the map $\Theta \nabla$ is not invertible. This difficulty is overcome by transferring the result stated in Theorem 2.3 to co-vector fields instead of scalar fields. Then instead of going back to scalar fields one goes back with the hamiltonian formulation to co-vector fields instead and from there, by integration, to the corresponding potentials. Of course, for doing that one requires that the co-vector fields generated this way are closed (a notion which we will introduce in the next section). This requirement of constructing only closed co-vector fields will lead to the notion of compatibility (treated in Section 6). Another difficulty with this concept arises already for the KdV when other boundary conditions at $\pm\infty$ are considered. Then we cannot write down the hamiltonians anymore for this equation since the integrals (3.10) to (3.12) then clearly diverge. In order to subsume even this case under a common theory we have to lift all our notions to a new level of abstraction. This new level of abstraction, which will be formulated in the next section, then provides a more transparent setup so that the necessary considerations can be carried out more easily.

4   Lie derivatives

In this section we would like to review the basics of symplectic geometry and Hamiltonian mechanics on an abstract level. This high degree of abstraction will enable us to represent the relevant results in a very concise way.

Let $\mathcal{F}$ be some commutative algebra (over $\mathbb{R}$ or $\mathbb{C}$) with identity. We now assume $(\mathcal{L}, [\ ], \mathcal{F})$ to be a Lie-module ([17]). Recall that being a **Lie module** means that $(\mathcal{L}, [\ ])$ is a Lie algebra such that a multiplication between elements of $\mathcal{L}$ and $\mathcal{F}$ is defined and that, furthermore, there is a canonical homomorphism

$$K \to L_K$$

from $\mathcal{L}$ into the derivations on $\mathcal{F}$. For $K, G \in \mathcal{L}$ and $f \in \mathcal{F}$ these derivations have to fulfill

$$[K, fG] = f[K, G] + L_K(f)G \tag{4.1}$$

$$L_K L_G - L_G L_K = L_{[K,G]} . \tag{4.2}$$

Of course, being a **derivation** on $\mathcal{F}$ means that the product rule

$$L_K(fg) = L_K(f)g + f L_K(g) \ for \ all \ f, g \in \mathcal{F} \tag{4.3}$$

holds. In the following we require, for technical reasons, that the map $K \to L_K$ is *injective*.

Remark 4.1: *Lie modules are the canonical extensions of Lie algebras admitting a Lie algebra homomorphism into the derivations of $\mathcal{F}$. To be precise: Let $\mathcal{L}_1$ be some Lie algebra contained in some $\mathcal{F}$-module $\mathcal{L}$ such that $\mathcal{L}$ is the linear hull of $\{fK | f \in \mathcal{F}, K \in \mathcal{L}_1\}$. Then if a Lie algebra homomorphism $K \to L_K$ from $\mathcal{L}_1$ into the derivations of $\mathcal{F}$ is given then there is a unique extension of $(\mathcal{L}_1, [\ ])$ into a Lie module structure $(\mathcal{L}, [\ ], \mathcal{F})$ such that (4.1) and (4.2) hold.*
The proof of this remark is a simple computation. One takes $K_1, K_2 \in \mathcal{L}_1$, then makes the obvious definition

$$[f_1 K_1, f_2 K_2] := f_1 f_2 [K_1, K_2] + f_1 L_{K_1}(f_2) K_2 - f_2 L_{K_2}(f_1) K_1 , \tag{4.4}$$

and the extension to all of $\mathcal{L}$ is obtained by taking sums. ■

If suitable topologies in $\mathcal{L}$ and $\mathcal{F}$ are given then we assume that all quantities introduced below are continuous. For $\mathcal{F}$-linear functionals $\gamma : \mathcal{L} \to \mathcal{F}$ we denote the application of γ to $K \in \mathcal{L}$ by $< \gamma, K >$. Such a functional γ is said to be **closed** if

$$L_K < \gamma, G > - L_G < \gamma, K > = < \gamma, [K, G] > \ for \ all \ G, K \in \mathcal{L} . \tag{4.5}$$

For $f \in \mathcal{F}$ we denote by ∇f the special $\mathcal{F}$-linear functional on $\mathcal{L}$ given by

$$< \nabla f, K >:= L_K f \; for \; all \; K \in \mathcal{L}. \tag{4.6}$$

Because of (4.2) all these ∇f are closed. A suitable $\mathcal{F}$-module of $\mathcal{F}$-linear maps $\mathcal{L} \to \mathcal{F}$ generated by closed $\mathcal{F}$-linear functionals $\gamma : \mathcal{L} \to \mathcal{F}$ is denoted by $\mathcal{L}^*$. We assume that $\mathcal{L}^*$ contains all ∇f, $f \in \mathcal{F}$. Elements in $\mathcal{L}^*$ which are of the form ∇f are called **gradients** and f is called the **potential** of ∇f. Observe that for $f, g \in \mathcal{F}$ elements of the form $g \nabla f$ are in general not gradients. Therefore the gradients do not form an $\mathcal{F}$-module.

An important observation is that the derivative L_K can be extended to all tensors, i.e. to all $\mathcal{F}$-multilinear forms on $\mathcal{L}^*$ and $\mathcal{L}$. This extension is obtained by defining first

$$L_K G := [K, G] \; for \; all \; G \in \mathcal{L} \tag{4.7}$$

and then by the requirement that for L_K the *product rule* holds for tensor products and for those quantities which come from inserting elements of $\mathcal{L}$ and $\mathcal{L}^*$ into $\mathcal{F}$-multilinear forms. This general extension of L_K is again called **Lie derivative** with respect to K.

Recall that $\mathcal{F}$–multilinear forms are maps from $(\otimes \mathcal{L}^*)^r \otimes (\otimes \mathcal{L})^n$, $n, r \subset \mathbb{N}$, into $\mathcal{F}$ which are $\mathcal{F}$-linear in each entry. These multilinear forms are called **tensors** (n-times **covariant** and r-times **contravariant**). Elements of $\mathcal{L}$ and $\mathcal{L}^*$ are special tensors which are 1-times contravariant and covariant, respectively. In the following we do not distinguish between an $\mathcal{F}$-linear operator $\Theta : \mathcal{L}^* \to \mathcal{L}$ and the tensor $\tilde{\Theta} : \mathcal{L}^* \otimes \mathcal{L}^* \to \mathcal{F}$ given by $\tilde{\Theta}(\gamma_1, \gamma_2) :=< \gamma_1, \Theta \gamma_2 >$. In the same way we identify operators $J : \mathcal{L} \to \mathcal{L}^*$ and $\Phi : \mathcal{L} \to \mathcal{L}$ with special tensors which are two-times covariant and once co-contravariant, respectively.

To illustrate the construction of L_K we compute Lie derivatives for 1-times covariant tensors and for 2-times contravariant tensors. First, we compute the Lie derivative for some fixed $\gamma \in \mathcal{L}^*$. We consider $< \gamma, G >$, $G \in \mathcal{L}$. The product rule applied to $< \gamma, G >$ yields

$$< L_K(\gamma), G >= L_K < \gamma, G > - < \gamma, [K, G] > \; .$$

i.e. the linear map $L_K(\gamma) : \mathcal{L} \to \mathcal{F}$ is

$$L_K(\gamma) = L_K \cdot \gamma - \gamma \cdot L_K \tag{4.8} .$$

For later use we note that for $f \in \mathcal{F}, K \in \mathcal{L}$ the following holds

$$L_{(fK)}(\gamma) = f L_K(\gamma) + < \gamma, K > \nabla f \; . \tag{4.9}$$

As an additional example we take some $\mathcal{F}$-linear operator $\Theta : \mathcal{L}^* \to \mathcal{L}$. Its Lie derivative L_K we compute again by the product rule applied to $< \gamma_1, \Theta \gamma_2 >$ where γ_1, γ_2 are arbitrary chosen elements in $\mathcal{L}^*$. This yields

$$< \gamma_1, L_K(\Theta) \gamma_2 >= L_K < \gamma_1, \Theta \gamma_2 > - < L_K(\gamma_1), \Theta \gamma_2 > - < \gamma_1, \Theta L_K(\gamma_2) > \tag{4.10}$$

On the right side, the Lie-derivative of the first term is given by definition of the Lie-module, and the Lie derivatives of the γ's were already determined by (4.8). Since γ_1 and γ_2 were arbitrary, (4.10) defines completely the Lie derivatives for the two-times contravariant tensor Θ. In the same way we can define, by induction, the derivative L_K for arbitrary tensors.

For purely covariant tensors α, i.e. multilinear forms on $(\otimes \mathcal{L})^n$ we can define a so called **exterior derivative** d, a notion which plays an important role for hamiltonian vector fields. On $\mathcal{F}$ we define this exterior derivative d to be the gradient

$$d_{|\mathcal{F}} = \nabla , \tag{4.11}$$

and when a tensor is r-times covariant $(r \geq 1)$ then we define this exterior derivative by

$$(d\alpha) \bullet K := L_K(\alpha) - d(\alpha \bullet K) \quad for \ all \ K \in \mathcal{L} . \tag{4.12}$$

Here $\alpha \bullet K$ means that K is inserted as the first entry in the form α, for example $\gamma \bullet K = \ < \gamma, K >$ when $\gamma \in \mathcal{L}^*$. One easily sees that (4.12) is an inductive definition over the order of covariance. In the following notation we use the convention that d and L_K are more binding than $\bullet$, i.e. $d\alpha \bullet K = (d\alpha) \bullet K \neq d(\alpha \bullet K)$, and similarly for the Lie derivative. Furthermore we observe that we may use (4.12) as the definition for the exterior derivative also in case of zero-forms $f \in \mathcal{F}$ if we adopt the formal notation that for zero-forms f the expression $f \bullet K$ is equal to zero. More generally, for n-**forms** (n-times covariant tensors) α we define $\alpha \bullet K_1 \bullet ... \bullet K_{n+1}$ (i.e. application of $n + 1$ $\bullet$'s) to be zero. This notation will considerably shorten subsequent proofs.

Observation 4.2:

(i) Exterior derivative and Lie-derivative commute.

$$L_K d = d L_K \tag{4.13}$$

(ii) As usual we obtain that $d \cdot d = 0$, a fact which is well known for concrete situations from differential geometry.

Proof:

(i): Consider arbitrary $G, K \in \mathcal{L}$ and covariant α, then by use of (4.12) we obtain:

$$
\begin{aligned}
((L_G d - d L_G)\alpha) \bullet K &= L_G(d\alpha \bullet K) - d\alpha \bullet L_G K - L_K L_G \alpha + d(L_G \alpha \bullet K) \\
&= L_G L_K \alpha - L_G d(\alpha \bullet K) - d\alpha \bullet L_G K - L_K L_G \alpha + d(L_G \alpha \bullet K) \\
&= L_{[G,K]}(\alpha) - (L_G d - d L_G)(\alpha \bullet K) - d(\alpha \bullet [G,K]) - d\alpha \bullet L_G K \\
&= d\alpha \bullet [G,K] - (d\alpha) \bullet L_G K - (L_G d - d L_G)(\alpha \bullet K) \\
&= (d L_G - L_G d)(\alpha \bullet K)
\end{aligned}
$$

Hence $((L_G d - d L_G)\alpha) \bullet K = (L_G d - d L_G)(\alpha \bullet K)$ and the claim follows by induction over the order of covariance. Observe that the necessary beginning of our induction argument is given by the fact that $\bullet$ applied to zero-forms gives zero.

(ii): We again use a repeated argument over the order of covariance. Consider an arbitrary covariant α and arbitrary vector fields K, G, then:

$$
\begin{aligned}
(d \cdot d\alpha) \bullet K \bullet G &= (L_K d\alpha \bullet G - d(d\alpha \bullet K) \bullet G \\
&= L_K(d\alpha \bullet G) - d\alpha \bullet L_K G - d(d\alpha \bullet K) \bullet G \\
&= L_K(d\alpha \bullet G) - d\alpha \bullet L_K G - L_G(d\alpha \bullet K) \bullet G + d(d\alpha \bullet K \bullet G) \\
&= L_K L_G \alpha - L_K d(\alpha \bullet G) + L_{[G,K]}\alpha + d(\alpha \bullet [K,G]) - L_G L_K \alpha \\
&\quad + L_G d(\alpha \bullet K) + d(L_K \alpha \bullet G) - d(d(\alpha \bullet K) \bullet G) \\
&= d(-L_K(\alpha \bullet G) - \alpha \bullet [G,K] + L_G(\alpha \bullet K) + (L_K \alpha) \bullet G - d(\alpha \bullet K) \bullet G) \\
&= d(L_G(\alpha \bullet K) - d(\alpha \bullet K) \bullet G) \\
&= d^2(\alpha \bullet K \bullet G)
\end{aligned}
$$

Again, the beginning of our induction argument is given by the fact that $\bullet$ applied to zero-forms gives zero. ∎

Definition 4.3:

(i) *A tensor T is said to be* **invariant with respect to the vector field** K *if* $L_K T = 0$.

(ii) *Observe that condition (4.5) for γ being closed can be written in terms of the exterior derivative as $d\gamma = 0$. Therefore we define a covariant tensor α to be* **closed** *if $d\alpha = 0$.*

Remark 4.4:

(i) *Gradients are closed because of $d \cdot d = 0$. If $\mathcal{L}$ is the vector field Lie algebra on some manifold M and $\mathcal{F}$ are the scalar fields, then locally the converse is also true (Poincaré Lemma ([25]).*

(ii) *Observe that (4.12) implies that any closed covariant tensor α is invariant with respect to K if and only if $\alpha \bullet K$ is again closed.*

(iii) *Let J be some invertible $\mathcal{F}$-linear operator $J : \mathcal{L} \to \mathcal{L}^*$. If $L_G J = 0$ for all those G with closed JG then J itself must be closed.*

Proof of (ii) and (iii):

(ii): Direct consequence of (4.12).

(iii): Because $\mathcal{L}^*$ is generated by its closed elements and since J is invertible we have that $\mathcal{L}$ is the $\mathcal{F}$-linear hull of those G in $\mathcal{L}$ such that JG is closed. For $f \in \mathcal{F}$ we obtain $L_{(fK)}J - d(JfK) = f(L_K J - d(JK))$. Therefore $L_K J - d(JK)$ vanishes for all K if it vanishes for the subset of those G such that JG is closed. So the assumption on J shows that the right side of (4.12) vanishes. Hence we have $d(J) \bullet K = 0$ for all K. ∎

Now, let $\Theta : \mathcal{L}^* \to \mathcal{L}$ be some antisymmetric $\mathcal{F}$ linear map. We define a **bracket** in $\mathcal{L}^*$ in the following way:

$$\{\gamma_1, \gamma_2\}_\Theta := L_{(\Theta\gamma_1)}\gamma_2 - L_{(\Theta\gamma_2)}\gamma_1 + d < \gamma_1, \Theta\gamma_2 > \qquad (4.14)$$

for $\gamma_1, \gamma_2 \in \mathcal{L}^*$.

Before presenting the basic result for these brackets we gather some useful identities. Using (4.12) we can rewrite (4.14) as

$$\{\gamma_1, \gamma_2\}_\Theta := L_{(\Theta\gamma_1)}\gamma_2 - (d\gamma_1) \bullet (\Theta\gamma_2) . \qquad (4.15)$$

So when γ_1 is *closed* then by application of (4.12) this bracket reduces to

$$\{\gamma_1, \gamma_2\}_\Theta := L_{(\Theta\gamma_1)}\gamma_2 \; for \; closed \; \gamma_1 \in \mathcal{L}^* . \qquad (4.16)$$

Furthermore we easily find with (4.9) how this bracket acts on multiplication with elements of $\mathcal{F}$

$$\{\gamma_1, f\gamma_2\}_\Theta := f\{\gamma_1, \gamma_2\}_\Theta + (L_{(\Theta\gamma_1)}f) \, \gamma_2 . \qquad (4.17)$$

Theorem 4.5: *Let $\Theta : \mathcal{L}^* \to \mathcal{L}$ be $\mathcal{F}$-linear and antisymmetric, then the following are equivalent:*

(i) $\{ , \}_\Theta$ defines a Lie algebra among the closed elements of $\mathcal{L}^*$

(ii) $\Theta\{\gamma_1, \gamma_2\}_\Theta = [\Theta d\gamma_1, \Theta d\gamma_2]$ for all closed $\gamma_1, \gamma_2 \in \mathcal{L}^*$.

(iii) $\{ , \}_\Theta$ defines a Lie algebra in $\mathcal{L}^*$

(iv) $\Theta\{\gamma_1, \gamma_2\}_\Theta = [\Theta\gamma_1, \Theta\gamma_2]$ for all $\gamma_1, \gamma_2 \in \mathcal{L}^*$

(v) $L_{\Theta\gamma}(\Theta) = 0$ for all closed $\gamma \in \mathcal{L}^*$.

(vi) For arbitrary $\gamma \in \mathcal{L}^*$ we have that Θ is invariant with respect to $\Theta\gamma$ if and only if $d\gamma \bullet (\Theta\gamma_1) \bullet (\Theta\gamma_2) = 0$ for all $\gamma_1, \gamma_2 \in \mathcal{L}^*$.

Proof:
In the following we omit, for simplicity, the sub-Θ in the brackets $\{ \}_\Theta$.
(i) $\Leftrightarrow$ (ii): Observe that by use of (4.14) then the antisymmetry of Θ implies for closed γ's that

$$\{\gamma_1, \gamma_2\} = L_{(\Theta d\gamma_1)}\gamma_2 = -L_{(\Theta d\gamma_2)}\gamma_1. \qquad (4.18)$$

So, $\{ , \}$ is antisymmetric anyway, and only the Jacobi identity has to be proved in order to show that this is a Lie algebra. Using (4.18) we find for the triple bracket

$$\begin{aligned}\{\gamma_3, \{\gamma_1, \gamma_2\}\} &= -L_{\Theta\{\gamma_1, \gamma_2\}}\gamma_3 \\ &= L_{(\Theta\gamma_3)}L_{(\Theta\gamma_1)}\gamma_2 = -L_{(\Theta\gamma_3)}L_{(\Theta\gamma_2)}\gamma_1 .\end{aligned}$$

Now, using for the cyclic sum suitable representations obtained from this formula we find with (4.2)

$$
\begin{aligned}
\{\gamma_3\{\gamma_1,\gamma_2\}\} + \text{ cyclic } &= L_{(\Theta d\{\gamma_1,\gamma_2\})}\gamma_3 + L_{(\Theta d\gamma_1)}L_{(\Theta d\gamma_1)}\gamma_3 - L_{(\Theta d\gamma_2)}L_{(\Theta d\gamma_1)}\gamma_3 \\
&= \{L_{(\Theta\{\gamma_1,\gamma_2\})} - [L_{(\Theta d\gamma_1)}, L_{(\Theta d\gamma_2)}]\}\gamma_3 \\
&= \{L_{(\Theta\{\gamma_1,\gamma_2\})} - L_{[\Theta d\gamma_1,\Theta d\gamma_2]}\}\gamma_3 \qquad (4.19)
\end{aligned}
$$

Since γ_3 was arbitrary, the Jacobi identity for the triple product can only hold if

$$
L_{(\Theta d\{\gamma_1,\gamma_2\})} - L_{[\Theta d\gamma_1,\Theta\gamma_2]} = 0 \qquad (4.20)
$$

which is equivalent to (ii) since $K \to L_K$ was assumed to be injective. On the other hand, whenever (ii) (and (4.20) as a consequence) holds then the cyclic sum obviously must be equal to zero, which implies the Jacobi identity.

The implications (iv) $\to$ (ii) and (iii) $\to$ (i) are obvious.

(ii) $\to$ (iv) and (i) $\to$ (iii) are either done by direct computation or by using remark 4.1 together with (4.17). To see this observe that (i) gives a Lie algebra $(\mathcal{L}^*_{cl}, \{\ \})$, where $\mathcal{L}^*_{cl}$ are the closed elements in $\mathcal{L}^*$ and that (4.17) gives a Lie algebra homomorphism $\gamma \to L^*_\gamma = L_{\Theta\gamma}$ from $\mathcal{L}^*_{cl}$ into the derivations of $\mathcal{F}$. Now taking the unique extension (described in remark 4.1) to the $\mathcal{F}$-module generated by $\mathcal{L}^*_{cl}$ (which by definition is equal to $\mathcal{L}^*$), one obtains the Lie algebra defined by (4.14). The fact that the homomorphism Θ extends from $\mathcal{L}^*_{cl}$ to $\mathcal{L}^*$ is due to the property that $\Theta L^*_\gamma = L_{\Theta\gamma}$.

(v) $\Leftrightarrow$ (ii): By (4.16) the condition

$$
\Theta L_{(\Theta\gamma)}\gamma_1 = L_{(\Theta\gamma)}(\Theta\gamma_1) \ \text{ for all closed } \gamma,\gamma_1 \in \mathcal{L}^* \qquad (4.21)
$$

is equivalent to (ii). And by the product rule this is equivalent to (v).

(vi) $\Leftrightarrow$ (iv): Using (4.15) and the antisymmetry of Θ we find for arbitrary $\gamma,\gamma_1,\gamma_2 \in \mathcal{L}^*$

$$
\begin{aligned}
< \gamma_1,(L_{(\Theta\gamma)}\Theta)\gamma_2 > &= < \gamma_1, L_{(\Theta\gamma)}(\Theta\gamma_2) > - < \gamma_1, \Theta L_{(\Theta\gamma)}\gamma_2 > \\
&= < \gamma_1,[\Theta\gamma,\Theta\gamma_2] > - < \gamma_1, \Theta\{\gamma,\gamma_2\} > + < \gamma_1, \Theta(d\gamma \bullet \Theta\gamma_2) > \\
&= < \gamma_1,[\Theta\gamma,\Theta\gamma_2] > - < \gamma_1, \Theta\{\gamma,\gamma_2\} > - d\gamma \bullet (\Theta\gamma_2) \bullet (\Theta\gamma_1) .
\end{aligned}
$$

Hence we obtain

$$
< \gamma_1,(L_{\Theta\gamma}\Theta)\gamma_2 > + d\gamma \bullet (\Theta\gamma_2) \bullet (\Theta\gamma_1) = < \gamma_1,[\Theta\gamma,\Theta\gamma_2] - \Theta\{\gamma,\gamma_2\} > \qquad (4.22) .
$$

Equating the right side of (4.22) to zero is equivalent to (iv) and equating its left side to zero is equivalent to (v). ∎

Let us have a closer look at condition (v): If Θ is invertible, then the condition therein imposed on γ is equivalent to $d\gamma = 0$, i.e. γ must be closed. Hence, for $J := \Theta^{-1}$ condition (v) means that J is K-invariant ($K \in \mathcal{L}$) if and only if JK is closed. Looking now at 4.4 (ii) and (iii) we see that for an invertible Θ one of the equivalent conditions of the theorem above is fulfilled if and only if $J = \Theta^{-1}$ is closed. In the finite dimensional theory the antisymmetric closed invertible J are called symplectic. So, loosely speaking, Θ has the algebraic behaviour of the inverse of a symplectic operator. Therefore, if one of the conditions of Theorem 4.5 is fulfilled, Θ is said to be **implectic**, a name which stands for inverse-symplectic. Sometimes instead of implectic, the name **Poisson tensor** is chosen. For reasons which will become obvious in the next section the $\{\ ,\ \}_\Theta$ are called the **Poisson brackets** with respect to Θ.

We decided not to insist on the invertibility condition because of the infinite dimensional nature of our manifolds, so we have to extend the notion *symplectic* for this more general situation: An operator $J : \mathcal{L} \to \mathcal{L}^*$ is called **symplectic** if it is antisymmetric and closed, and if in addition its kernel $ker(J) = \{G \in \mathcal{L} | JG = O\}$ is a Lie ideal in $\mathcal{L}$. Being a Lie ideal means of course that $[K, G] \in ker(J)$ for all $G \in ker(J)$ and $K \in \mathcal{L}$. This additional ideal-condition is automatically fulfilled if J is invertible because then $ker(J) = 0$.

In a analogy to implectic operators one can use symplectic operators J to construct in $J\mathcal{L}$ suitable Poisson brackets: Take $\gamma_1, \gamma_2 \in J\mathcal{L}$ and choose G_1, G_2 such that $\gamma_i = JG_i$, $i = 1, 2$. Then we define

$$\{\gamma_1, \gamma_2\}^{(J)} := L_{G_1}(\gamma_2) - L_{G_2}(\gamma_1) + <\gamma_1, G_2> \ . \tag{4.23}$$

Rewriting that with (4.12) and using $d(J) = 0$ we obtain

$$\{\gamma_1, \gamma_2\}^{(J)} = L_{G_1}(G_2) - d(JG_1) \bullet G_2 = JL_{G_1}G_2 + d(J) \bullet G_1 \bullet G_2 = J[G_1, G_2] \ .$$

Since $ker(J)$ is an ideal with respect to $[\ ,\ ]$ the bracket $\{\ ,\ \}^{(J)}$ does not depend on the choice of G_1, G_2. Furthermore, because J is one-to-one from the equivalence classes modulo $ker(J)$ to $J\mathcal{L}$, the bracket $\{\ ,\ \}^{(J)}$ must be a Lie algebra such that J is a homomorphism from $\{\ ,\ \}^{(J)}$ into $[\ ,\ ]$

We may summarize this section: *An implectic operator makes out of $\mathcal{L}^*$ a Lie algebra module $(\mathcal{L}^*, \{\ ,\ \}_\Theta, \mathcal{F})$, with corresponding $\mathcal{F}$-derivations L_γ^*, $\gamma \in \mathcal{L}^*$ such that Θ is a homomorphism from this Lie algebra module into $(\mathcal{L}, [\ ,\ ], \mathcal{F})$. Here homomorphism means that for all $\gamma_1, \gamma_2 \in \mathcal{L}^*$ we have*

$$\Theta\{\gamma_1, \gamma_2\}_\Theta = [\Theta\gamma_1, \Theta\gamma_2] \tag{4.24}$$

and

$$\Theta L_{\gamma_1}^* = L_{\Theta\gamma_1} \ . \tag{4.25}$$

In general, it is highly desirable to construct for given tensors suitable Lie algebra elements which leave these tensors invariant. Indeed, as we will see, the search for symmetries, conservation laws, invariant spectral problems and the like may be subsumed under this general theme. In view of that problem it is obvious that symplectic and implectic tensors must play a fundamental role because for them the closed elements in $\mathcal{L}^*$ imediatly give rise to such invariances.

5   Hamiltonian and Bi-Hamiltonian Fields

In this section we always assume that Θ is an implectic operator. For some of the results presented here the reader is referred to the fundamental papers ([10,] [11,] [12,] [14,] [15]).

Comparing the different Poisson brackets defined in Sections 3 and 4, one discovers that they are defined on rather different spaces. However, these two notions are easily connected if, as before in the concrete situation, a *bracket* among the elements of $\mathcal{F}$ is defined as follows:

$$\{f_1, f_2\}_\Theta := < df_2, \Theta df_1 > = L_{(\Theta df_1)} f_2 \ for \ f_1, f_2 \in \mathcal{F} . \tag{5.1}$$

One easily sees that ∇ maps the $\mathcal{F}$-brackets into the $\mathcal{L}^*$-brackets:

$$\nabla\{f_1, f_2\}_\Theta = \{\nabla f_1, \nabla f_2\}_\Theta . \tag{5.2}$$

This suggests that these $\mathcal{F}$-brackets also form a Lie algebra. Indeed this is the case, the proof is literally almost the same as the proof (i) $\Leftrightarrow$ (ii) in the last theorem[7].

Definition 5.1:

   (i) *Elements $K \in \mathcal{L}$ which are of the form $K = \Theta\gamma$ with closed γ and implectic Θ are called* **hamiltonian** *(with respect to Θ).*

   (ii) *Given some symplectic J, then elements $K \in \mathcal{L}$ such that JK is closed are called* **inverse-hamiltonian**[8] *(with respect to J).*

   (iii) *Let Θ be implectic and J be closed, and assume that J is not the inverse of Θ. Then some K is called a* **bi-hamiltonian field** *(with respect to Θ and J) if JK is closed and if there is some closed γ such that $K = \Theta\gamma$.*

The power of bi-hamiltonian fields is seen from:

Observation 5.2: *Consider a bi-hamiltonian field*

$$K = \Theta\gamma, \ JK = \gamma_1, \ \gamma \ and \ \gamma_1 \ being \ closed$$

as described above, and define $K_{n+1} = (\Theta J)^n K$. Then all the K_n and $\gamma_n := JK_n$ are invariant with respect to K.

Proof:

---

[7]An interesting question is whether or not the Jacobi identity for the $\mathcal{F}$-brackets is a sufficient condition to guarantee that Θ is implectic. In most situations this is indeed the case. One only needs some kind of Poincaré Lemma to show that.

[8]Observe that for invertible Θ or J the notions hamiltonian and inverse-hamiltonian do coincide. This is the reason why in the finite dimensional theory the notion inverse-hamiltonian does not appear, since there symplectic forms are usually assumed to be non-degenerate. In the infinite dimensional situation however, the assumption of nondegeneracy is not advised because invertibility of operators usually involves topological considerations and depends very much on the spaces under consideration.

Using the bi-hamiltonian nature of K we see from Remark 4.4 (ii) and from Theorem 4.5 (v) that Θ and J are invariant with respect to K. Hence, by the product rule, all K_n have to be invariant with respect to K. ∎

Let us now illustrate the notions and techniques of the last chapter in the light of the concrete situation which was considered in Sections 2 and 3.

Standard situation :
Let M be a C^∞-manifold, we consider C^∞-tensor fields on M. In particular let $\mathcal{L}$ be the vector fields, $\mathcal{F}$ be the scalar fields and let $\mathcal{L}^*$ be those C^∞-maps from M into the cotangent-bundle which are given by assigning to $u \in M$ a continuous linear functional on the tangent space $T_u M$ at u.

In order to carry out computations with Lie derivatives we should show what these look like in charts. As we already know, for a scalar field f the Lie derivative is the usual gradient

$$L_A f = <\nabla f, A> = f'[A] \tag{5.3}$$

and the application of L_A to a co-vector field γ is found to be

$$L_A \gamma = \gamma'[A] + A'^+ \gamma \tag{5.4}$$

where A'^+ denotes the transpose of the operator A' with respect to the duality between tangent and cotangent space.

Furthermore, if $\Theta : T^*M \to TM$ and $J : TM \to T^*M$ are two-times contravariant and two-times covariant, respectively, then their Lie derivatives are

$$L_A \Theta = \Theta'[A] - \Theta A'^+ - A'\Theta \tag{5.5}$$

$$L_A J = J'[A] + A'^+ J + J A' \tag{5.6}$$

Finally the Lie derivative for an operator $\Phi : TM \to TM$ is equal to

$$L_A \Phi = \Phi'[A] - A'\Phi + \Phi A' . \tag{5.7}$$

Fix some $K \in \mathcal{L}$, then $f \in \mathcal{F}$ and $G \in \mathcal{L}$ are invariant with respect to K if and only if f is a conserved quantity and G a symmetry group generator, respectively. Using the Lie derivatives, we see that Θ is implectic if and only if condition 3 (ii) of Theorem 3.2 is fulfilled. In the same way we obtain as equivalent condition for J being closed that

$$< J(u)'[G_1]G_2, G_3 > + < J(u)'[G_2]G_3, G_1 > + < J(u)'[G_3]G_1, G_2 > = 0 \tag{5.8}$$

must hold for all $G_1, G_2, G_3 \in \mathcal{L}$. Hence Section 4 generalizes the situation considered in Section 3.

Now, we consider again the dynamical system

$$u_t = K(u) . \tag{1.1}$$

In analogy to Definition 5.1 we call (1.1) a **bi-hamiltonian system** if there are closed $J(u)$, implectic $\Theta(u)$ and *closed* γ_0, γ_1 such that

$$K = \Theta\gamma_0 \text{ and } JK = \gamma_1 . \tag{5.9}$$

If that is the case then we obtain as consequence of 5.2
Observation 5.3: *Define inductively*

$$K_1 = K \text{ and } K_{n+1} = \Theta J K_n \tag{5.10}$$

$$\gamma_{n+1} = J\Theta\gamma_n \tag{5.11}$$

then all the K_n and γ_n are invariant with respect to K. Hence the K_n are symmetry group generators for (1.1) and, if the γ_n are gradients, then they are gradients of conserved quantities for the evolution equation (1.1).

Example 5.4: Korteweg-de Vries equation
Consider the situation as in Example 3.5. We observe that the inverse of D

$$(D^{-1}f)(x) := \int_{-\infty}^{x} f(\xi)d\xi$$

is a well defined operator $D^{-1} : S \to S^*$. The Korteweg-de Vries equation

$$u_t = K(u) := 6uu_x + u_{xxx} \tag{3.8}$$

is a bi-hamiltonian system since for the implectic $\Theta = D^3 + 2(Du + uD)$ and the symplectic $J = D^{-1}$ we have that (5.9) is fulfilled when γ_0 and γ_1 are taken to be the gradients of I_0 and I_1 as given in (3.10) and (3.11), respectively. Hence putting

$$K_{n+1} := (\Phi)^n K, \text{ where } \Phi = D^2 + 2DuD^{-1} + 2u , \tag{5.12}$$

then all these K_n are symmetry group generators. Since Φ is recursively generating symmetry group generators, this usually is called a **recursion operator** [20(]sometimes also a **strong symmetry**, for example in [21,] [5,][3)]. □
Although Observation 5.3 is very useful for constructing symmetry groups, there still remains a major problem is the analysis of the system (1.1). Ultimately, we are interested in constructing suitable coordinates (action variables) for the flow, so we need scalar quantities which are invariant under the flow. Certainly, action variables give rise to symmetry group generators, however the converse is not always true, because the corresponding co-vector fields may not be closed. Therefore the question whether or not the fields γ_n are closed is of great importance. If that happens then, by the Poincaré lemma, one can, at least locally, construct suitable coordinates. And if, as in the example, the manifold under consideration is a vector space then even the construction of global potentials is a simple excercise:

Lemma 5.5: *Let the manifold of all u's be a vector space. Then a co-vector field $\gamma(u)$ is the gradient of a scalar field $F(u)$ if and only if γ is closed.*

Proof:

Since gradients are closed we only have to prove the existence of F for closed γ. So let $\gamma(u)$ be closed. Consider the scalar field

$$F(u) \;=\; \int_0^1 <\gamma(\lambda u), u> d\lambda. \tag{5.13}$$

Take some arbitrary $v(u)$ in the tangent bundle. Then by (5.4) we obtain

$$
\begin{aligned}
<\nabla F(u), v> \;&=\; \int_0^1 \{\lambda <\gamma'(\lambda u)[v], u> \;+\; <\gamma(\lambda u), v>\} d\lambda \\
&=\; \int_0^1 \{<\lambda \gamma'(\lambda u)[u], v> \;+\; <\gamma(\lambda u), v>\} d\lambda \tag{5.14} \\
&=\; \int_0^1 \frac{d}{d\lambda}\{<\lambda\gamma(\lambda u), v>\} d\lambda \;=\; <\gamma(u), v>
\end{aligned}
$$

This proves $\nabla F(u) = \gamma(u)$ since $v(u)$ was arbitrary ∎
Observe that this proof can be generalized to the situation where the manifold can be parametrized by a star-shaped subset of a vector space.

6   Compatibility

In this section we introduce the essential structure which will be responsible for the fact that in most cases of bi-hamiltonian systems the invariant co-vector fields constructed by use of Observation 5.2 are indeed closed. Again we will consider the necessary arguments at a rather general level. This is not done in order to introduce an unreasonable level of abstraction but rather to get rid of unnecessary ballast. Further, it turns out that with a general formulation of **compatibility** in Lie algebras one is more flexible with respect to applications. In our presentation we follow closely [9.]

Let, $(\mathcal{L},[\ ,\ ])$ be some Lie algebra over $\mathbb{R}$ or $\mathbb{C}$. We call $(\mathcal{L},[\ ,\ ])$ the **reference algebra**. Let furthermore Λ be a vector space over the same scalars. Now, consider a linear transformation

$$T : \Lambda \rightarrow \mathcal{L} \ .$$

from Λ into the reference algebra $\mathcal{L}$. We call some (bilinear) product $\{\ ,\ \}$ in Λ (not necessarily assumed to be a Lie product) a T-**product** if T is a homomorphism into $(\mathcal{L},[\ ,\ ])$, i.e. if

$$T\{a,b\} = [Ta, Tb] \ for \ all \ a,b \in \Lambda \ . \tag{6.1}$$

To emphasize that some product is a T-product we write $[\ ,\ ]_T$ instead of $\{\ \}$. A linear $\Phi : \mathcal{L} \rightarrow \mathcal{L}$ is said to be **hereditary** if

$$[a,b]_\Phi := [\Phi a, b] + [a, \Phi b] - \Phi[a,b] \tag{6.2}$$

defines a Φ-product in $\mathcal{L}$. Using then $\Phi[a,b]_\Phi = [\Phi a, \Phi b]$, we see that Φ is hereditary if and only if:

$$\Phi^2[a,b] + [\Phi a, \Phi b] = \Phi\{[\Phi a, b] + [a, \Phi b]\} \ for \ all \ a,b \in \mathcal{L} \ . \tag{6.3}$$

Rephrasing a notion introduced earlier we call a linear map $\Phi : \mathcal{L} \to \mathcal{L}$ **invariant** with respect to $k \in \mathcal{L}$ if

$$\Phi[k, b] = [k, \Phi b] \ for \ all \ b \in \mathcal{L} \tag{6.4}$$

Theorem 6.1:   *Let Φ be a hereditary map which is invariant with respect to k. Then $\{\Phi^n k | n \in I\!N_0\}$ is an abelian subset of $(\mathcal{L}, \ [\ ])$. If Φ is invertible then $\{\Phi^n k | n \in Z\!\!\!Z \}$ is abelian as well.*

For the proof we need:

Lemma 6.2:   *Let Φ be hereditary and let it be invariant with respect to k. Then Φ is invariant with respect to Φk. If Φ is invertible then it is invariant with respect to $\Phi^{-1} k$ as well. Thus the set $\{k | \Phi \ invariant \ with \ respect \ to \ k\}$ of all elements which leave Φ invariant is a subalgebra of $\mathcal{L}$ which is invariant under the application of Φ (and of Φ^{-1} if Φ is invertible).*

Proof:

Replace a by k in (6.3). Since Φ is invariant with respect to k the first and fourth term cancel and the equality reads

$$[\Phi k, \Phi b] = \Phi[\Phi k, b] \ for \ all \ b \ . \tag{6.5}$$

This clearly implies that Φ is invariant with respect to Φk. If Φ is invertible, we replace a in (6.3) by $\Phi^{-1} k$ and apply Φ^{-1} to the remaining two terms. ∎

Proof of Theorem 6.1:

From the Lemma 6.2 we obtain by induction that Φ is invariant with respect to any $\Phi^m k$ and $\Phi^n k$. Hence using antisymmetry we find

$$[\Phi^m k, \Phi^n k] = \Phi^{m+n}[k, k] = 0$$

for all m, n. For invertible Φ, in this argument Φ^{-1} has to replace Φ. ∎

Remark 6.3:   *Let Φ be hereditary and let a_1 and a_2 be eigenvectors of Φ , i.e.*

$$\Phi a_i = \lambda_i a_i \ , \quad \lambda_i = scalar, \ i = 1, 2 \ .$$

Then for these a_i relation (6.3) is equivalent to

$$(\Phi - \lambda_1)(\Phi - \lambda_2)[a_1, a_2] = 0 \ . \tag{6.6}$$

Hence, when an operator Φ has a spectral resolution, this operator is hereditary if and only if all the corresponding spectral projections are algebra homomorphisms.

Now, let us return to the general situation of maps from Λ into $\mathcal{L}$. Assume that in Λ we have $T-$ and $\Psi-$products $[\ , \ ]_T$ and $[\ , \ ]_\Psi$, respectively. These products are said to be **compatible** if

$$\{a, b\} := [a, b]_T + [a, b]_\Psi \tag{6.7}$$

defines a $(T + \Psi)$-product in Λ.

Lemma 6.4:   *Let $[\ , \ ]_T$ and $[\ , \ ]_\Psi$ be $T-$ and $\Psi-$products, respectively. These products are compatible if and only if*

$$T[a, b]_\Psi + \Psi[a, b]_T = [Ta, \Psi b] + [\Psi a, Tb] \ for \ all \ a, b \in \Lambda \tag{6.8}$$

Proof:
Observe that $T[a,b]_T = [Ta, Tb]$ and $\Psi[a,b]_\Psi = [\Psi a, \Psi b]$. So, (6.8) is obviously equivalent
to

$$(T + \Psi)\{[a,b]_T + [a,b]_\Psi\} = [(T + \Psi)a, (T + \Psi)b] \, , \tag{6.9}$$

which proves the claim. ∎

Observation 6.5: *Let λ be a scalar. Obviously, $[\ ,\ ]_{\lambda T}$ defined by $[a,b]_{\lambda T} := \lambda[a,b]_T$ is a
(λT)-product whenever $[\ ,\ ]_T$ is a T-product. Now, replacing in (6.8) Ψ and $[\ ,\ ]_\Psi$ by $\lambda\Psi$
and $[\ ,\ ]_{\lambda\Psi}$, respectively, we see that (6.8) remains valid. In other words, (6.8) is linear in
Ψ (as well as in T). Hence, if $[\ ,\ ]_{T_1}$ and $[\ ,\ ]_{T_2}$ are compatible , and if both are compatible
with $[\ ,\ ]_T$ then $[\ ,\ ]_{\lambda T_1} + [\ ,\ ]_{\sigma T_2}$ is always compatible with $[\ ,\ ]_T$.*

Observation 6.6: *Consider the case when the reference algebra is equal to Λ, i.e. $\Lambda = \mathcal{L}$
and put $T = I, \Psi = \Phi$. Furthermore, assume that $[\ ,\ ]_T$ is the given product in $(\mathcal{L}, [\ ])$, and
that $[\ ,\ ]_\Psi = \{\ \}$ is a second product such that $\Phi : (\mathcal{L}, \{\ \}) \to (\mathcal{L}, [\ ])$ is a homomorphism.
Then (6.8) holds if and only if $\{\ \}$ is the product defined in (6.3). Hence, Φ is hereditary
if and only if $(\mathcal{L}, \{\ ,\ \})$ and $(\mathcal{L}, [\ ,\ ])$ are compatible.*

In order to shorten our notions we call Ψ and T compatible if their Ψ- and T- products,
$[\ ,\ ]_\Psi$ and $[\ ,\ ]_T$ are compatible. By application of this notion to the special case of hereditary operators Φ_1, Φ_2 we see that Φ_1 and Φ_2 are compatible if and only $\Phi_1 + \Phi_2$ is again
hereditary.

Theorem 6.7: *Consider maps $T, \Psi : \Lambda \to \mathcal{L}$ and their corresponding products $[\ ,\ ]_T$ and
$[\ ,\ ]_\Psi$. Assume that Ψ is invertible. Then T and Ψ are compatible if and only if $\Phi = T\Psi^{-1}$
is hereditary.*

Proof:
Define a second product in $\mathcal{L}$ by

$$\{a, b\} := \Psi[\Psi^{-1}a, \Psi^{-1}b]_T \; for \; a, b \in \mathcal{L} \, .$$

Then $\Phi : (\mathcal{L}, \{\ \}) \to (\mathcal{L}, [\ ,\ ])$ is a homomorphism. Using the definition of $\{\ ,\ \}$ and (6.1)
(for Ψ instead of T) we obtain

$$
\begin{aligned}
(I + \Phi)([a,b] + \{a,b\}) &= (T + \Psi)\Psi^{-1}([a,b] + \{a,b\}) \\
&= (T + \Psi)([\Psi^{-1}a, \Psi^{-1}b]_\Psi + [\Psi^{-1}a, \Psi^{-1}b]_T) \, .
\end{aligned}
$$

For general $a, b \in \mathcal{L}$ the right side is equal to $[(T + \Psi)\Psi^{-1}a, (T + \Psi)\Psi^{-1}b]$ if and only if T
and Ψ are compatible, but this expression is equal to $[(I + \Phi)a, (I + \Phi)b]$ and equal to the
left side if and only if I and Φ are compatible. Hence the compatibility of T and Ψ and
that of I and Φ are equivalent. Now using Observation 6.6 we see that this is equivalent to
Φ being hereditary. ∎

By similar arguments we find:

Observation 6.8: *Let Φ, Ψ be compatible hereditary operators and assume that Φ and
Ψ do commute. Then $\Phi\Psi$ is hereditary. As a consequence, if Φ be hereditary, then any
polynomial in Φ is hereditary.*

Now consider again the
Lie-module situation:
Let $(\mathcal{L}, [\ ], \mathcal{F})$ be a Lie module as considered in Section 4 and define $\Lambda = \mathcal{L}^*$. If $\Phi : \mathcal{L} \to \mathcal{L}$ is a tensor, then condition (6.3) is easily rephrased as

$$\Phi L_A(\Phi) = L_{(\Phi A)}(\Phi) \ \text{for all } A \in \mathcal{L} \ . \tag{6.10}$$

So Φ is hereditary if and only if (6.10) is fulfilled. In case of the *standard situation* this can be expressed in charts as

$$\Phi \Phi'[A]B - \Phi[\Phi' A]B = \Phi \Phi'[B]A - \Phi[\Phi' B]A \ \text{for all } A, B \in \mathcal{L} \tag{6.11}$$

a condition which appeared in [3.] Because of Theorem 4.5 (4.14) defines a Θ-product in $\mathcal{L}^* = \Lambda$ if and only if Θ is implectic. Since Θ enters the definition (4.14) linearly we have:

Observation 6.9: *Two implectic operators Θ_1, Θ_2 are compatible if and only if $\Theta_1 + \Theta_2$ is again implectic.*

From Theorem 6.7 we obtain:
Corollary 6.10: *Let Θ_1, Θ_2 be implectic and assume that Θ_1 is invertible. Then $\Theta_1 + \Theta_2$ is implectic if and only if $\Phi = \Theta_2 \Theta_1^{-1}$ is hereditary.*

These results we apply to the
Bi-hamiltonian case: Let Θ_1, Θ_2 be compatible implectic operators such that Θ_1 is invertible. Let K be a bi-hamiltonian vector field

$$K = \Theta_1 \gamma_1 = \Theta_2 \gamma_2, \ \gamma_1, \gamma_2 \ closed \ .$$

Then define

$$\begin{aligned} \Phi &:= \Theta_2 \Theta_1^{-1}, \ \Phi^+ := \Theta_1^{-1} \Theta_2 \\ J_n &:= \Theta_1^{-1} \Phi^n, \ \Theta_{n+1} = \Phi^n \Theta_1 \end{aligned} \tag{6.12}$$

Theorem 6.11:

(i) *All J_n and all γ_n are closed.*

(ii) *All tensors $\Phi, \Phi^+, K_n, \gamma_n, J_n, \Theta_n$ are invariant with respect to every K_m, in particular all K_n, K_m commute.*

(iii) *If, in addition, Θ_2 is invertible, then the Θ_n are implectic.*

Before we can prove this we need to introduce a canonical extension $\tilde{\mathcal{L}}$ of $\mathcal{L}$.

The affine extension of $\mathcal{L}$:

Let $\tilde{\mathcal{L}} = \mathcal{L} \otimes \mathbb{C}[\xi]$ be the Lie algebra of formal power series in the indeterminate ξ with coefficients in $\mathcal{L}$. Extend in the same way $\mathcal{F}$ and $\mathcal{L}^*$

$$\tilde{\mathcal{F}} = \mathcal{F} \otimes \mathbb{C}[\xi], \quad \tilde{\mathcal{L}}^* = \mathcal{L}^* \otimes \mathbb{C}[\xi] .$$

Define that all Lie derivatives, and consequently the exterior derivative, ignore the variable ξ, i.e.

$$[\sum_n A_n \xi^n, \sum_m B_m \xi^m] = \sum_{n,m} [A_n, B_m] \xi^{n+m} . \tag{6.13}$$

Obviously $(\tilde{\mathcal{L}}, [\ ], \tilde{\mathcal{F}})$ is again a Lie module. We can embed the tensor structure of $\mathcal{L}$ into that of $\tilde{\mathcal{L}}$ by treating ξ as scalar. By comparison of coefficients we then obtain that a covariant tensor $T[\xi]$ with respect to $\tilde{\mathcal{L}}$

$$T[\xi] = \sum T_n \xi^n ,$$

(given by a formal power series in ξ with tensors in $\mathcal{L}$ as coefficients) is closed if and only if all the T_n are closed. Now having this additional structure we present as a simple excercise:

Proof of Theorem 6.11:
Consider the affine extension $\tilde{\mathcal{L}}$ of $\mathcal{L}$. Since Θ_1, Θ_2 are compatible implectic operators we have that $\tilde{\Theta} : \tilde{\mathcal{L}}^* \to \tilde{\mathcal{L}}$ defined by $\tilde{\Theta} = \Theta_1 + \xi\Theta_2$ is again implectic (Observation 6.5). Observe that $\tilde{\Theta}$ has an inverse $\tilde{J}$

$$\tilde{J} = \tilde{\Theta}_1^{-1} = \Theta_1^{-1} \sum_{n=0}^{\infty} (-\xi)^n (\Theta_2 \Theta_1^{-1})^n . \tag{6.14}$$

Hence $\tilde{J}$ must be closed in $\tilde{\mathcal{L}}$. Therefore every

$$J_n = \Theta_1^{-1} (\Theta_2 \Theta_1^{-1})^n = \Theta_1^{-1} (\Phi)^n$$

must be closed in $\mathcal{L}$. This is the essential step where we needed the extension of $\mathcal{L}$. From now on we argue in $\mathcal{L}$.

From the bi-hamiltonian formulation we know (Observation 5.2) that the Θ_1, Θ_2, Φ and all the K_n are K-invariant (product rule).
(i): We already know that J_n is closed and K-invariant. Hence by Remark 4.4 (ii) the $\gamma_n = J_n K$ must be closed. This holds for all n.
(ii): We know that Φ is invariant with respect to all the K_m (consequence of Lemma 6.2). Since Θ_1^{-1} and the γ_n are closed, Θ_1^{-1} must be invariant with respect to $K_n = \Theta_1 \gamma_n$ (Remark 4.4 (ii)). By the product rule we then find that $J_m = \Theta_1^{-1} \Phi^m$ and $\Theta_{m+1} = \Phi^m \Theta_1$ are K_n-invariant.
(iii): If Θ_2 is invertible as well we may interchange the role of Θ_1 and Θ_2 in order to see that $\tilde{J}_n = \Theta_2^{-1} \Phi^{-n}$ is closed. So its inverse Θ_n^{-1} must be implectic. ∎

7   Examples and Applications

Observe that when the duality between tangent and co-tangent space is represented as in Example 3.5 then the differential operator D is trivially implectic. In this section we show how, from this knowledge, new pairs of implectic operators can be constructed.

We first point out the relation between Lei-module-isomorphisms and variable transformations. This connection allows efficient use of the invariant manner in which we introduced the main notions.

Consider Lie modules $(\mathcal{L}, \mathcal{F})$ and $(\tilde{\mathcal{L}}, \tilde{\mathcal{F}})$. In both modules we denote the Lie product by [ ] since no confusion arises. A pair (S, σ) of maps $S : \mathcal{L} \to \tilde{\mathcal{L}}$ and $\sigma : \mathcal{F} \to \tilde{\mathcal{F}}$ is said to be a **Lie-module-homomorphism** if

- S and σ are homomorphisms with respect to the algebraic structures in $\mathcal{L}$ and $\mathcal{F}$, respectively,

- $S(fK) = \sigma(f)S(K)$ for all $f \in \mathcal{F}$ and $K \in \mathcal{L}$.

- $S \cdot L_K = L_{(S(K))}$ for all $K \in \mathcal{L}$.

If S and σ are invertible then this is called a **Lie-module-isomorphism**. Isomorphisms allow us to carry over the whole tensor structure from $(\mathcal{L}, \mathcal{F})$ to $(\tilde{\mathcal{L}}, \tilde{\mathcal{F}})$. To do this we first define for $\gamma \in \mathcal{L}^*$ the corresponding $\tilde{\gamma} \in \tilde{\mathcal{L}}^*$ by

$$< \tilde{\gamma}, \tilde{K} > := \sigma(< \gamma, S^{-1}\tilde{K} >) . \tag{7.1}$$

The map $S^* : \gamma \to \tilde{\gamma}$ is called the **reciprocal image**. Let Ψ be some $\mathcal{L}$-tensor (r-times contravariant and n-times covariant), then the corresponding $\tilde{\mathcal{L}}$-tensor $\tilde{\Psi}$ is defined by

$$\tilde{\Psi}(S^*\gamma_1, ..., S^*\gamma_r, SK_1, ..., SK_n) := \sigma \cdot \Psi(\gamma_1, ..., \gamma_r, K_1, ..., K_n) \; for \; \gamma_i \in \mathcal{L}^* \; and \; K_i \in \mathcal{L} . \tag{7.2}$$

For example, if a two-times contravariant tensor Θ is taken in operator notation then (7.2) means that

$$< S^*\gamma_1, \tilde{\Theta}S^*\gamma_2 > = \sigma(< \gamma_1, \Theta\gamma_2 >) \; for \; all \; \gamma_1, \gamma_2 \in \mathcal{L}^* \tag{7.3}$$

where $< , >$ denote the respective dualities in $\mathcal{L}$ and $\tilde{\mathcal{L}}$. This yields

$$\tilde{\Theta} = S\Theta S^T \tag{7.4}$$

where $S^T : \tilde{\mathcal{L}}^* \to \mathcal{L}^*$ is the transpose of S given by

$$\sigma(< S^*\tilde{\gamma}, K >) = < \tilde{\gamma}, S^*K > .$$

Observe that in this formula we used $S^{*T} = S^{-1}$. With the same ease other transformation formulas may be explicitly determined (see [4)]. As a consequence of our invariant definitions we find

Remark 7.1: *The notions implectic, symplectic, closed and hereditary are invariant under Lie-module isomorphisms.*

The most important Lie-module isomorphisms are given by variable transformations and these constitute an efficient tool for the construction of new compatible pairs of implectic operators.

Variable transformations:

Let M and $\tilde{M}$ be C^∞ manifolds and denote the respective manifold variables by u and $\tilde{u}$. Assume that there is a function $u \to \tilde{u} = T(u)$ such that T and its inverse (denoted by $\tilde{T}$) are C^∞. Let the meaning of $\mathcal{L}$ and $\mathcal{F}$ be as in the *standard situation* and consider the corresponding Lie module with respect to M. Then vector fields from M to $\tilde{M}$ are transformed by the variational derivative of T:

$$K(u) \to \tilde{K}(\tilde{u}) := T'(\tilde{T}(\tilde{u}))[K(\tilde{T}(\tilde{u}))] \tag{7.5}$$

and for scalars we define

$$f(u) \to \tilde{f}(\tilde{u}) := f(\tilde{T}(\tilde{u})) . \tag{7.6}$$

These transformations define a Lie-module isomorphism.

The most simple example for that isomorphism is when M is a vector space and

$$\tilde{u} = \lambda u + a , \tag{7.7}$$

where λ is some scalar and a some constant vector in M. Then we have that $S := T' = \lambda \, Id$ and we obtain

Remark 7.2: *Under the substitution $u \to \lambda u + a$ (a a constant vector and λ some scalar) in each tensor the properties: implectic, symplectic, closed and hereditary are preserved.*

Example 7.3: Construction of compatible pairs

For technical reasons, we now consider the space $\mathcal{S}_-$ of C^∞-functions f in the real variable $x \in \mathbb{R}$ having the property that f and all its derivatives vanish at $-\infty$ faster than any rational function and that at $+\infty$ all derivatives grow at most polynomially. As $\mathcal{S}_+$ we denote the corresponding space where the role of $-\infty$ and $+\infty$ has been interchanged. Between $\mathcal{S}_+$ and $\mathcal{S}_-$ we introduce an L^2 scalar product as in (3.9)

$$< U, u > = \int U(x)u(x)dx, \quad U \in \mathcal{S}_+, u \in \mathcal{S}_- . \tag{7.8}$$

As before, D denotes differentiation with respect to x and D^{-1} denotes integration from $-\infty$ to x. Let the manifold under consideration be $M = \mathcal{S}_-$. Observe that D is implectic and invertible. We consider the variable transformation $\tilde{u} = T(u) := u^2 + u_x$ on M. Observe that, because of the boundary conditions we have chosen, this is one to one by the Implicit Function Theorem. To see that, compute for $T'(u) = 2u + D$ the inverse of $T'(u)$ by solving the linear differential equation $T'(u)z = g$ for given g and unknown z. On $\mathcal{S}_-$ this has a unique solution and the operator $T'(u)^{-1}$ mapping g into z is

$$T'(u)^{-1} = \exp\left(-2(D^{-1}u)\right) D^{-1} \exp\left(+2(D^{-1}u)\right) . \tag{7.9}$$

Starting with the implectic operator D (with respect to the manifold variable u) we find by variable transformation (formula (7.4)) that

$$\begin{aligned}
\tilde{\Theta}(\tilde{u}) &= T'(u)DT'(u)^T \\
&= (2u + D)D(2u - D) \\
&= D^3 + 2D(u^2 + u_x) + 2(u^2 + u_x)D
\end{aligned} \tag{7.10}$$

Using the relation between u and $\tilde{u}$ we see that

$$\tilde{\Theta}(\tilde{u}) = D^3 + 2D\tilde{u} + 2\tilde{u}D \qquad (7.11)$$

is implectic as it was claimed already in Example 3.5. Now performing the substitution (for $a(x) = 1$) as described in Remark 7.1 we find that

$$\tilde{\Theta}(\tilde{u} + 1) = D^3 + 2(D\tilde{u} + \tilde{u}D) + 4D = \tilde{\Theta}(\tilde{u}) + 4D \qquad (7.12)$$

is again implectic. Because D is already known to be implectic we have that D and $\tilde{\Theta}$ are compatible. Hence by Corollary 6.10

$$\tilde{\Phi} = \tilde{\Theta}D^{-1} = D^2 + 2D\tilde{u}D^{-1} + 2\tilde{u} , \qquad (7.13)$$

must be hereditary.

Using Theorem 6.11 we find that

$$\tilde{\Theta}_2(\tilde{u}) = \tilde{\Phi}(\tilde{u})\tilde{\Theta}(\tilde{u}) \qquad (7.14)$$

again is implectic. We transform that back from the $\tilde{u}$-variable to the u-variable in order to obtain with (7.10) the following implectic operator:

$$
\begin{aligned}
\Theta(u) &= T'^{-1}\tilde{\Theta}_2(T'^{-1})^T \\
&= T'^{-1}\tilde{\Theta}D^{-1}\tilde{\Theta}(T'^{-1})^T \\
&= T'^{-1}T'D(T')^T D^{-1}T'D(T')^T(T'^{-1})^T \\
&= D(T')^T D^{-1}T'D \\
&= D(2u - D)D^{-1}(2u + D)D \\
&= -D^3 + 4DuD^{-1}uD
\end{aligned}
$$

Now using $u \to iu$ we find with Remark 7.1 that

$$\Theta(u)_{mKdV} = D^3 + 4DuD^{-1}uD \qquad (7.15)$$

is implectic. Using Remark 7.2 and $u \to (1/4)\sqrt{2}(2u + 1)$ we find that

$$\Theta_{Gardner} = \frac{1}{2}(\Theta(u)_{mKdV} + \Theta(u) + D)$$

is also implectic. here $\Theta(u)$ is the implectic operator given in (7.11) (only $\tilde{u}$ replaced by u). Since $\Theta(u)$ and D are compatible $\Theta(u) + D$ is again implectic and Θ_{mKdV} and $\Theta(u) + D$ must be compatible. Taking now $1/4\Theta_{mKdV} + 3/4(\Theta(u) + D)$ (which is implectic by Observation 6.5) and substituting $u \to (2u - 3)$ we find that $\Theta_{mKdV} + (3/4)D$ is implectic. Hence Θ_{mKdV} and D are compatible.

Observation 7.4: *If the duality between function spaces is taken to be (3.9) then the differential operators*

$$\Theta_0 = D$$

$$\Theta_{KdV} = D^3 + 2Du + 2uD$$

$$\Theta_{mKdV} = D^3 + 4DuD^{-1}uD$$

are implectic. Every two of these are compatible.

Example 7.5: Conserved quantities for the KdV

Consider the situation of Example 3.5. where the operator

$$\Theta_{KdV} = D^3 + 2Du + 2uD$$

was introduced. Taking into account the compatibility between the implectic operators D and Θ_{KdV} (as just proved) we get from Corollary 6.10 that

$$\Phi = \Theta D^{-1} = D^2 + 2DuD^{-1} + 2u , \qquad (bihamKdV)$$

is hereditary. As a consequence (Theorem 6.11) we have that the

$$\gamma_{n+1} = D^{-1}K_{n+1} = D^{-1}\Phi(u)^n K(u), \text{ where } K(u) = 6uu_x + u_{xxx} \qquad (7.16)$$

are closed co-vector fields. These fields are invariant for any of the flows $u_t = K_n(u)$. Hence (Lemma 5.5) all

$$I_n(u) = \int_0^1 < \gamma_n(\lambda u), u > d\lambda = \int_0^1 \int_{\mathbb{R}} \gamma_n(\lambda u(x))u(x)d\lambda dx \qquad (7.17)$$

are conserved quantities for every one of these flows, especially for the KdV. In addition, one easily sees that all these quantities commute with respect to the Poisson brackets defined by either Θ_{KdV} or D. $\square$

Example 7.6: Further systems

Using the compatible pairs we have up to now, and more which we generate easily by variable transformations, we can generate new and nontrivial hereditary operators. For example

$$\Phi_{mKdV} = \Theta_{mKdV} D^{-1}$$
$$\Phi_{Gardner} = \Theta_{Gardner} D^{-1}$$
$$\Phi_{sineGordon} = 2\Phi_{mKdV}^{-1}$$
$$\Phi_{BBM-like}(v) = (I - D^2)^{-1}\Phi_{KdV}(v - v_{xx})$$
$$\Phi_{sG-KdV} = \Phi_{KdV}^{-1}\Phi_{sineGordon}$$

The hereditary nature of these operators is easily seen from compatibility of known pairs. For example, take Φ_{sG-KdV} then this can be written as $\Theta_{KdV}\Theta_{mKdV}^{-1}$, and must be hereditary since these two are compatible. For $\Phi_{BBM-like}(v)$ we use the compatibility of Θ_{KdV} and of $D - D^3$, and then we performed a variable transformation $u = v - v_{xx}$.

Since none of these operators depend explicitly on x we find that they are all invariant with respect to the special vector field u_x. So for the equations

$$u_t = \Phi(u)u_x$$

we find (by application of Theorem 6.1) infinitely many symmetry group generators

$$K_n(u) = \Phi^n(u)u_x \tag{7.18}$$

Let us list these equations (following the above order)

$$u_t = u_{xxx} + 6u^2 u_x \qquad\qquad (modified\ KdV)$$

$$u_t = u_{xxx} + 6u^2 u_x + 6uu_x + u_x \qquad\qquad (Gardner\ eq.)$$

$$u_t = \sin\left(2\int_{-\infty}^{x} u(\xi)d\xi\right) \qquad\qquad (potential\_sG\ eq.)$$

$$v_t - v_{xxt} = v_{xxx} - 2vv_{xxx} - 4v_x v_{xx} + 6vv_x \qquad\qquad (BBM\_like\ eq.)$$

$$v_{xt} = 2v_{xx}\cos(2v) + 4(v_x - v_x^2)\sin(2v) + 2v_{xx}\int_{-\infty}^{x}\sin(2v(\xi))d\xi \qquad\qquad (KdV\_sG\ eq.)$$

In the case of the last equation (KdV_sG), we have performed an additional variable transformation $u = v_x$. The equation (potential_sG) is connected to so called **sine_Gordon** equation since substituting first $u = v_x$ and performing then a 45-degree rotation in the space of independent variables yields

$$v_{\xi\xi} - v_{\eta\eta} = \sin(2v)\ . \qquad\qquad (sine\_Gordon\ equation)$$

Most of these equations are well known from the literature. They all have a bi-hamiltonian formulation if the solution manifold is suitably chosen such that these operators are well defined. The invariant co-vector fields generated by this recursion mechanism are all closed because they are then generated by a hereditary operator, which stems from a compatible pair of impleetic operators. For example, in case of the modified Korteweg-de Vries equation (modifiedKdV) the compatible pair of implectic operators is D and Θ_{mKdV}. The equation itself has the form

$$u_t = u_{xxx} + 6u^2 u_x = \Theta_{mKdV}(u)\nabla\frac{1}{2}\int_{\mathbb{R}} u^2(\xi)d\xi\ . \tag{7.19}$$

Furthermore

$$D^{-1}(u_{xxx} + 6u^2 u_x) = \nabla\int_{\mathbb{R}}(-u_x^2(\xi) + \frac{1}{2}u^4(\xi))d\xi\ . \tag{7.20}$$

Application of Theorem 6.11 then shows that

$$I_n(u) = \int_0^1 \int_{\mathbb{R}} \gamma_n(\lambda u(x))u(x)dxd\lambda = \int_0^1 \int_{\mathbb{R}} (D^{-1}\Theta_{mKdV}(\lambda u(x)))^n u(x)d\lambda dx \tag{7.21}$$

are conserved quantities for the modified Korteweg-de Vries equation. Observe that these are also conserved quantities for the potential sine-Gordon equation (potential_sG eq.) since that equation is generated by the inverse of the operator Φ_{mKdV}. Similar arguments go through for the other equations given above.

We like to mention that the applications of the theory presented in this paper are in no way exhausted by these examples. There are many more such as: the nonlinear Schrödinger equation, two-component systems, Spin chains, and the bi-hamiltonian formulations given by Fokas and Santini ([2,] [24,] [23,] [22]) for equations in two independent variables. These last examples are interesting in so far as they require the full generality of notions and methods as introduced in Sections 4,5 and 6.

8   Integrability and Solitons

Let us briefly review:

Complete integrability in the finite dimensional case:
The hamiltonian flow

$$u_t = \Theta(u)\nabla H(u). \tag{3.3}$$

on a $2N$-dimensional manifold M with invertible implectic operator Θ is called completely integrable if it admits N conserved quantities $I_1 := H, I_2, ..., I_N$ such that the corresponding symmetry group generators $\Theta\nabla I_1, \Theta\nabla I_2, ..., \Theta\nabla I_N$ commute. Furthermore these fields are required to be linearly independent at each manifold point. These $I_1, I_2, ..., I_N$ are called **action variables**.

Observation 8.1:
In this case one can find N closed and pairwise commuting vector fields $A_1, ..., A_N$ such that

$$L_{A_i} I_j = \delta_{ij} I_i \tag{8.1}$$

or, by use of (4.16) and Theorem 4.5

$$[A_i, \Theta\nabla I_j] = \delta_{ij}\Theta\nabla I_i . \tag{8.2}$$

*The $\Theta\nabla I_i$ are called **action fields** and the A_i are called the conjugate **angle fields**).*

The proof of this statement is technically involved, so we will only give a brief sketch. The arguments are an adaption of [16, page 28].

Proof :
STEP 1:
First one shows that around each manifold point u_0 coordinates $\{I_1, .., I_N, Q_1, ..., Q_N\}$ can be chosen such that $J(u) := \Theta(u)^{-1}$ is constant in that chart. This is done in the following way:
Represent the manifold around u_0 by some open ball in a vector space E with coordinates $\{I_1, .., I_N, \tilde{Q}_1, ...\tilde{Q}_N\}$. Then consider the operators $J(u)$ and $J_0(u) := J(u_0)$ and take a deformation $J_t(u) := J(u) + t(J_0(u) - J(u))$, $1 \geq t \geq 0$ from J to J_0. Observe that $J_t(u_0) = J_0(u_0)$ is invertible for all t with $1 \geq t \geq 0$. The openness of the set of isomorphisms shows that there is a ball B around u_0 such that $J_t(u)$ is invertible for all $u \in B$ and $1 \geq t \geq 0$. The inverse of $J_t(u)$ we denote by $\Theta_t(u)$. Since $(J_0(u) - J(u))$ is closed we find by the Poincaré lemma a 1-form γ in B such that $(J_0(u) - J(u)) = d\gamma$ and $\gamma(u_0) = 0$. Now take the t-dependent vector field $K(t, u) = \Theta_t(u)\gamma$ and consider the equation

$$v_t = K(t, v) \tag{8.3}$$

in B. Since $K(t = 0, u) = 0$ we can assume (by eventually restricting B again) that (8.3) has a unique solution for all $1 \geq t \geq 0$ and all $u \in B$. Define $\varphi(u)$ to be the solution of (8.3) for $t = 1$ and initial condition $v(t = 0) := u$ and let $\tilde{I}_i(u) = I_i(\varphi(u))$, $\tilde{Q}_i(u) = Q_i(\varphi(u))$, $i =$

$1, ..., N$. Observe that $\tilde{I} = I$ and that $J(u)$ now is constant in the chart given by these new coordinates.

STEP 2:

Consider the constant $J(u)$ as constructed above in the vector space given by the coordinates $\{I_1, .., I_N, Q_1, ..., Q_N\}$ and endow that space with the usual Euclidean metric of $\mathbb{R}^{2N}$. Since $J(u)$ is invertible and antisymmetric its matrix representation must be of the form

$$\begin{pmatrix} 0 & S \\ -S & 0 \end{pmatrix}$$

where S is invertible and symmetric. Hence by a change of basis among the Q's we can assume that S is the $N \times N$ identity matrix. Finally, taking $A_i = -\Theta\nabla(I_iQ_i)$ we locally find the desired vector fields.

STEP 3:

We observe that different local realizations of (8.1) differ on the overlap of their domain only by a suitable combination of the $\Theta\nabla I_i$, hence by globally defined vector fields. This property allows us to patch the A_i from one chart to the next so that they coincide on the overlap. Hence we can define them globally. ∎

Observe that locally potentials for the A_i exist, let us call them Q_i. Then (8.1) implies that for every of the flows $u_t = \Theta(u)\nabla I_n$ the $Q_m, n \neq m$ are conserved quantities, whereas Q_n changes with t but is the absolute part of the time-dependent conserved quantity $Q_n - tI_n$. So taking these coordinates we arive at

Observation 8.2: *On some $2N$-dimensional manifold, endowed with the invertible implectic operator Θ, let there be given N pairwise commuting scalar fields $I_1, ..., I_N$ (commuting with respect to the corresponding Poisson brackets). Then around each point there are local coordinates $\{I_1, ..., I_N, Q_1, ..., Q_N\}$ such that the hamiltonian flows $u_t = \Theta(u)\nabla I_n$ are linear in these coordinates so that all but Q_n are invariant and that the action on the Q_n is such that this grows linear with t.*

Now we use the vector fields from Observation 8.1 to define the following operator $J_2 \mathcal{L} \to \mathcal{L}^*$

$$J_2 = \sum_{i=1}^{N} \Theta^{-1}A_i \otimes dI_i - dI_i \otimes \Theta^{-1}A_i . \tag{8.4}$$

Obviously this is an antisymmetric tensor and it is closed because both dI_i and $\Theta^{-1}A_i$ are closed. So it may serve as a symplectic operator and one easily finds that

$$\Phi := \Theta J_2 \tag{8.5}$$

is hereditary. Using the relations (8.1) we see right away that

$$\begin{aligned} J_2\Theta\nabla I_i &= I_i\nabla I_i \\ J_2 A_i &= -I_i\Theta^{-1}A_i \end{aligned} \tag{8.6}$$

and that therefore the eigenvalues of Φ are given by the action variables. So we have

Observation 8.3: *For a finite dimensional completely integrable system in* $2N$*-dimensional space, with action varaibles* $\{I_1, ..., I_N\}$ *there always exists a hereditary operator* Φ *such that its spectrum is doubly degenerate. The eigenvalues are given by the action variables and the corresponding eigenvectors are the action fields and their conjugate angle fields.*

Of course, the converse is also true: Whenever on a $2N$*-dimensional manifold we have a hereditary operator* Φ *with doubly degenerate spectrum and some implectic operator* Θ *such that* $\Theta\Phi$ *is closed, then when the the gradients of the eigenvalues are mapped with* Θ *onto vector fields they form a commuting algebra of vector fields (consequence of Remark 6.3). Hence any dynamic of system given by any linear combination of these vector fields must be completely integrable.*

Now let us return to the general situation of infinitely dimensional manifolds. Here the notion of complete integrability has not yet been definitely defined in the literature. Often such a system is called completely integrable if it admits an infinite dimensional abelian symmetry group of hamiltonian fields. This of course, is a somehow loose definition since it is easy to construct situations where such a symmetry group does not suffice to guarantee a parametrization by action and angle variables. Instead of attempting here a general definition we shall show that, under reasonable conditions, the existence of a compatible bi-hamiltonian pair leads to complete integrability on finite dimensional reductions given by the corresponding symmetry group generators. We will give a survey on the results which can be obtained in that direction, for details the reader is referred to [8].

Let us recall the situation we considered before. On a suitable manifold M the evolution equation $u_t = K_1(u)$ where $u = u(x,t) \in M$ was considered. We assume that there is a hereditary recursion operator $\Phi(u)$ generated out of a compatible hamiltonian pair

$$\Phi(u) = \Theta_2(u)\,\Theta_1^{-1}(u) =: \Theta_2(u)\,J(u) \ .$$

As shown the operator Φ then generates a hierarchy of pairwise commuting infinitesimal symmetry group generators

$$K_{n+1}(u) := \Phi^n(u)K_1(u)$$

for the evolution equation under consideration.

In addition to what we assumed until now we require furthermore the existence of a **scaling symmetry** $\tau_0(u)$. By that we mean:

$$[\tau_0, K_1] = (\varrho + 1)K_1 \tag{8.7}$$

and

$$L_{\tau_0}\Phi = \Phi \ . \tag{8.8}$$

As a consequence of the scaling property the recursive application of Φ on τ_0 produces a second hierarchy of vector fields, the so-called **mastersymmetries** [6] $\tau_n = \Phi^n\tau_0$ such that the following commutator relations hold between the symmetries K_n and the mastersymmetries τ_n

$$[K_n, K_m] = 0 \ , \quad [\tau_n, K_m] = (m + \varrho)K_{n+m} \ , \quad [\tau_n, \tau_m] = (m - n)\tau_{n+m} \ . \tag{8.9}$$

Indeed, these commutator relations are a simple consequence of the hereditary property of Φ. One should observe that the relation $[\tau_n, K_m] = (m + \varrho)K_{n+m}$ is equivalent to the fact that $\tau_n + (m + \varrho)K_{n+m}$ is a time-dependent symmetry group generator of $u_t = K_m(u)$. A Lie algebra consisting of τ's and K's fulfilling (8.9) is called a **hereditary algebra**. These scaling symmetries exist for almost all popular soliton equations, and even in those cases where a scaling symmetry or a hereditary cannot be found one can nevertheless construct a suitable hereditary algebra. For example, in the KdV case $\tau_0(u) = \frac{1}{2}xu_x + u$ is the scaling symmetry, and for the mKdV and the potential sine-Gordon one finds $\tau_0(u) = xu_x + u$.

From the invariance of the symmetry group generators one finds that the submanifold (see for example [19])

$$M_N = \{ u \mid \text{there exists } \alpha_n \text{ such that } \sum_{n=0}^{N} \alpha_n K_n = 0\} \qquad (8.10)$$

is invariant under any of the flows $u_t = K_n(u)$, in particular under $u_t = K_1(u)$. This manifold is called the manifold of N-**soliton solutions**. For the KdV typical two- and three-solitons are given in figures 3 and 4[9] .

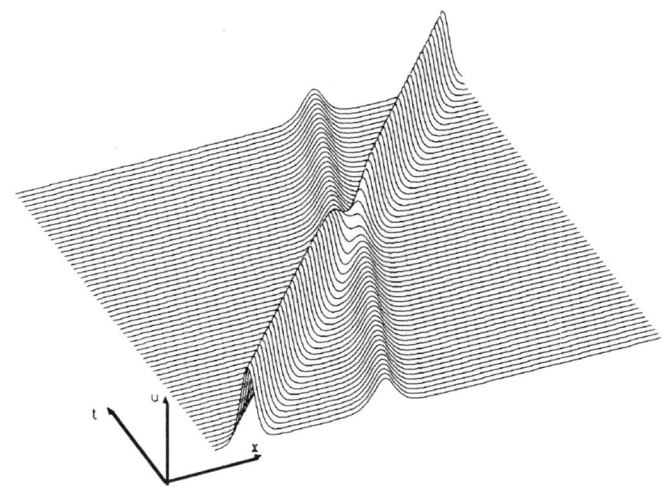

Fig. 3: Two-soliton of the KdV

---

[9]I am indebted to Thorsten Schulze for plotting these figures.

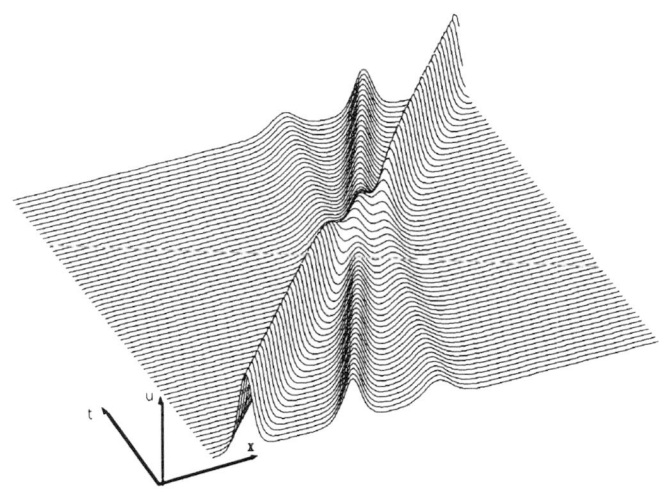

Fig. 4: Two-soliton of the KdV

In case when the boundary conditions at infinity for possible solutions u are chosen in such a way that the resulting manifold M_N has dimension $2N$ then by a lengthy but simple analysis [8] (mainly of the hereditary structure of Φ) one obtains from this structure the following result

Theorem 8.4:

(1) For all $r, p \in \mathbb{N}_0$ we have the following representation of the tangent space $T_u M_N$ of M_N at the point u

$$T_u M_N = \text{span} \{ K_r, K_{r+1}, ..., K_{r+N-1}, \tau_p, \tau_{p+1}, ..., \tau_{p+N-1} \} \ .$$

(2) Whenever the α_n are the coefficients given by (8.10) (to define the manifold point u) then the following hold

    (i) For all $r \in \mathbb{N}_0$ we have the following identities on M_N:

$$\sum_{n=0}^{N} \alpha_n K_{n+r} = 0 \quad \text{and} \quad \sum_{n=0}^{N} \alpha_n \tau_{n+r} = 0 \ .$$

    (ii) The discrete eigenvalues $c_1, ..., c_N$ of Φ are given as the zeros of the characteristic polynominal $P(\xi) = \sum_{n=0}^{N} \alpha_n \xi^n$.

    (iii) The corresponding eigenstates are $\tilde{V}_i = \Pi_i(\Phi) K_0$ and $\tilde{W}_i = \Pi_i(\Phi) \tau_0$, where $\Pi_i(\xi) = P(\xi)/(\xi - c_i)$.

As a direct consequence of Theorem 8.4 we obtain that the recursion operator Φ leaves the tangent space $T_u M_N$ of the reduced manifold invariant. Hence, the restriction $\bar{\Phi} := \Phi_{|red}$ of Φ to M_N is a linear operator on a finite dimensional space. This operator $\bar{\Phi}$ has the properties listed below in:

Observation 8.5:

(1) $\bar{\Phi}$ is invertible and can be written as $\bar{\Phi} = \bar{\Theta}_2 \bar{\Theta}_1^{-1}$ where $\bar{\Theta}_1, \bar{\Theta}_2$ are a compatible pair of implectic operators.

(2) The eigenvalues $c_1, ..., c_N$ of $\bar{\Phi}$ are doubly degenerated.

(3) Renorming the eigenstates $\tilde{V}_i, \tilde{W}_i$ leads to eigenstates V_i and W_i which are hamiltonian vector fields w.r.t. $\bar{\Theta}_1$ and $\bar{\Theta}_2$.

(4) The eigenstates V_i and W_i fulfil the commutator relations

$$[V_i, V_j] = 0 = [W_i, W_j] \quad , \quad [V_i, W_j] = \delta_{ij} V_j \ . \tag{8.11}$$

These last two results show that the finite dimensional reductions, given by those members of the abelian symmetry group which is generated by the bi-hamiltonian formulation is, under suitable boundary conditions at infinity, the same situation as we found in the completely integrable finite dimensional case.

Since the eigenstates V_i, W_i are hamiltonian vector fields and since they fulfill the canonical commutator relations (8.9), their potentials can be interpreted as action/angle variables for the flow induced by (1.1) on M_N.

Although all our considerations were of a purely algebraic nature we should remark that in most cases which are relevant from the physical viewpoint the N-soliton solutions (with vanishing boundary conditions at infinity) are those solutions which decompose into N single waves for $t \to \pm\infty$

$$u_N \cong \sum_{i=1}^{N} s_i(x + c_i t + q_i) \ .$$

This can be seen for the KdV from Figures 4 and 5. By comparison with the asymptotic data we get in these case we get a simple method for finding the eigenstates of the recursion operator.

Observation 8.6: *Taking the partial derivatives of u_N w.r.t. the asymptotic data*

$$\frac{\partial u_N}{\partial q_i} \quad \text{and} \quad \frac{\partial u_N}{\partial c_i}$$

one obtains eigenstates of the recursion operator $\bar{\Phi}$ for the eigenvalue c_i. The function $\partial u_N / \partial q_i$ then is the vector field corresponding to the action variable and this is called the interacting soliton *[7]*.

In case of the KdV equation a plot of such quantities is easily obtained (see Figure 5)

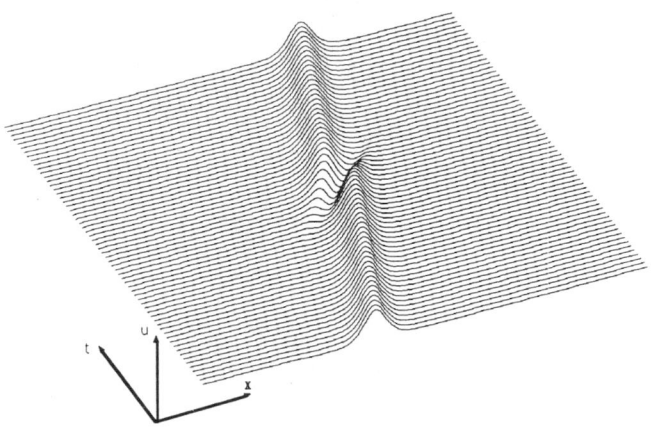

Fig. 5: Interacting soliton of the KdV

A corresponding conjugate eigenstate for the KdV is obtained by taking the derivative of the field function u_N with respect to the parameters given by an eigenvalue of the recursion operator.

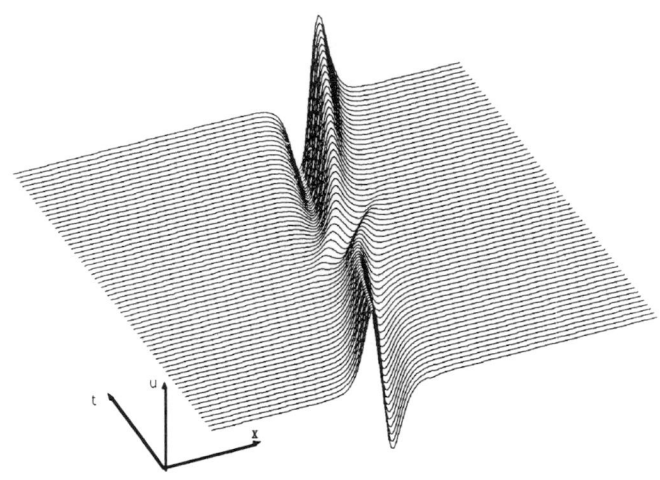

Fig. 6: Derivative of Angle-variable density of the KdV

It should be remarked that the field functions given by these plots themselves satisfy nonlinear equations which have a compatible bi-hamiltonian formulation [7]

References

[1] A. Dick: *Emmy Noether*, Birkhäuser-Verlag, Basel, 1970

[2] A.S. Fokas and P.M. Santini: *The Recursion Operator of the Kadomtsev-Petviashvili equation and the squared eigenfunctions of the Schrödinger Operator*, Studies in Appl. Math., 75, p.179-186, 1986

[3] B. Fuchssteiner: *Application of Hereditary Symmetries to Nonlinear Evolution equations*, Nonlinear Analysis TMA, 3, p.849-862, 1979

[4] A. S. Fokas and B. Fuchssteiner: *Bäcklund Transformations for Hereditary symmetries*, Nonlinear Analysis TMA, 5, p.423-432, 1981

[5] B. Fuchssteiner and A. S. Fokas: *Symplectic Structures, Their Bäcklund Transformations and Hereditary Symmetries*, Physica, 4 D, p.47-66, 1981

[6] B. Fuchssteiner: *Mastersymmetries, Higher-order Time-dependent symmetries and conserved Densities of Nonlinear Evolution Equations*, Progr. Theor.Phys., 70, p.1508-1522, 1983

[7] B. Fuchssteiner: *Solitons in Interaction*, Progress of Theoretical Physics, 78, p.1022-1050, 1987

[8] B. Fuchssteiner and G. Oevel: *Geometry and action-angle variables of multisoliton systems*, Reviews in Mathematical Physics, 1, p.415-479, 1990

[9] B. Fuchssteiner: *Compatibility in abstract algebraic Structures: Application to nonlinear systems*, Paderborn, preprint, 1990

[10] I.M.Gelfand and I.Y. Dorfman: Funktsional'nyi Analiz i Ego Prilozheniya, *Hamiltonian Operators and Algebraic Structures related to them*, 13, p.13-30, 1974

[11] I.M.Gelfand and I.Y. Dorfman: Funktsional'nyi Analiz i Ego Prilozheniya, *The Schouten bracket and Hamiltonian operators*, 14, p.71-74, 1980

[12] I.M.Gelfand and I.Y. Dorfman: Funktsional'nyi Analiz i Ego Prilozheniya, *Hamiltonian Operators and Infinite-Dimensional Lie-Algebras*, 15, p.23-40, 1981

[13] D. J. Korteweg and G. De Vries: *On the change of form of long waves advancing in a rectangular canal, and a new type of long stationary waves*, Philos.Mag.Ser. 5, 39, p.422-443, 1895

[14] F. Magri: *A simple model of the integrable Hamiltonian equation*, J.Math.Phys., 19, p.1156-1162, 1978

[15] F. Magri: *A Geometrical Approach to the Nonlinear Solvable equations*, Nonlinear Evolution equations and Dynamical Systems (M. Boiti, F. Pempinelli, G. Soliani eds.) Springer Verlag, p.233-263, 1980

[16] J. Marsden: *Applications of Global Analysis in Mathematical Physics*, Publish or Perish, Inc., 2, Boston, 1970

[17] E. Nelson: *Tensor Analysis*, Princeton University Press, Princeton NJ, 1967

[18] E. Noether: *Invariante Variationsprobleme*, Nachr.v.d. Gesellschaft d.Wiss. zu Göttingen, p.235-257, 1918

[19] S.P. Novikov, S. V. Manakov, L. P. Pitaevskii and V. E. Zakharov: *Theory of Solitons, The Inverse Scattering Method*, Consultants Bureau, New York-London, 1984

[20] P.J. Olver: *Evolution Equations possessing infinitely many symmetries*, J.Math.Phys., 18, 1977 p.1212-1215,

[21] C. Rebbi and G. Soliani: *Solitons and Particles*, World Scientific, Singapore, 1984

[22] P.M. Santini and A.S. Fokas: *Recursion Operators and Bi-Hamiltonian Structures of 2+1 dimensional systems*, in: Topics in Soliton Theory and Exactly solvable Nonlinear equations (eds: M. Ablowitz, B. Fuchssteiner, M. Kruskal) World Scientific Publ., Singapore, 1987 p.1-19,

[23] P.M. Santini and A.S. Fokas: *Recursion Operators and Bi-Hamiltonian structures in Multidimensions. I*, Commun. Math. Phys., 115, p.375-419, 1988

[24] P.M. Santini, A.S. Fokas: *Recursion Operators and Bi-Hamiltonian structures in Multidimensions. II*, Comm. Math. Phys., 116, p.449-474, 1988

[25] B. Schutz: *Geometrical Methods of Mathematical Physics*, Cambridge University Press, Cambridge-London-La Rochelle-New York-Melbourne-Sidney, 1980

[26] S. Yamamuro: *Differential Calculus in Topological Linear Spaces*, 374, Springer Verlag, Berlin-Heidelberg-New York, 1974

[27] S. Yamamuro: *A theory of differentiation in locally convex spaces*, AMS, Providence, Rhode Island, 1979

Symmetric Chaos

Greg King and Ian Stewart
Nonlinear Systems Laboratory
Mathematics Institute
University of Warwick,
Coventry CV4 7AL, England

Abstract

Chaos is the occurrence of highly irregular behaviour in a deterministic dynamical system. It is well known that both ordered and chaotic states can be obtained in such systems by varying parameters. In contrast we here consider circumstances in which aspects of both order and chaos may be present in the same behaviour and at the same parameter values. Namely, in a dynamical system with symmetry, chaotic attractors may themselves possess a degree of symmetry. The non-chaotic dynamics of symmetric systems has attracted increasing attention of late, and the time appears to be ripe to extend the investigation into the chaotic regime.

We describe some experimental situations in which this type of behaviour appears to be implicated, including patterned turbulence in fluids and nonlinear oscillations in electronic circuits. We describe sample results from various numerical experiments, to illustrate some of the phenomena that can occur, and describe some simple but fundamental mathematical results that go some way towards explaining them. The field has many open questions, some of which are indicated.

1   Introduction

Recent years have seen a tremendous flowering of the theory of deterministic chaos: the occurrence in non-random dynamics of highly irregular and apparently stochastic behaviour: see Cvitanović [13], Devaney [14], Guckenheimer and Holmes [22]. Another development of some interest has been the growing recognition of symmetry-breaking in equivariant (symmetrical) dynamical systems as a mechanism for pattern-formation, Golubitsky *et al.* [19]. The aim of this paper is to describe recent attempts to combine the two ideas by considering chaotic dynamics in equivariant dynamical systems, thereby creating a framework for the study of patterned chaos. Our treatment is not exhaustive: material not discussed here can be found in Crowe *et al.* [12], Franjioni *et al.* [17], Milnor [27], and Piña and Cantoral [29].

As motivation we have selected the phenomenon of turbulent Taylor vortices in the Taylor-Couette system (section 2). We choose this not because it has been rigorously explained — indeed the resolution of its paradoxical features reported here and due to Golubitsky remains conjectural — but because it embodies the precise combination of symmetry and chaos that we aim to capture. General background on symmetry in dynamics is sketched in section 3. In section 4 we exemplify some of the phenomena that can occur in a series of numerical experiments: the cubic logistic map on the line; mappings of the plane equivariant under the dihedral group D_3; and mappings of the 2-torus equivariant under an action of D_4.

Section 5 collects together some simple but useful theorems from the literature and folklore of the subject. The first addresses the issue of the occurrence of 'interesting' dynamics in equivariant diffeomorphisms as opposed to mappings; the second 'explains' the occurrence of symmetry-increasing crises; and the third demonstrates that the symmetries of strange attractors may differ from those possible for fixed points. Section 6 enters more deeply into the mathematics of symmetric chaos and sketches some recent results of Krupa and Roberts [25] on symmetry-locking in equivariant circle maps.

In section 7 we return to applications with a survey of several experiments on symmetric chaos in electronic circuits: the Van der Pol-Duffing oscillator with cubic or quintic characteristic, and systems of coupled identical oscillators. These systems exhibit many of the types of behaviour predicted theoretically: in particular the occurrence of symmetry-increasing crises. The general issue of extracting symmetries of strange attractors from time series is addressed in section 8; and in section 9 we return to the motivating problem of turbulent Taylor vortices to describe a plausible scenario for their formation, due to Chossat and Golubitsky, and to discuss a possible method for testing this scenario experimentally.

2 Patterned Chaos

Interest in chaotic dynamics of symmetric systems arises from the occurrence in experiments of 'coherent structures': dynamical states that combine local chaos with global pattern. In this section we describe one example of this phenomenon, to provide a degree of motivation for the more theoretical work to be described below, whose connections with applications are currently stimulating but speculative. The classic example of what we have in mind is the formation of turbulent Taylor vortices in Taylor-Couette flow, see Fenstermacher *et al.* [15]. The experimental apparatus comprises two coaxial cylinders, with fluid confined between them, see Figure 1.

In early experiments the outer cylinder is fixed and the inner one rotates; more recently both cylinders are free to rotate. If the outer cylinder is held fixed and the inner cylinder is slowly speeded up, then a variety of patterned flows is observed. For suitable gap widths the 'main sequence' of bifurcations takes the following form:

1. *Couette flow.* Laminar flow depending only on the radial position: featureless.

2. *Taylor vortices.* The flow stratifies into horizontal vortex layers, with the axial velocity components in neighbouring vortices acting in opposite directions.

3. *Wavy vortices.* The interface between neighbouring vortices develops a wavelike structure, which travels azimuthally without changing form.

4. *Modulated wavy vortices.* The wavy interface begins to oscillate axially as well as rotating azimuthally.

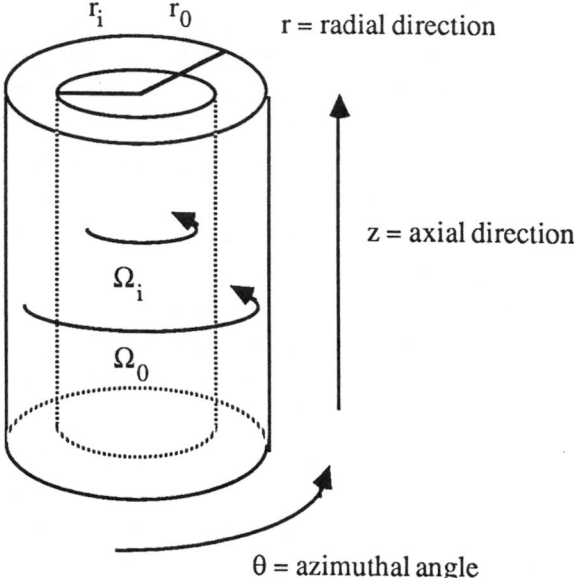

Figure 1: The Taylor-Couette apparatus. Here Ω_i, Ω_o denote the angular velocities of the inner and outer cylinders respectively, and r_i, r_o denote their radii.

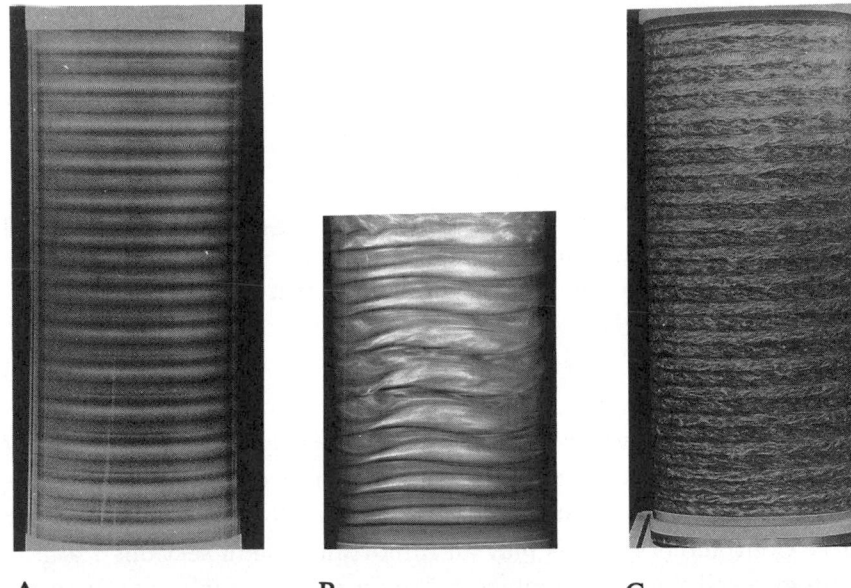

A   B   C

Figure 2: (a) Taylor vortices. (b) Modulated wavy vortices (near ends) and wavy turbulence (middle). (c) Turbulent Taylor vortices (b). [Pictures (a) and (c) courtesy of Harry Swinney and Randy Tagg.]

5. *Wavy turbulence.* The wavy interface becomes ever more complicated and turbulence sets in.

6. *Turbulent Taylor vortices.* The stratified pattern of Taylor vortices reasserts itself, but now the overall flow is turbulent and the interface between vortices is 'flat' only 'on the average', in the following sense: if the velocity of the flow is averaged over a period of time then along the 'boundary' between turbulent vortices, the average axial velocity is zero. The comparison between Taylor vortices and Turbulent Taylor vortices is shown in Figure 2.

3   Symmetry in Dynamics

Throughout most of this paper we concentrate on discrete dynamical systems:

$$x_{t+1} = f(x_t, \lambda). \tag{1}$$

Here $f : \mathbf{R}^n \times \mathbf{R}^r \to \mathbf{R}^n$ is a C^∞ mapping, and $\lambda \in \mathbf{R}^r$ is an r-dimensional bifurcation parameter. Usually either $r = 0$ (no bifurcation parameter) or $r = 1$. Continuous systems

$$\frac{\mathrm{d}x}{\mathrm{d}t} = f(x, \lambda) \tag{2}$$

have been widely studied, and the results of this section apply equally to them. Discrete systems arise as Poincaré sections of continuous ones, and continuous systems can be obtained by suspending discrete ones. Continuous systems play an important role in sections 7 and 8. More generally we may replace $\mathbf{R}^n$ by a C^∞ manifold, for example a circle or a torus.

A subset A of $\mathbf{R}^n$ (or more generally of the ambient manifold) is said to be an *attractor* for (1) if

(a) $f(A) = A$.

(b) A has a dense orbit.

(c) There exists a neighbourhood U of A such that for every $u \in U$ we have $\omega(u) \subset A$, where $\omega(u)$ is the ω-limit set of u.

A similar definition holds for the continuous case. For the definition of ω-limit set see Guckenheimer and Holmes [22].

Suppose that a compact Lie group G acts linearly on $\mathbf{R}^n$. Let $\gamma.x$ denote the image of $x \in \mathbf{R}^n$ under the action of $\gamma \in G$. We say that f is *G-equivariant* if

$$f(\gamma.x) = \gamma.f(x)$$

for all $x \in \mathbf{R}^n, \gamma \in G$. (More generally the actions of G on source and target may differ, but we encounter this case only briefly in section 6). We require the following terminology:

(a) The *G-orbit* of $x \in \mathbf{R}^n$ is

$$G.x = \{\gamma.x \mid \gamma \in G\}.$$

The concept of a G-orbit is important, because if $x(t)$ is any solution of the dynamical system 1 or 2 then so is $\gamma.x(t)$. That is, *all dynamical phenomena come as group orbits*. In particular if A is an attractor for the dynamics then so is $\gamma.A$ for all $\gamma \in G$. We say that $\gamma.A$ is *conjugate to A*.

(b) The *isotropy subgroup* of $x \in \mathbf{R}^n$ is

$$\Sigma_x = \{\gamma \in G \mid \gamma.x = x\}.$$

This measures the degree of symmetry of a point x. We also require a generalisation. If $x \in \mathbf{R}^n$ we let $O(x)$ be the orbit of x, and define the *(orbital) symmetry group of x* to be

$$\Delta_x = \{\gamma \in G \mid \gamma \overline{O(x)} = \overline{O(x)}\}.$$

If A is an attractor having a dense orbit $f^t(x)$, then the *symmetry group* of A, written Δ_A, is defined to equal Δ_x.

(c) If Σ is a subgroup of G (which henceforth we write as $\Sigma \le G$) then the *fixed-point space* of Σ is

$$\text{Fix}(\Sigma) = \{x \in \mathbf{R}^n \mid \sigma.x = x \text{ for all } \sigma \in \Sigma\}.$$

Fixed-point spaces are useful because of the following result. It is trivial to prove but fundamental and remarkably powerful:

Lemma 3.1
If f is G-equivariant and $\Sigma \le G$ then

$$f(\text{Fix}(\Sigma)) \subset \text{Fix}(\Sigma).$$

Proof
Let $x = \sigma.x$ for all $\sigma \in \Sigma$. Then $\sigma.f(x) = f(\sigma.x) = f(x)$.

Corollary 3.2
For every subgroup $\Sigma \le G$, the fixed-point space $\text{Fix}(\Sigma)$ is invariant under the dynamics of f in equations (1, 2).

Thus equivariant dynamical systems are naturally equipped with a set of invariant linear subspaces for the dynamics. Often their study can be simplified (at least in some respects) by restricting them to such subspaces. For example, finding solutions that break symmetry to a subgroup $\Sigma \leq G$ is equivalent to finding solutions that lie inside Fix(Σ).

If $g : \mathbf{R}^n \to \mathbf{R}$ is such that $f(\gamma.x) = f(x)$ for all $x \in \mathbf{R}^n, \gamma \in G$, we say that g is *G-invariant*. Invariance and equivariance impose strong restrictions on the form of mappings and functions, and these restrictions are basic to the theory of dynamics with symmetry. The main general result for C^∞ maps is due to the combined efforts of Hilbert, Schwarz, and Poénaru:

Theorem 3.3
(a) There exist finitely many G-invariants $\rho_1, \ldots, \rho_r$ such that every C^∞ G-invariant is of the form $g(\rho_1, \ldots, \rho_r)$ for a C^∞ map $g : \mathbf{R}^r \to \mathbf{R}$.
(b) There exist finitely many polynomial G-equivariant maps $h_1, \ldots, h_s :$ $\mathbf{R}^n \to \mathbf{R}^n$ such that every G-equivariant is of the form $g_1 h_1 + \ldots + g_s h_s$ where the g_j are C^∞ G-invariants.

We call $\rho_1, \ldots, \rho_r$ a *Hilbert basis* for the invariants and $h_1, \ldots, h_s$ a *system of generators* for the (module of) equivariants (over the invariants).

Warning
Computing the functions ρ_i and mappings h_j is not always easy, and is normally done on a case by case basis. There is no simple direct general method.

Example
Let $G = \mathbf{D}_n$ acting on $\mathbf{C}$. The group is generated by ζ and κ, where $\zeta.z = e^{2\pi i/n} z$ and $\kappa.z = \bar{z}$. A Hilbert basis for the invariants is given by the functions
$$z\bar{z}, \quad \mathrm{Re}(z^n),$$

and a system of generators for the equivariants is

$$z, \ \bar{z}^{n-1}.$$

See Golubitsky *et al.* [19] p.52.

4 Numerical Examples

In order to set the scene and exhibit some of the new phenomena that arise in the chaotic dynamics of symmetric systems, we discuss a variety of numerical experiments.

4.1 Cubic Logistic Map

The simplest nontrivial group action is $\mathbf{Z}_2$ acting on the line by $x \mapsto -x$. The *logistic map* $x \mapsto kx(1-x)$ has been widely studied, but is not equivariant for this $\mathbf{Z}_2$-action. However, a minor variant, the *cubic logistic map*, is. It takes the form $x \mapsto kx(1-x^2)$. Because this is odd in x, equivariance is immediate. See also Rogers and Whitley [30] and Chossat and Golubitsky [9].

Figure 3a shows a numerical bifurcation diagram for the cubic logistic map. It is obtained by fixing the value of k, iterating the map for long enough to remove transients, and then plotting the next 100 iterates of x. After that, k is increased slightly and the process is repeated.

We see an initial bifurcation from a $\mathbf{Z}_2$-symmetric fixed point $x = 0$ to an asymmetric fixed point (unlike the initial bifurcation in the logistic map, which is to a period-2 point). Then there is a period-doubling cascade (of asymmetric points), leading to chaos, with the usual periodic windows interspersed. A period-3 window is clearly visible.

After that, however, there is a new phenomenon: a sudden 'explosion' of the attractor, from an asymmetric interval contained in the positive half-line into a symmetric interval containing both negative and positive values of x. This is an example of a phenomenon that Grebogi *et al.* [21] call a *crisis*. What has happened is that two disjoint strange attractors have collided and 'fused' into a single

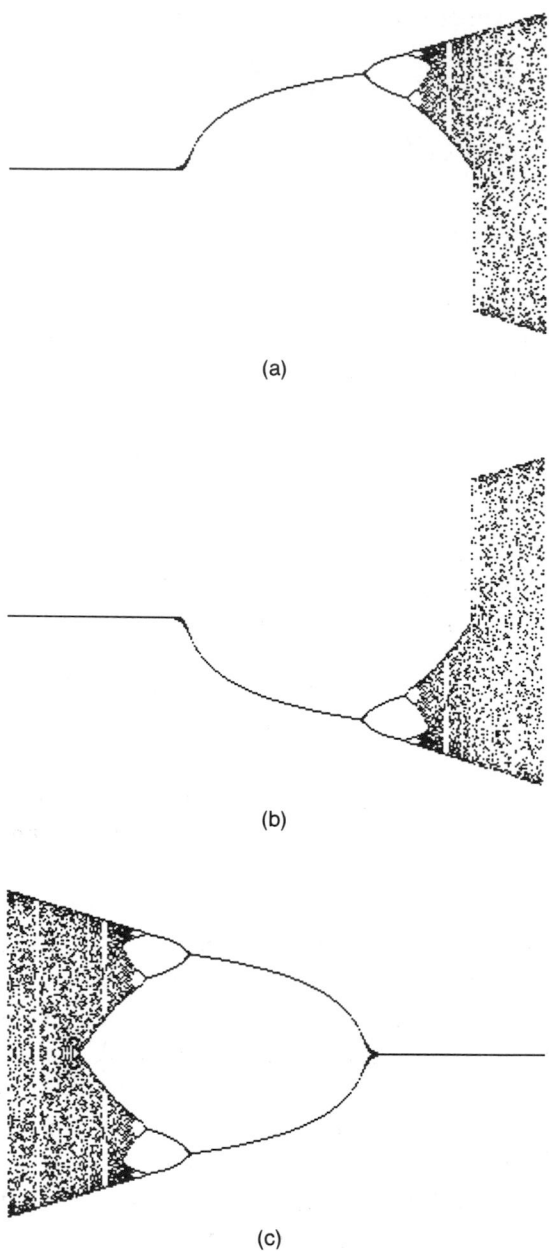

(a)

(b)

(c)

Figure 3: Bifurcation diagrams for cubic logistic map. (a) Start-ing from positive x. (b) Starting from negative x. (c) Showing all attractors.

attractor. Recall that an attractor, by definition, is *indecomposable*, that is, it cannot be split into two components, each of which is itself an attractor. This is a consequence of the 'dense orbit' condition (b) of section 3. It is important to realise that the symmetric attractor that exists after the crisis is not just a union of two separate asymmetric attractors: it is indecomposable and has its own dense orbit. To prove this we must use symbolic dynamics: see section 6 below; but it is easy to convince oneself numerically, as follows. Figure 3(a) is obtained by starting each iteration (for each k) at a positive value of x. If instead we start at a negative value, we obtain Figure 3(b). This is Figure 3(a) upside down. Indeed this is a consequence of the equivariance: if for some value of k we know that a susbset $A \subset \mathbf{R}$ is an attractor, then so is its conjugate $-A$. If we now combine the two bifurcation diagrams into one, Figure 3(c), we see that the sequence prior to the crisis actually involves pairs of symmetrically related attractors, and that these do indeed merge at the crisis point.

The curious feature of such an event is that it *increases* the symmetry of the attractor. Normally, bifurcations tend to *break* symmetry. It is true that reversing the bifurcation parameter trivially turns a symmetry-breaking bifurcation into a symmetry-increasing one, but generally there is a 'natural' direction for the bifurcation parameter, in which the degree of nonlinearity increases, often leading from order into chaos; and for non-chaotic states the bifurcations in that direction normally lead to less and less symmetry. However, when strange attractors are present, symmetry-increasing crises are common. They have been reported several times in the literature: see Kitano *et al.* [24], Sakaguchi and Tomita [31], Szabó and Tél [33], Chossat and Golubitsky [9],[10]. We postpone detailed discussion until section 5.2, and instead seek further examples.

4.2   Dihedral Symmetry

A rich source of examples is afforded by mappings of the plane equivariant under the natural action of the dihedral group $\mathbf{D}_n$. The equivariants for this group action are stated in section 3 above. Chossat and Golubitsky [10] have studied the family of $\mathbf{D}_n$-equivariant mappings

$$f(z, \lambda) = (\alpha u + \beta v + \lambda)z + \gamma \bar{z}^{n-1}. \qquad (3)$$

Field and Golubitsky [16] have produced high-resolution computer pictures of the attractors of 3 in which pixels are colour-coded according to how many times the point lands on them. This gives an impression of the invariant measure on the attractor. One of Field's pictures was used as the cover for the SERC Nonlinear Systems Initiative brochure [32], and a second such picture drawn by Andrew Cliffe appears inside.

We here fix $n = 3$. To set the scene we first give some samples of numerically plotted attractors in Figure 4. These (and most other attractors in this paper) were plotted using *Kaos*, a dynamical systems package for SUN workstations written by Swan Kim and John Guckenheimer.

Symmetry-increasing crises readily occur in these maps. An example is shown in Figure 5, where three copies of a $\mathbf{Z}_2$-symmetric attractor collide to create a $\mathbf{D}_3$- symmetric attractor.

As observed by Grebogi *et al.* [21], when a crisis occurs, there is a short period after the collision during which time the state x flips intermittently between the remnants of the original three attractors. This intermittency occurs here and also in other examples below.

Another example includes some regular dynamics as well, Figure 6(a-f). Here (a) shows not two invariant circles, as might appear from a static view, but a single $\mathbf{Z}_2$-symmetric quasiperiodic orbit. Successive points hop alternately between the two loops. In (b) this breaks down (in what appears to be the usual manner for invariant circles, but symmetrized) to form a curious attractor that appears to consist of thin curves but actually has extra fine structure not shown here. The conjugate attractors to this, shown in (d), merge in (e) to from a single fully symmetric attractor. By stage (f) this has acquired considerable extra structure.

To some extent the structure visible in these pictures is controlled by the fact that the mapping (3) is in general not a diffeomorphism. There is a tendency for the gross structure to be closely related to the *critical values* of the mapping, that is, the points x at which its Jacobian is singular. For example we show in Figure 7 the set of critical

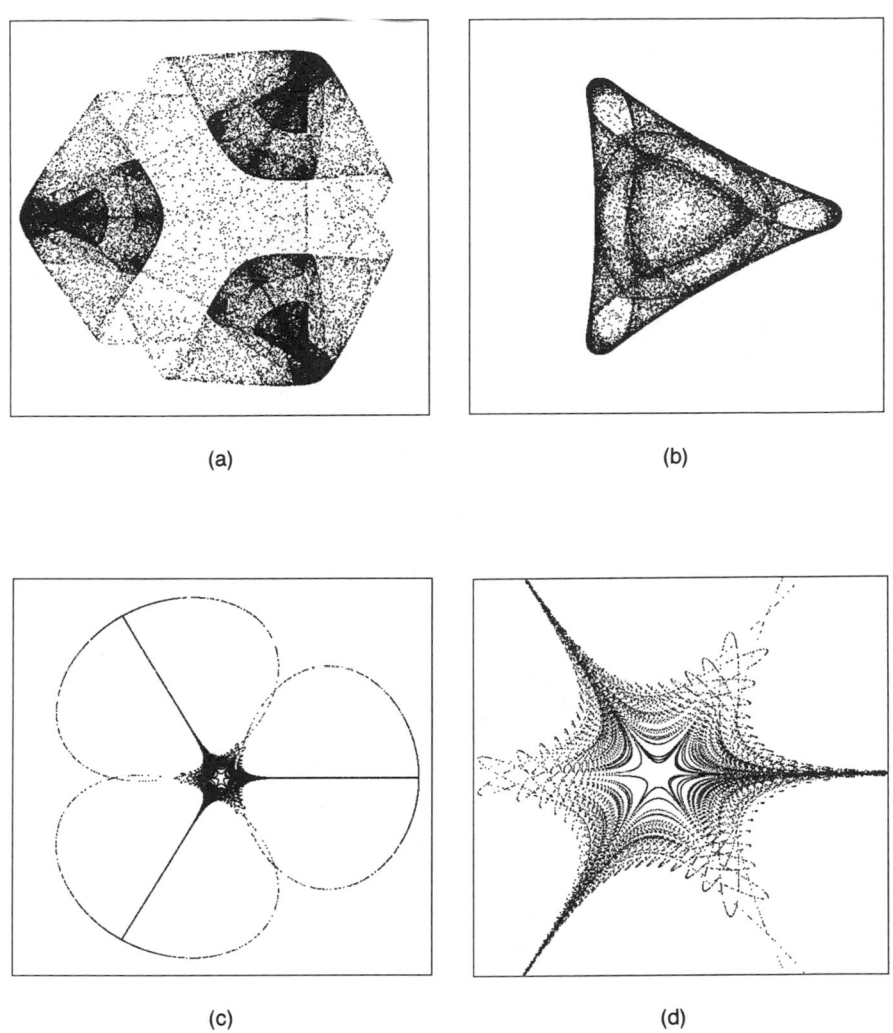

(a)

(b)

(c)

(d)

Figure 4: Attractors for the mapping (3) with the following parameter values: (a) $\alpha = -1.1, \beta = .17, \gamma = -.79, \lambda = 1.89$. (b) $\alpha = 1.8, \beta = 0, \gamma = 1.34164, \lambda = -1.89$. (c) $\alpha = -1, \beta = 0, \gamma = .944911, \lambda = 1$. (d) $\alpha = 1, \beta = 0, \gamma = .944911, \lambda = 1$. (Close up of centre of Figure 6(c).)

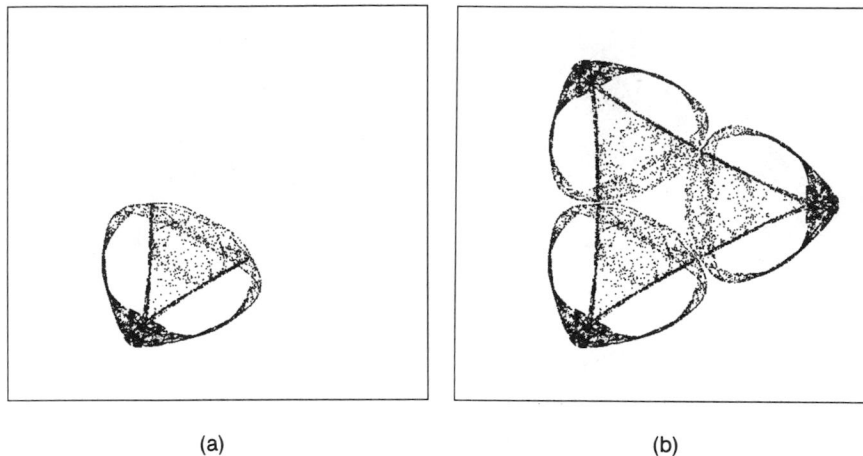

(a)   (b)

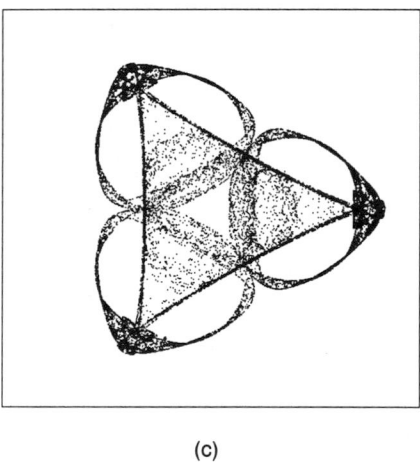

(c)

Figure 5: Symmetry-increasing crisis in the mapping (3). (a) $\alpha =$ $-1.1, \beta = .212, \gamma = .6, \lambda = 1.89$. (One attractor shown.) (b) $\alpha =$ $-1.1, \beta = .212, \gamma = .6, \lambda = 1.89$. (Three conjugate attractors shown.) (c) $\alpha = -1.1, \beta = .213, \gamma = .6, \lambda = 1.89$. (One attractor shown.)

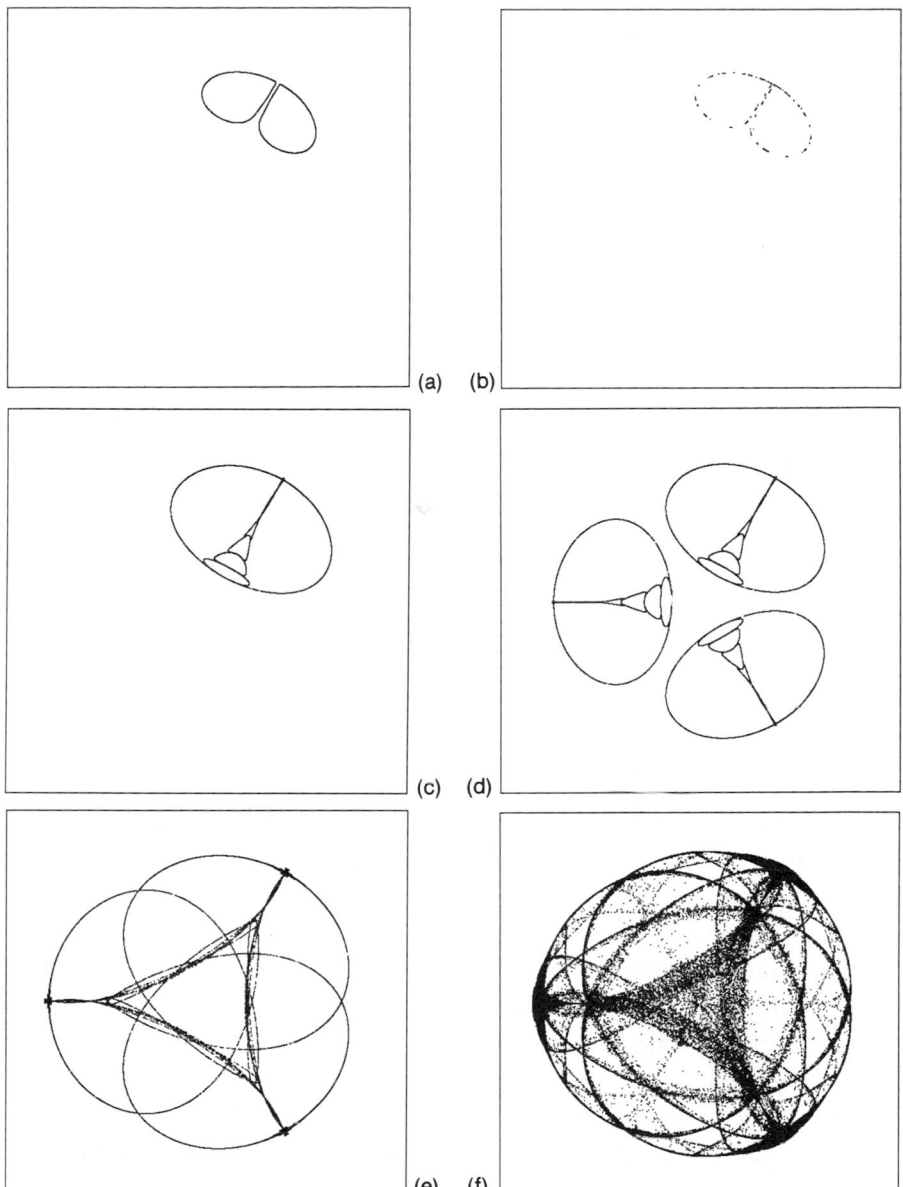

Figure 6: Bifurcations of attractors for the mapping (3) with param-
eter values (a) $\alpha = -.9, \beta = 0, \gamma = -.8$, and varying λ. (a) $\lambda = 1.15$.
(b) $\lambda = 1.155$.(c) $\lambda = 1.22$. (One attractor shown.) (d) $\lambda = 1.22$.
(Three conjugate attractors shown.) (e) $\lambda = 1.3$.(f) $\lambda = 1.35$.

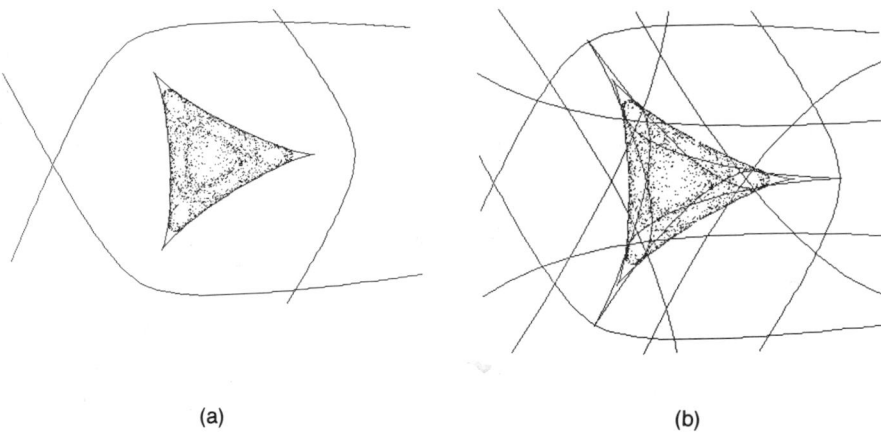

(a)   (b)

Figure 7: (a) Critical set, (b) critical set together with its first iterate, for the mapping (3) with parameter values as in Figure 4(b).

values, and its image under the map f, for the case corresponding to Figure 4(b).

Moreover, Golubitsky has given a simple argument to show that $\mathbf{D}_n$-symmetric *diffeomorphisms* of the plane cannot have fully symmetric strange attractors. (By Lemma 3.1 they must leave the axes of all reflections in $\mathbf{D}_n$ invariant; hence, being diffeomorphisms, they must permute the regions between them. The linear term close to the origin shows that this permutation must be either the identity or it must transpose opposite pairs. Neither can create a fully symmetric attractor.)

However, the critical values cannot be the whole story. As we show in section 5.1, much of the structure observed above for mappings also occurs for equivariant *diffeomorphisms* on $\mathbf{R}^4$, which do not *have* critical values! There is something of a mystery here.

Figure 7 illustrates a feature of the critical values that appears to be closely related to this mystery. Namely, the fold lines that determine the critical values can form trapping regions, confining the dynamics. (Observe how the cusped triangle is folded up and laid down within itself.) If no stable regular dynamics is possible within such a region, then a strange attractor must be present; moreover, it is likely to encounter the fold lines along parts of the edge of that region. So critical values provide a plausible mechanisms for the formation of attractors, especially strange attractors.

For mappings in the plane it often seems to be the case that the critical set, or its first or second iterate, lie very close to any attractors that occur. One question that this raises is the following: when can we obtain the attractor by iterating the critical set? Clearly this cannot happen for diffeomorphisms; but the numerical evidence is that it happens for a very broad class of maps.

4.3 Torus Maps

This section describes preliminary stages of some joint work with Golubitsky, which was stimulated by a morning television programme in Minneapolis on computer design of textiles. An especially rich source of symmetric strange attractors arises from equivariant torus maps $f : \mathbf{T}^2 \to \mathbf{T}^2$ where $\mathbf{T}^2$ is the 2-torus. Here we discuss the $\mathbf{D}_4$-equivariant family of maps

$$\begin{pmatrix} x \\ y \end{pmatrix} \mapsto \begin{pmatrix} (\lambda + \alpha\cos(2\pi y))\sin(2\pi x) + \beta\sin(4\pi x) + \gamma\sin(6\pi x)\cos(4\pi y) + kx \\ (\lambda + \alpha\cos(2\pi x))\sin(2\pi y) + \beta\sin(4\pi y) + \gamma\sin(6\pi y)\cos(4\pi x) + ky \end{pmatrix}$$

$$(4)$$

where $\alpha, \beta, \gamma, \lambda$ are real parameters and k is an integer parameter. The form of this mapping is deduced from a Fourier expansion, together with linear terms, by applying the generators of the symmetry group $\mathbf{D}_4$ of the unit square defining a fundamental domain for the torus. Only 'low order' terms of the Fourier series are retained here.

We content ourselves with some examples revealing the extraordinary richness of these maps.

The sequence in Figure 8 illustrates a symmetry-increasing crisis. In (a) we see one attractor, symmetric about a diagonal. In (b) we see its four conjugates. In (c) not all of the attractor has yet appeared, because of intermittency. The figure shows three pairs of sets, each

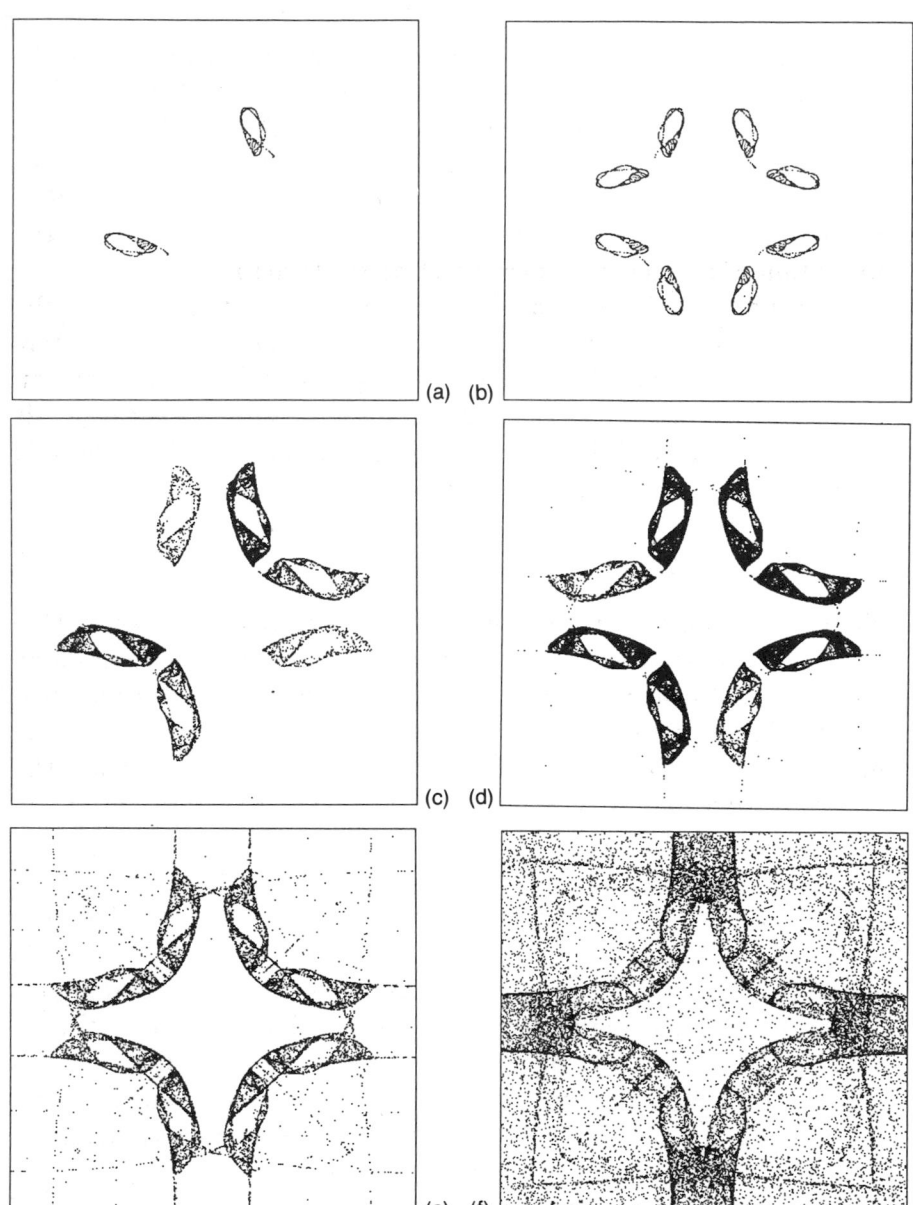

Figure 8: Bifurcating sequence of equivariant torus maps, equation (4). Here $\beta = .22, \gamma = .05, \lambda = .25, k = 2$ throughout. (a) $\alpha = .15$, one attractor shown. (b) $\alpha = .15$, four conjugate attractors shown. (c) $\alpha = .168$. (d) $\alpha = .168$, more iterations than (c). (e) $\alpha = .17$. (f) $\alpha = .19$.

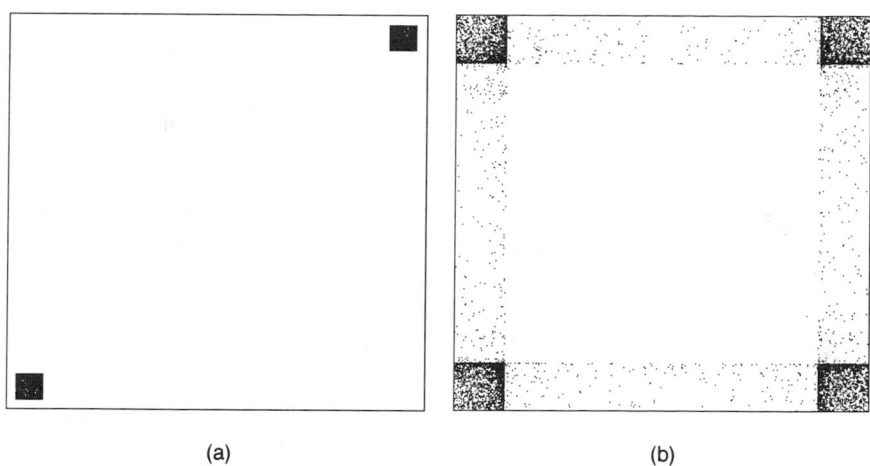

(a)   (b)

Figure 9: Symmetry-increasing crisis in equivariant torus maps. Here $\alpha = 0, \gamma = 0, \lambda = 2, k = 0$ throughout. (a) $\beta = -1.25$. (b) $\beta = -1.2502$.

with a different density of points; the fourth is missing. If the plot is continued for long enough we obtain (d), and the fourth pair of sets appears. The density still appears non- uniform and it takes a long time to even out. By (e) and (f) the attractor has merged into an apparently connected set.

A different symmetry-increasing collision is shown in Figure 9. Here the initial symmetries are reflection in the diaognal and rotation through π; the final symmetries are the whole of $\mathbf{D}_4$.

The rich variety of attractors possessed by these torus maps is illustrated in Figure 10.

We do not study the dynamics of these maps in any detail here. However, it would be interesting to try to extend some of the ideas of section 6 below from circle maps to torus maps.

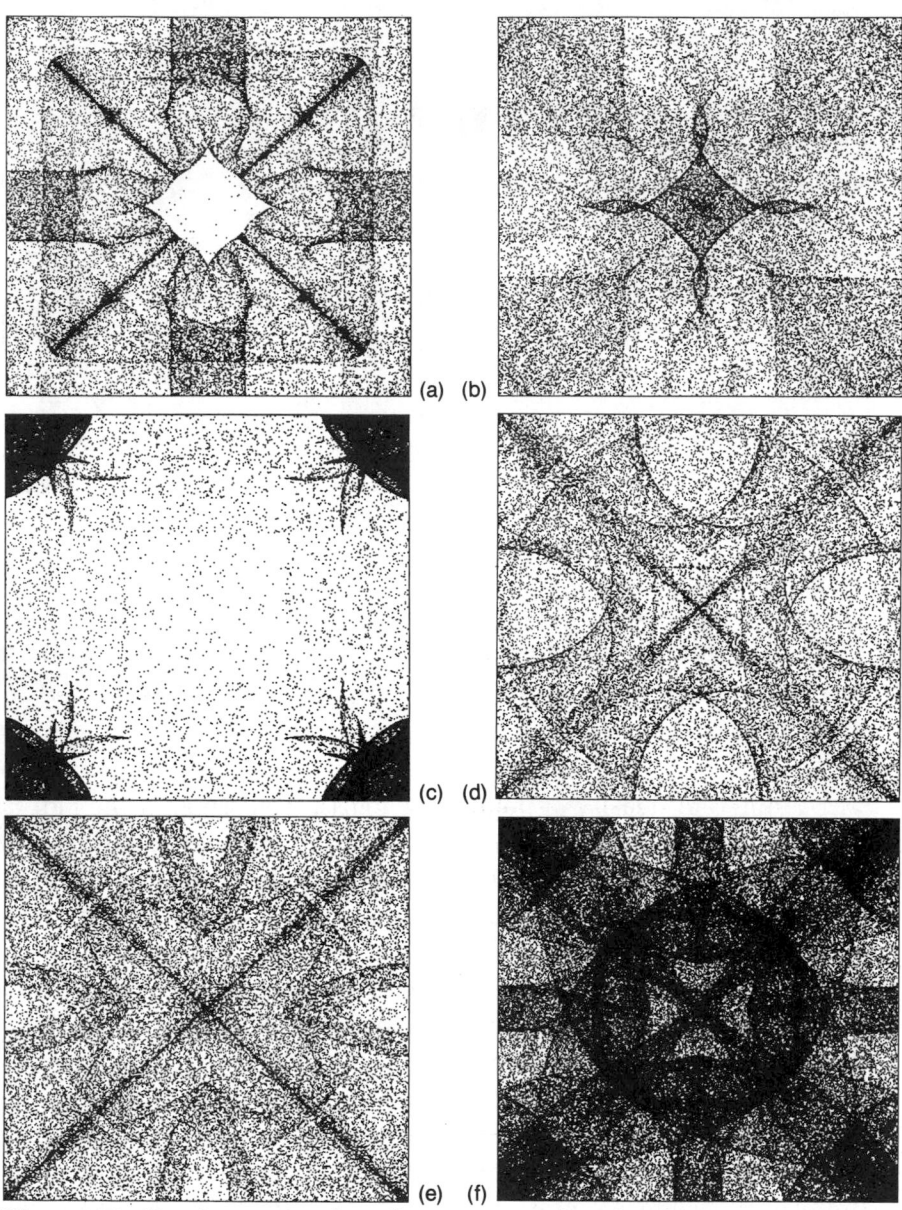

(a) (b)

(c) (d)

(e) (f)

Figure 10: Further examples of attractors of equivariant torus maps.
(a) $\alpha = .2, \beta = .23, \gamma = .1, \lambda = .25$, $k = 2$. (b) $\alpha = -.7$, $\beta = .4$,
$\gamma = .1$, $\lambda = .25$, $k = 1$. (c) $\alpha = 0$, $\beta = -.13$, $\gamma = .26$, $\lambda = -.7$, $k = 5$.
(d) $\alpha = .55$, $\beta = .33$, $\gamma = .19$, $\lambda = .15$, $k = 1$. (e) $\alpha = .55$, $\beta = .33$,
$\gamma = .19$, $\lambda = .05$, $k = 1$. (f) $\alpha = .42$, $\beta = .33$, $\gamma = .19$, $\lambda = .05$,
$k = 1$.

5   Useful Theorems

In this section we collect together various useful results from the
'folklore' of symmetric chaos. The first concerns the question of
obtaining *diffeomorphisms* that exhibit 'exotic' behaviour. The second explains how symmetry-increasing crises come about. The third
relates to the possible isotropy subgroups of non- point attractors,
especially strange ones.

5.1   Shadow Lift Lemma

Most of the examples that we discuss in this paper are mappings
rather than diffeomorphisms, that is, they are usually non- invertible. However, Poincaré maps derived from continuous systems are
always diffeomorphisms. Inasmuch as much of the observed structure in strange attractors of maps seems to be related to the critical
values (see section 4.2) one may ask whether mappings are really the
appropriate objects to study.

   In fact they are: we now show that structure observed in mappings in dimension n can be found in diffeomorphisms in dimension
$2n$. We begin with some definitions.

   Suppose that $f : \mathbf{R}^n \to \mathbf{R}^n$ is a mapping. Let (y_t) be a sequence
of points in $\mathbf{R}^n$ and let $\epsilon > 0$. We say that (y_t) $\epsilon-shadows$ f for
$t_0 \leq t \leq t_1$, or is an $\epsilon-$ *pseudoorbit* of f, if for all t with $t_0 \leq t \leq t_1$
we have $\|y_{t+1} - f(y_t)\| < \epsilon$. That is, the points of the sequence (y_t)
for t in the interval considered look like an orbit of f, except for a
perturbation of size $< \epsilon$ at each iteration. The *shadowing lemma*
of Bowen [4], see also Guckenheimer and Holmes [22] p.251, states
that under such conditions, then for any $\delta > 0$ we can choose ϵ small
enough such that there exists an orbit (x_t) of f with $\|y_t - x_t\| < \delta$.
That is, an $\epsilon-$pseudoorbit lies within δ of a true orbit. This result
is often invoked to justify the use of computer graphics to display
the structure of strange attractors, when by definition trajectories
on such attractors are subject to exponential divergence and hence
to numerical instability.

   Our next result explains the relevance of phenomena observed in
mappings to phenomena that can occur in diffeomorphisms.

Lemma 5.1 *(Shadow- Lifting)*
Suppose that $f : \mathbf{R}^n \to \mathbf{R}^n$ is a G−equivariant mapping, having a compact attractor A. Let and $t_0 < t_1$ be given. For any $\epsilon > 0$ define the mapping $F : \mathbf{R}^n \times \mathbf{R}^n \to \mathbf{R}^n \times \mathbf{R}^n$ by

$$F(x,y) = (\epsilon y + f(x), x).$$

Then:
(a) F is a diffeomorphism.
(b) F is equivariant for the diagonal action $\gamma.(x,y) = (\gamma.x, \gamma.y)$ of G on $\mathbf{R}^n$ of $\mathbf{R}^n \times \mathbf{R}^n$.
(c) If A lies within a ball centre the origin of radius B, then for any $y_0 \in A$ the sequence $y_t = \pi_1(F^t(y_0))$ ϵB−shadows f, where π_1 is projection onto the first factor $\mathbf{R}^n$ of $\mathbf{R}^n \times \mathbf{R}^n$.

Proof
(a) The inverse of F is G where

$$G(x,y) = (y, \frac{x - f(y)}{\epsilon}).$$

(b) Let $\gamma \in G$. Then $\gamma.F(x,y) = \gamma.(\epsilon y + f(x), x) = (\gamma.(\epsilon y + f(x)), \gamma.x) = (\epsilon \gamma.y + \gamma.f(x), \gamma.x) = (\epsilon \gamma.y + f(\gamma.x), \gamma.x) = F(\gamma.x, \gamma.y)$.
(c) By definition we have $\|\pi_1(F(x,y) - f(x)\| = \|\epsilon y\|$, where $y \in A$. The result is now immediate.

We call F the ϵ−*lift* of f. Applied to the logistic map it produces the Hénon map; applied to the odd logistic map it produces an 'approximate Poincaré map' used by Holmes [23]. If we take ϵ small enough, then the shadow lifting lemma shows that projections onto the first factor of attractors of F are arbitrarily close to attractors of f. In practice, even when ϵ is quite large a lot of the 'strange' structure of the attractors of f can be observed, although not at precisely the same parameter values for $\alpha, \beta, \gamma, \lambda$: see Figure 11.

Moreover, it is easier to find interesting behaviour for mappings in low dimensions than it is for diffeomorphisms. In particular, although as we have already observed no $\mathbf{D}_3$−equivariant mapping of

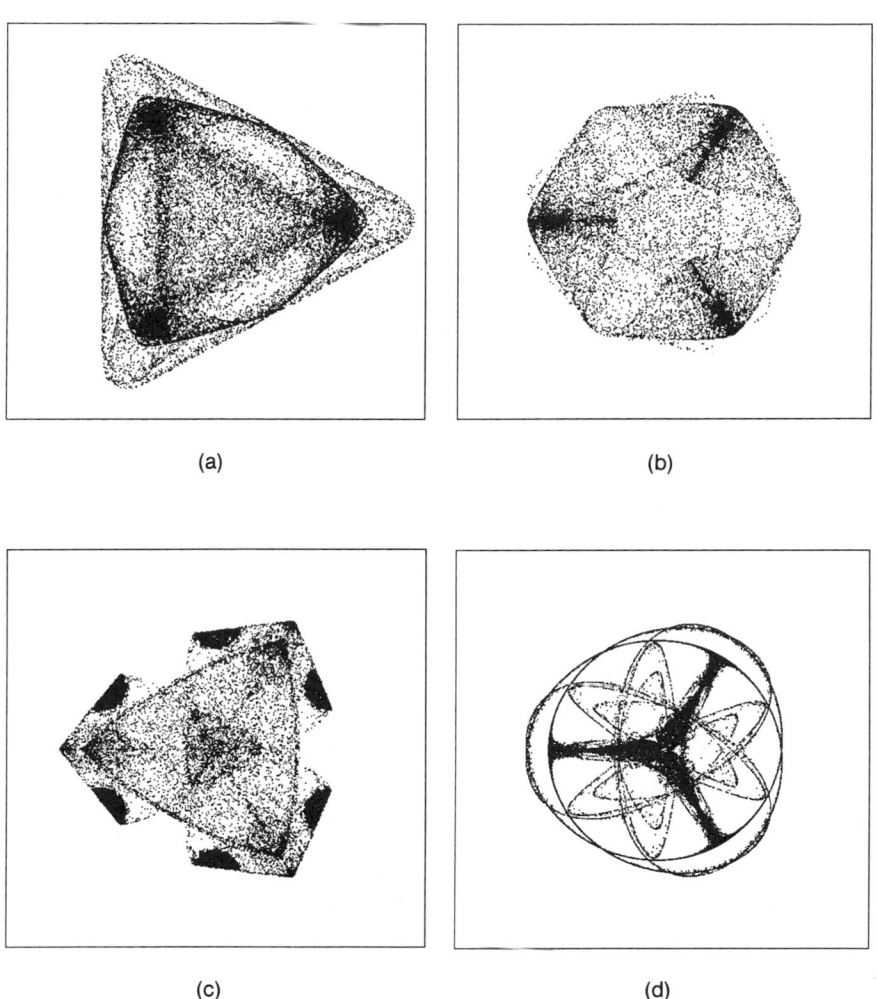

(a)

(b)

(c)

(d)

Figure 11: Projections into $\mathbf{R}^2$ of ϵ−lifts to $\mathbf{R}^4$ of $\mathbf{D}_3$−equivariant mappings (3). (a) $\alpha = -1.1, \beta = .213, \gamma = .6, \lambda = 1.7, \epsilon = .2$. (b) $\alpha = -1.1, \beta = .17, \gamma = -.79, \lambda = 1.5, \epsilon = .2$. (c) $\alpha = -1.1, \beta = .213, \gamma = -.6, \lambda = 1.8, \epsilon = .25$. (d) $\alpha = -.91, \beta = 0, \gamma = -.8, \lambda = 1, \epsilon = .205$.

the plane can have a fully symmetric attractor, this can occur for $\mathbf{D}_3$−equivariant mappings on $\mathbf{R}^4$ (modulo the conjecture that what appear on the computer screen to be strange attractors really *are* strange attractors).

5.2   Symmetry-increasing Crises

In this section we 'explain' why collisions of strange attractors in equivariant systems tend to increase symmetry, following arguments of Chossat and Golubitsky [10]. Let $f : \mathbf{R}^n \to \mathbf{R}^n$ be a G-equivariant mapping and let A be an attractor for f, satisfying properties (a-c) of section 3.

Proposition 5.2 *(Symmetry-Increasing Crises)*
Let $f : \mathbf{R}^n \to \mathbf{R}^n$ be a G-equivariant mapping and let A be an attractor for f in the above sense. Suppose that for some $\gamma \in G$ we have

$$A \cap \gamma(A) \neq \emptyset$$

Then $\gamma(A) = A$.

Proof
Let $T = \{f^n(z)\}$ be a dense orbit in A, and suppose we can show that

$$\omega(x) = A \text{ for every } x \in T. \tag{5}$$

Let U be an open neighbourhood of A satisfying (c) above and define $Q = \gamma(U) \cap U$. By assumption U is open and non- empty. Since T is dense in A and $Q \cap A \neq \emptyset$, we have $Q \cap T \neq \emptyset$. Let $x \in Q \cap T$. Then $\omega(x) \in A$. Since $x \in \gamma(U)$ we also have $\omega(x) \subset \gamma(A)$. Therefore $A \subset \gamma(A)$. Reversing the roles of A and $\gamma(A)$ we obtain $\gamma(A) \subset A$. Thus $A = \gamma(A)$.

It remains to prove 5. Let $x \in T$ and let $a \in A$. We show $a \in \omega(x)$. If $a \notin T$ then by density of T, $a \in \omega(x)$. To deal with the case when $a \in T$ we choose u so that the orbit of u is T, and then it is enough to show that $\omega(u)$ includes T, because $\omega(x) = \omega(f^k(u)) = f^k(\omega(u))$ for some k. But this follows from the recurrence property $f(A) = A$, which implies that there is a sequence of iterates $f^{t_n}(u) \to u$.

This proposition implies that when two conjugate attractors A and $\gamma(A)$ collide, then the result includes γ as a symmetry. Moreover, if A already has symmetry Δ_A then it is easy to check that the new combined attractor has symmetry $\langle \Delta_A, \gamma \rangle$ generated by Δ_A and γ. Thus it is *not* a surprise to find that crises increase symmetry.

5.3 Isotropy

For a given group G acting on $\mathbf{R}^n$ by a given representation, it is known that in general the isotropy subgroups of single points do not comprise all possible (closed) subgroups of G. For example the isotropy subgroups of $\mathbf{D}_n$ acting in its standard representation on $\mathbf{R}^2$ are (up to conjugacy) $\mathbf{1}$, $\mathbf{Z}_2$, and $\mathbf{D}_3$; the subgroup $\mathbf{Z}_3$ cannot occur.

What about isotropy subgroups of *subsets* of $\mathbf{R}^n$? In particular, can the isotropy subgroups of non-point attractors differ from those of single points?

Until recently the only isotropy groups of attractors encountered in numerical experiments on the family (3) were $\mathbf{1}$, $\mathbf{Z}_2$, $\mathbf{D}_3$ — the same as for points. It was unknown whether $\mathbf{Z}_3$ can occur. The answer (numerically, at least) is provided by Figures 12, 13, which we stumbled upon by accident. This attractor emerges only after a long transient, which looks very similar to the attractors that occur at nearby parameter values. It seems to be rather fragile.

We shed a little more light on the question by sketching a construction showing that when G acts faithfully, every finite cyclic subgroup can be an isotropy group of an attractor of some G-equivariant mapping.

Let G be compact, acting linearly and (without loss of generality) orthogonally. Consider a cyclic group $H = \mathbf{Z}_m = \langle 1, \gamma, \ldots, \gamma^{m-1} \rangle$. Suppose that D_0 is a fundamental domain for the H-action on the unit sphere $\mathbf{S}^{n-1}$, so that the cone D on D_0 is a fundamental domain for H acting on $\mathbf{R}^n$. That is,

$$\mathbf{R}^n = \bigcup_{k=0}^{m-1} \gamma^k(D), \tag{6}$$

the union is disjoint except for the origin, and D contains a non-empty open set U.

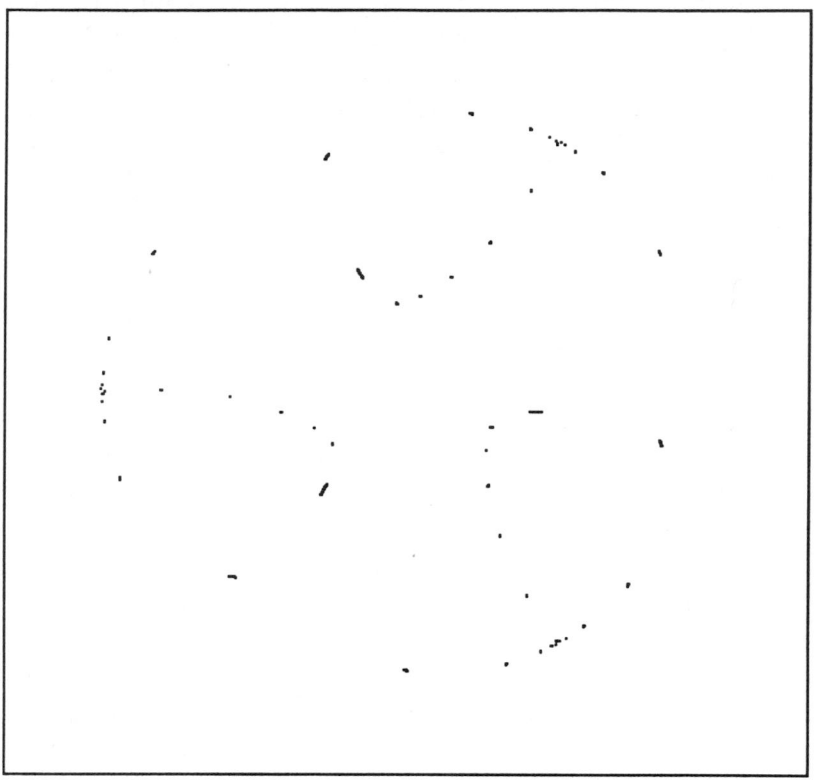

Figure 12: $\mathbf{Z}_3$-symmetric attractor for the mapping (3) with parameter values $\alpha = -.9, \beta = 0, \gamma = .8, \lambda = 1.25002$. Although the attractor here looks periodic (with fairly large period) close-ups of individual specks (Figure 13) reveal a structure similar to that of the Hénon attractor.

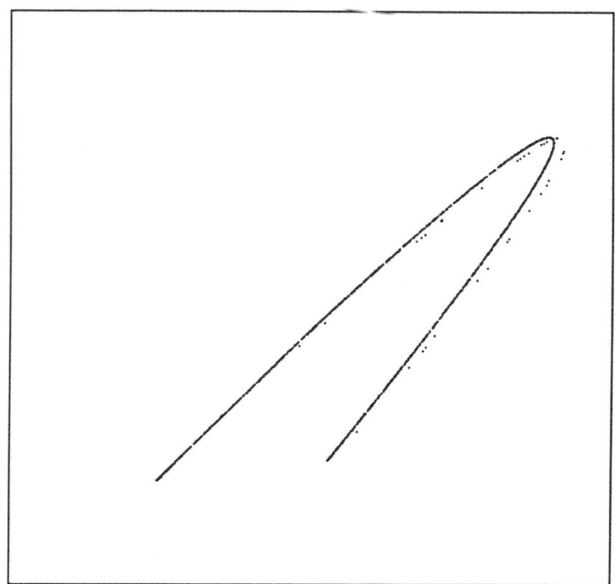

Figure 13: Close-up of speck in Figure 12.

Choose a map $\phi : U \to U$ such that ϕ^m has an attractor A, for example the Hénon map restricted to a suitable disc. (The recurrence of the Hénon map together with its lack of periodicity presumably implies that this attractor is the same as that of ϕ, but we do not need to know this here.) Define

$$V = \bigcup_{k=0}^{m-1} \gamma^k(U) \tag{7}$$

and define $\hat{\phi} : V \to V$ as follows. If $x \in \gamma^k(U)$ then $\hat{\phi}(x) = \gamma.\phi(x) \in \gamma^{k+1}(U)$. Then extend $\hat{\phi}$ to a continuous mapping on $\mathbf{R}^n$ and average over G to obtain an equivariant extension, which we also call $\hat{\phi}$. The map $\hat{\phi}$ is similar to ϕ except that it moves 'one place on' from $\gamma^k(U)$ to $\gamma^{k+1}(U)$. Thus $\hat{\phi}^m(u) = \phi^m(u)$ for all $u \in U$. Therefore $\hat{\phi}$ has an attractor B formed by m disjoint copies of A,

$$B = \bigcup_{k=0}^{m-1} \gamma^k(A), \tag{8}$$

and $\gamma(B) = B$. We conclude that the isotropy subgroup of B contains H.

6   Symmetry-locking and Equivariant Kneading Theory

We now describe some initial steps towards a rigorous theory of equivariant chaos. Krupa and Roberts [25] have developed equivariant versions of symbolic dynamics and kneading theory to study circle maps that are equivariant under the action of the dihedral group $\mathbf{D}_n$. In this section we summarize some of their ideas, which are among the most substantial rigorously proved results on the dynamics of maps with symmetry.

Following standard practice they employ circle maps as models for the following dynamical situation. Consider a dynamical system with $\mathbf{Z}_n$ symmetry, in which there is a rotating wave with $\tilde{\mathbf{Z}}_n$ symmetry. Assume that this undergoes Hopf bifurcation to create a torus with $\mathbf{Z}_n$ symmetry, and that this torus subsequently breaks up. If we consider a Poincaré section then we have a $\mathbf{Z}_n$-equivariant Poincaré mapping ϕ defined on an annulus (a neighbourhood of the cross-section of the torus, which is a circle). Assume that the torus is strongly stable, so that ϕ is strongly contracting from the annulus to the circle. Then we may approximate the important dynamics of ϕ by its restriction to the circle. After the torus breaks down this approximation remains valid, and we can expect to observe chaotic dynamics in the circle map.

To keep the discussion specific, consider the family of *Arnold maps*

$$f_{\omega\mu} : \mathbf{S}^1 \to \mathbf{S}^1$$

defined by

$$f_{\omega\mu}(x) = x + \omega - \frac{\mu}{2\pi k} \sin 2\pi k x \ (\text{mod } 1). \tag{9}$$

This has a cyclic symmetry group $G = \mathbf{Z}_k$ defined by $x \mapsto x + \frac{s}{k}, s = 0, \ldots, k - 1$. We study how the orbital symmetry group Δ_x varies with the parameters ω and μ for x belonging to attractors of f. Here $S^1 = R/Z$ is the circle.

For certain values of ω the map (9) has a further symmetry. For example, if $\omega = 0, \frac{1}{2}$, then $f_{\omega\mu}$ is equivariant under $x \mapsto -x$. More generally, if $\omega = \frac{r}{2k}$ then

$$f_{\omega\mu}(-x) = -f_{\omega\mu}(x) + \frac{r}{k} \tag{10}$$

for $r = 0, \ldots, 2k - 1$. This can be thought of as D_k- symmetry in which the group actions on source and target are different. In particular it implies that all points $x = \frac{s}{2k}$ are periodic. These periodic orbits have symmetry group $D_{(k,q)}$ where $\frac{p}{q} = \frac{r}{2k}$ in lowest terms.

In Figure 14 we give some examples of how the maps $f_{\omega\mu}$ bifurcate for various fixed rational ω, as μ varies. In cases (a) and (b) we have chosen ω and k to produce the D_k symmetry just mentioned. In contrast, in case (c) there is only Z_k symmetry. The figures are drawn in pairs. In the first of each pair, μ runs horizontally and the coordinate x on the circle runs vertically from 0 to 1, so top and bottom are identified. In the second polar coordinate version, μ is the radial coordinate and the angular variable is $2\pi x$. The figures are drawn by very slowly 'ramping' the value of μ, that is, increasing it by 0.0001 at each iteration; meanwhile the new value of x is given by formula (9). The entire bifurcation diagram is obtained as a trajectory of the resulting dynamical system.

In (a) we have $k = 5$ and the overall symmetry is D_5. We first see a period-10 point whose orbit has D_5 symmetry. The first bifurcation is period-doubling to a period-20 point, still with D_5 symmetry. The symmetry breaks to Z_5 at the next bifurcation (though this is hard to see because of the small scale). It is followed by a period-doubling cascade that preserves the remaining symmetry. Then a mini-crisis restores the symmetry to D_5, and the attractor grows without further crises until it occupies the entire circle. This chaotic band is punctuated by periodic windows, in which the symmetries also change. Then the chaos vanishes leaving a period-5 point with Z_5 symmetry: this undergoes its own period-doubling cascade, still with Z_5 symmetry. Another crisis again restores D_5 symmetry. The second chaotic band, also containing periodic windows, terminates

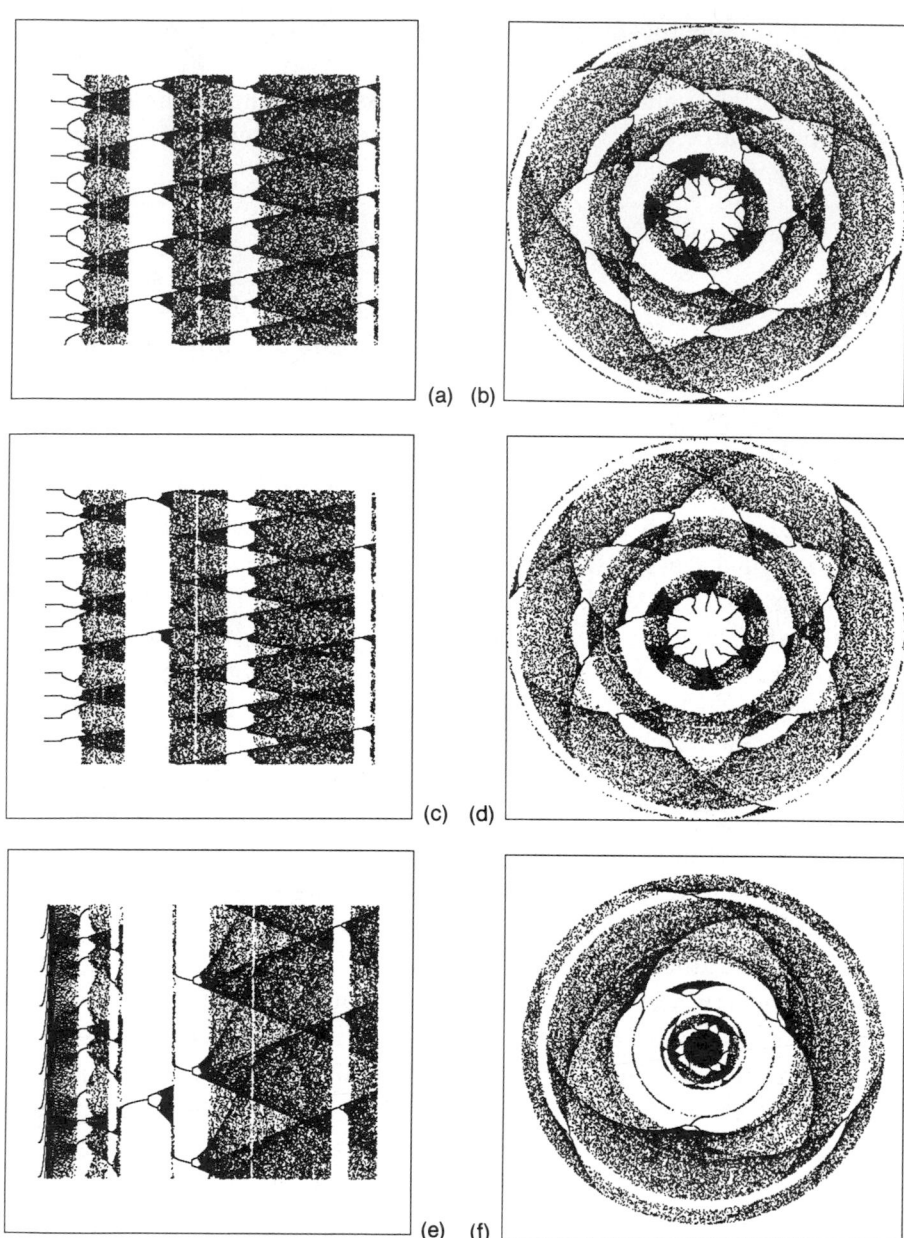

Figure 14: Examples of symmetry-locking in torus maps, shown in cartesian and polar bifurcation diagrams. Here μ runs from 0 to 10. (a) $k = 5, \omega = \frac{3}{10}$. (b) $k = 6, \omega = \frac{5}{12}$. (c) $k = 3, \omega = \frac{7}{8}$.

by the creation of a period-10 point with $\mathbf{D}_5$ symmetry, followed by a symmetry-breaking pitchfork bifurcation to $\mathbf{Z}_5$ symmetry... and so on.

Case (b), in which $k = 6$, is slightly different. Here the initial bifurcation is a symmetry-breaking pitchfork, starting a period- doubling cascade, and at this stage the orbit has $\mathbf{D}_3$ symmetry. (As well as the obvious translations there is a combination of a reflection and a translation.) There is a tiny local crisis that partially restores the symmetry, and a second crisis to restore full $\mathbf{D}_6$ symmetry. The chaotic band terminates in a fixed point with $\mathbf{Z}_2$ symmetry. It undergoes a symmetry-breaking pitchfork to trival symmetry and then a period- doubling cascade, still with trivial symmetry, until two successive crises restore $\mathbf{Z}_2$ and then $\mathbf{D}_6$ symmetry. The second chaotic band terminates in a $\mathbf{D}_6$-symmetric periodic point, which undergoes a pitchfork to $\mathbf{Z}_6$ symmetry and period-doubles; again two successive crises restore the symmetries. The third chaotic band terminates in a periodic point with $\mathbf{Z}_3$ symmetry... and so on.

Case (c) is included merely to show that when $\omega = p/q$ and $q \neq 2k$, new phenomena occur as well as those just described. Some of the structure visible is an artefact of the drawing procedure: in particular the dark band towards the left-hand end mainly corresponds to quasiperiodic motion in which the whole circle becomes the attractor. But again we see broken symmetries and symmetry-increasing crises.

Krupa and Roberts [25] systematize these observations and prove theorems that limit or pin down the possible phenomena. First, lift $f_{\omega\mu}$ to a map

$$F_{\omega\mu} : \mathbf{R} \to \mathbf{R}$$

with the same formula as $f_{\omega\mu}$ (but not taken (mod 1)). Define the *rotation interval* $\rho(f)$ as follows. If $x \in \mathbf{S}^1$ then

$$\rho(x, f) = \{z \in \mathbf{R} \mid z \text{ is a limit point of } [\frac{F^n(y) - y}{n}]_{n=1}^{\infty}\}. \quad (11)$$

where y is a lift of x to $\mathbf{R}$, and then set

$$\rho(f) = \bigcup_{x \in \mathbf{S}^1} \rho(x, f) \quad (12)$$

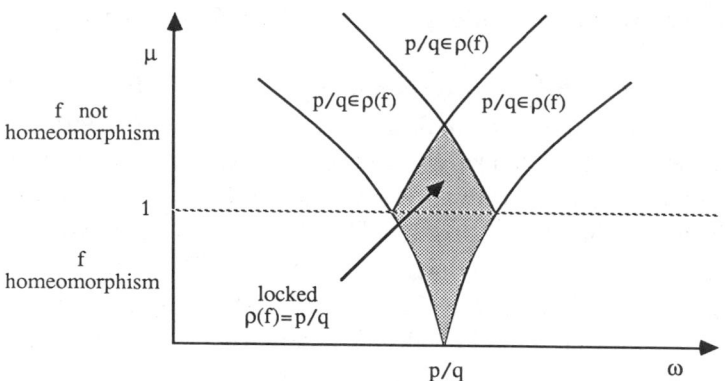

Figure 15: Schematic of an Arnold tongue for the torus map (9).

which is known to be a closed interval. The end-points of $\rho(f)$ vary continuously with f in the C^0 topology. If f is monotonic, that is, a homeomorphism, then the rotation intervals are points and there is a unique rotation number. In general, if $\rho(f)$ is a single point, the we say that f is *frequency- locked.*

Figure 15 shows the familiar 'Arnold tongue' picture for frequency-locking. Figures 14(a-d) are obtained by increasing μ vertically from a rational value p/q of ω.

The following theorem of Krupa and Roberts is based on ideas of Mackay and Tresser [26].

Theorem 6.1 *(Symmetry- Locking)*
Suppose that f is $\mathbf{Z}_n$- (respectively $\mathbf{D}_n$-) equivariant, and frequency-locked with $\rho(f) = \frac{p}{q}$ in lowest terms. Then for all $x \in \mathbf{S}^1$ the orbital symmetry group Δ_x is a subgroup of $\mathbf{Z}_{(k,q)}$ (respectively $\mathbf{D}_{(k,q)}$).

In particular, symmetry-increasing crises cannot occur when f is frequency-locked. For this reason we say that f is *symmetry-locked* when the hypotheses of the theorem hold. More generally f is symmetry-locked in some region of parameter space if there exists a subgroup Δ of G such that $\Delta_x \subset \Delta$ for all x when the parameters lie in that region. However, even when f is symmetry-locked there may exist windows in which symmetry is first broken and then restored via a crisis.

Crises happen on the boundary of the shaded region in Figure 15. The theorem is sharp in the sense that if ω is irrational and $\omega \in \rho(f)$, then there exists x such that $\Sigma_x = \mathbf{Z}_k$ (respectively $\mathbf{D}_k$).

Now we sketch how the existing theory of symbolic dynamics of one- humped maps (such as the logistic map) can be applied to this equivariant case. (See Devaney [14] for an introduction to symbolic dynamics.) Consider a parameter window with ω constant and μ increasing, such that there is a saddle-node bifurcation to a period q point. This occurs when the graph of f^q becomes tangent to the diagonal, as in Figure 16(a). This bifurcation creates an invariant interval I on which the graph of f^q resembles a one-humped map. The dynamics of the system is then largely controlled by that of the one-humped map; however, in addition we must take account of the manner in which f permutes the intervals $I, f(I), \ldots, f^{k-1}(I)$ and the symmetries. These latter are essentially combinatorial issues and do not pose any conceptual difficulties (although keeping the details straight is far from easy!). Figures 16(b-e) show the 'typical' progression of the bifurcation as μ increases: appropriate 'semi-global' monotonicity conditions are required to ensure that these pictures are actually valid, but in examples those conditions commonly do hold. The result is that f^q passes through an entire bifurcation sequence for the one-humped map: initial period-doublings in a Feigenbaum cascade, followed by chaos, with periodic windows, Šarkovsii ordering, and so forth. See Collet and Eckmann [11] or Devaney [14].

Again we observe the occurrence of symmetry-locking. Throughout Figures 16(a-d) the interval that defines the horizontal edge of the boxes is invariant, ceasing to be invariant only at the stage shown in Figure 16(d). While that interval remains invariant it is not possible to regain (through crises) any symmetries lost at the initial saddle-node bifurcation. Thus we have a *saddle-node symmetry-locking window*. The location of such windows can be found using symbolic

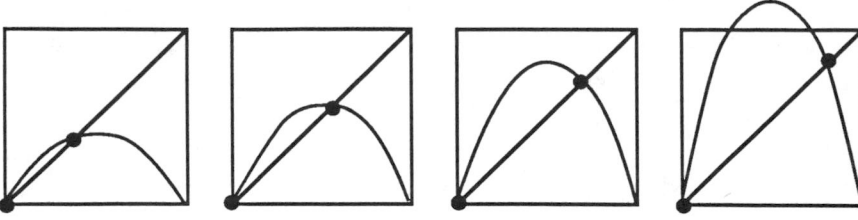

Figure 16: A saddle-node bifurcation in a circle map.

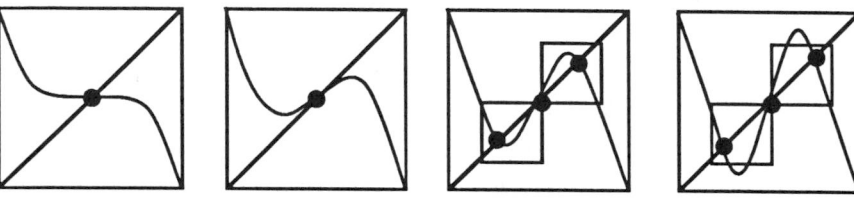

Figure 17: Pitchfork bifurcation.

dynamics, exactly as for the one-humped map. The symbolic dynamics tracks the itinerary of the critical point of the one-humped map, leading to the concept of a *kneading sequence*. The symmetries can be inserted into the symbolic dynamics to provide a rigorous check on how much symmetry exists within the window. It is conjectured that there always exist subwindows in which all possible broken symmetries occur.

Similar ideas apply to pitchfork bifurcations, where again we can find invariant intervals for f^q, as in Figure 17. Now, however, the behaviour within the parameter window is controlled by a two-humped map with $\mathbf{Z}_2$ symmetry; that is, it follows the model of the cubic logistic map, with period-doubling cascades, chaos, and a symmetry-increasing crisis. The crisis occurs at the stage of Figure 17(d), when the humps poke out of their confining boxes. Provided the humps remain within the outer box, a symmetry-increasing crisis occurs.

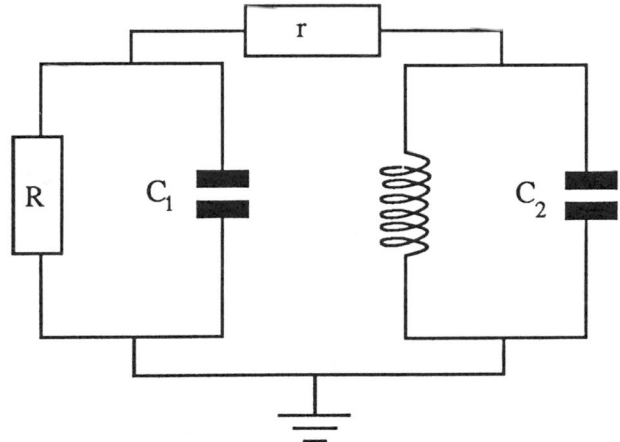

Figure 18: Circuit for a Van der Pol-Duffing Oscillator.

Considerations of this kind explain the majority of features actually observed in numerical experiments with the Arnold maps: see Krupa and Roberts [25] for details. Clearly there are many additional questions yet to be tackled.

7 Electronic Circuits

What connection do these theoretical ideas have with experiments? Can symmetric chaos be observed in ther laboratory? One area in which a number of experiments on symmetric chaos have been done is electronic circuits. We report here on some of this work: the systems discussed are a Van der Pol-Duffing Oscillator, a Van der Pol-Duffing Oscillator with quintic characteristic, and coupled identical oscillators. They are modelled by continuous dynamical systems, and studied by means of equivariant Poincaré maps.

7.1 Van der Pol-Duffing Oscillator

The circuit for a Van der Pol-Duffing Oscillator is shown in Figure 18. It consists of an RC-oscillator resistively coupled to an LC oscillator.

One resistor R is nonlinear with characteristic

$$I_R(V) = \nu + aV + bV^3.$$

When ν is zero, this has $\mathbf{Z}_2$ symmetry $V \mapsto -V$. Here we concentrate on this symmetric case. For the effect of the symmetry-breaking term ν see Gomes and King [20]. The dynamics of the circuit is determined by the dynamical system

$$\dot{V}_1 = -(\frac{1}{C_1})[bV_1^3 + (a + \frac{1}{r})V_1 - \frac{V_2}{r}] \tag{13}$$

$$\dot{V}_2 = \frac{1}{C_2}[\frac{(V_1 - V_2)}{r} - I_L] \tag{14}$$

$$\dot{I}_L = \frac{1}{L}V_2. \tag{15}$$

To nondimensionalize the variables we rescale, setting

$$X = \sqrt{br}V_1 \quad Y = \sqrt{br}V_2 \quad Z = \sqrt{br^3}I_L \quad t \mapsto \frac{1}{rC_2}t. \tag{16}$$

This yields the system

$$\dot{X} = -\gamma(X^3 - \alpha X - Y + \mu) \tag{17}$$

$$\dot{Y} = X - Y - Z \tag{18}$$

$$\dot{Z} = \beta Y \tag{19}$$

where $\alpha = -1 - ar$, $\beta = \frac{r^2 C_2}{L}$, $\mu = \sqrt{br^3}\nu$, and $\gamma = \frac{C_2}{C_1}$.

The dynamics of this system is complicated with numerous different types of bifurcation, [20]. In general there is a close resemblance to the bifurcations of the cubic logistic map, with asymmetric period-doubling sequences, followed by a transition to asymmetric chaos (of Rössler attractor type), followed by a symmetry-increasing crisis. Gaito and King [18] report experimental determinations of Poincaré sections which include this crisis, showing the expected intermittency immediately after collision. Figure 19 shows the flows, Poincaré sections and partial return maps (between appropriate segments of the

Poincaré section) in the two cases. The Poincaré sections are essentially 1-dimensional (though what appear to be curves presumably have the usual thin Cantor set cross-sectional structure, which we ignore). They involve several distinct 'curves', and it is a nontrivial matter to select the appropriate segments of the set for drawing return maps.

Observe that the return map has the expected Z_2 symmetry, with the group acting as rotation through π, and that the crisis occurs when the peaks in the mapping break through the zero level, permitting the state of the system to escape from the positive or negative interval and explore the entire line. This is similar to the behaviour of the cubic logistic map. The analogy can to some extent be explained by transposing the left and right halves of the interval $(-\pi, \pi)$ in Figure 19(c,d), after which transformation the Poincaré map qualitatively resembles a cubic curve. However, this is a discontinuous transformation which introduces some (not insoluble) interpretational problems.

The complete dynamics is complicated. It can be visualised using a *knot-holder*, a concept introduced — though not under that name — by Birman and Williams [3], which effectively represents the thin but intricately layered structure of the attractor as a surface. This takes the form of Figure 20. It can also be modelled in terms of a fast/slow decomposition of the vector field, in which the phase point makes chaotic transitions between two layers of a 3-sheeted surface, or alternatively between two adjacent potential wells. The spiral motion that occurs as the point jumps from one sheet to the other is responsible for some of the structure in the Poincaré sections, and we must restrict attention to suitable subsets, as mentioned above.

7.2   Quintic Van der Pol-Duffing Oscillator

Castro [8] has extended some of this work to an oscillator with a quintic characteristic

$$I_R(V) = \nu + aV + bV^3 + cV^5,$$

which can be thought of as modelling a particle moving in a symmetric system of three adjacent potential wells. Two experimentally

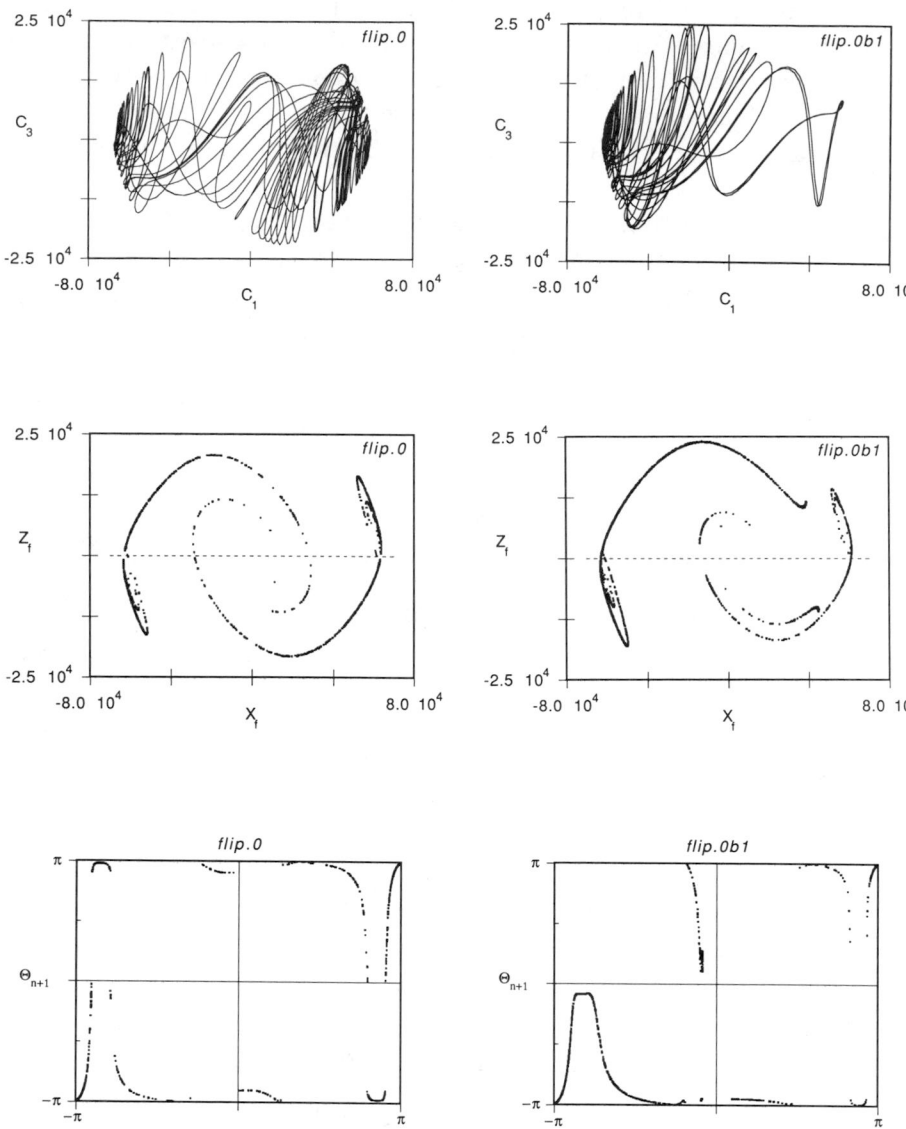

Figure 19: Experimentally determined flow, Poincaré section, and partial return map for the Van der Pol-Duffing oscillator. Left: symmetric case $\nu = 0$. Right: asymmetric case $\nu \neq 0$.

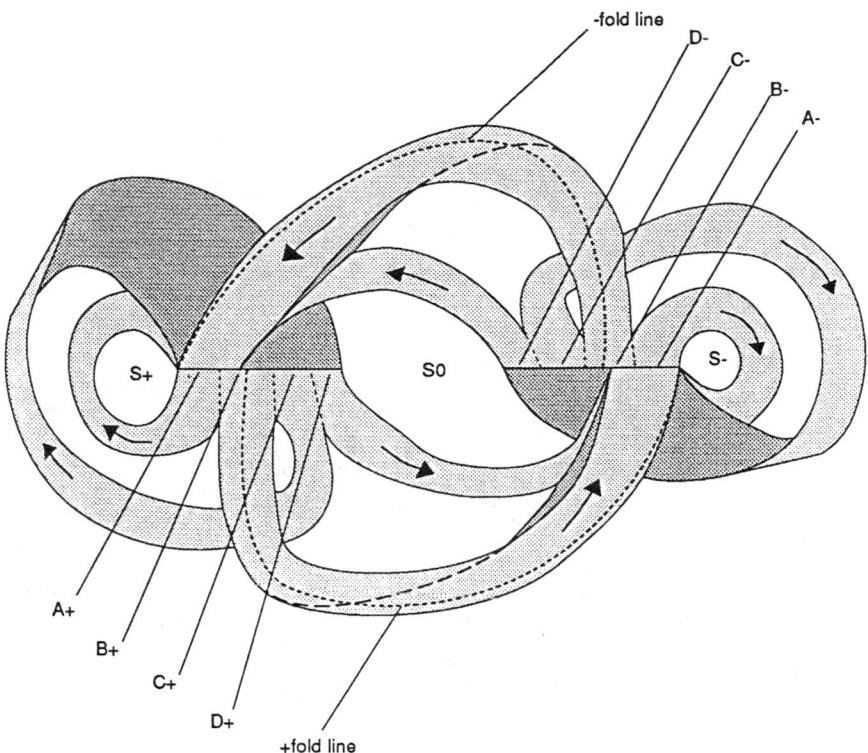

Figure 20: Knotholder for the fully symmetric strange attractor of the Van der Pol-Duffing oscillator.

determined Poincaré sections are shown in Figure 21. These figures
in many ways resemble an ϵ-lift of a quintic analogue of the logistic
map, but with ϵ quite large, taking the form

$$
\begin{pmatrix} x \\ y \end{pmatrix} \mapsto \begin{pmatrix} \epsilon y + Ax + Bx^3 + Cx^5 \\ x \end{pmatrix} \tag{20}
$$

See Figure 22.

This resemblance is related to a fast/slow decomposition of the
dynamics which corresponds to a physical picture of the oscillator as
a particle moving either in the central well or in all three. See [8] for
more details.

7.3   Coupled Identical Oscillators

Richer symmetries than $\mathbf{Z}_2$ occur in systems of n symmetrically coupled oscillators. Ashwin, King, and Swift [2] study systems of this
type having symmetry group $\mathbf{S}_n$, the symmetric group of degree
n. The symmetries are induced by permutations of the oscillators.
In particular they focus on the case $n = 3$. The cited paper concerns non-chaotic dynamics, which can be reduced to a $\mathbf{S}_3$- equivariant dynamical system on a 2-torus. They predict a new type of
codimension-1 bifurcation, and find it experimentally in a network
of coupled Van der Pol oscillators.

Ashwin [1] develops these ideas further, and studies chaos in such
systems when $n = 3$ or 4. He uses the circuit of Figure 23, which
is a system of coupled identical Van der Pol oscillators forced by
a sinusoidal signal. The results are plotted by assigning one of n
unit vectors, arranged in the plane at angles of $\frac{2\pi}{n}$, to the voltage
in each oscillator, and plotting a Poincaré section synchronised with
the times at which the forcing signal generator passes through zero
from negative to positive. This is a $\mathbf{D}_n$-equivariant projection of
the voltage data, in the sense of section 8.2 below. Because it is a
projection, the images of attractors may appear to overlap.

We describe two particular bifurcation sequences observed in
these experiments. Both begin with a $\mathbf{Z}_3$-symmetric 2-torus (a).
In the first, Figure 24, this symmetry-locks (b) to a 3:2 resonance,

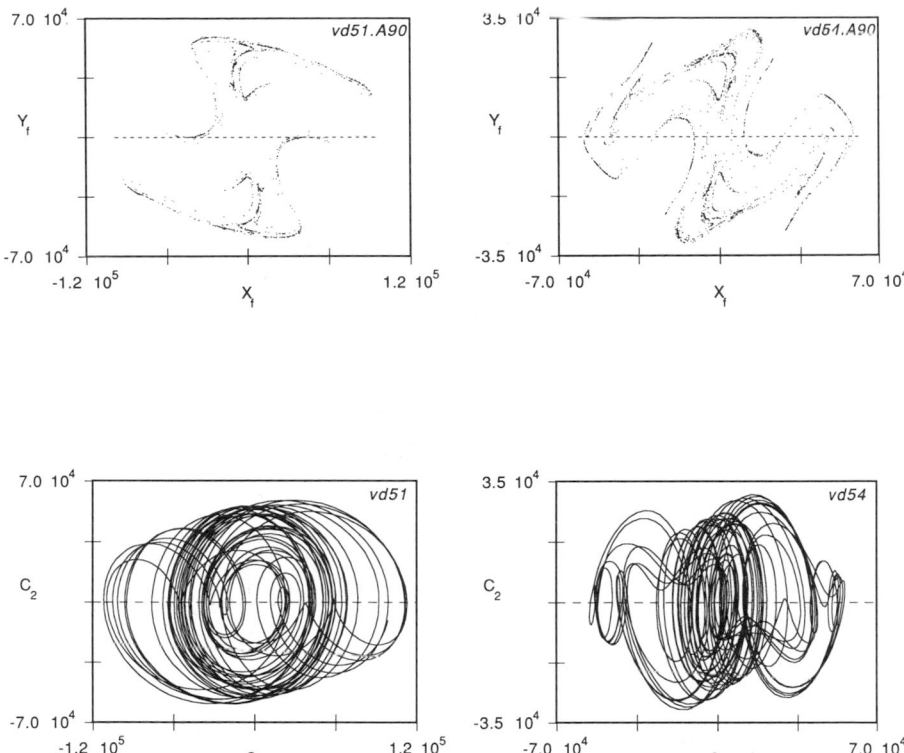

Figure 21: Experimental flows (above) and Poincaré sections (below) for the quintic Van der Pol-Duffing oscillator. (a) Confined to the central well. (b) Moving in all three wells.

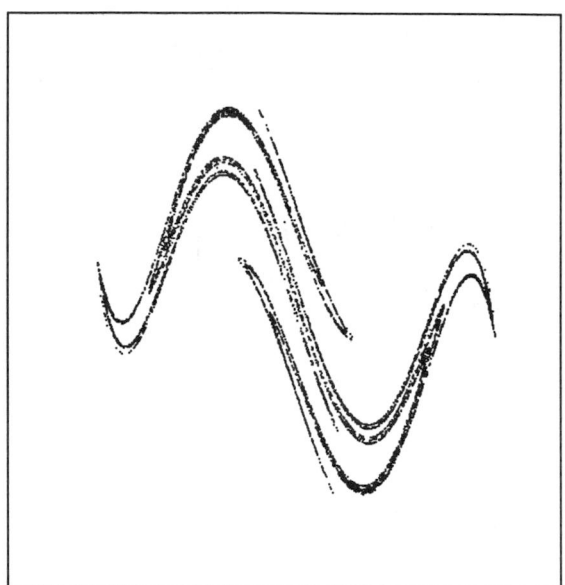

Figure 22: Attractor in a lift (20) of a quintic logistic map.

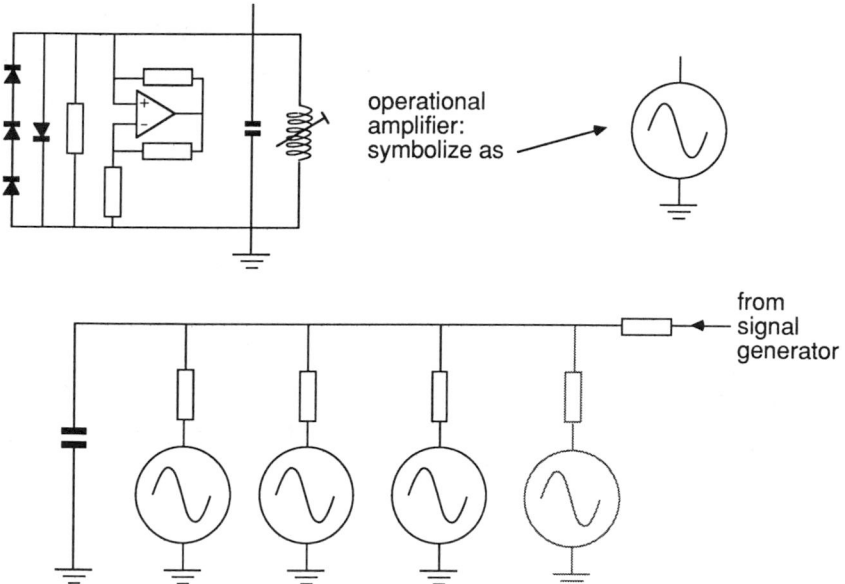

Figure 23: System of three or four forced coupled identical oscillators.

which period-doubles to a 6:4 resonance (c) and then undergoes a Sacher-Naimark torus bifurcation (d) to a 2-torus at 6:4 resonance. In (e) the 2-torus becomes a chaotic attractor with Z_3 symmetry. By (f) there has been a symmetry- increasing crisis to S_3 symmetry.

In the second, Figure 25, the initial bifurcation (b) is a Sacher-Naimark torus bifurcation to a 3-torus. The amplitude of the 3-torus increases, and we see (c) a bunching of trajectories where it is conjectured to approach a hyperbolic 2-torus. By (d) the two tori have collided in a heteroclinic bifurcation, destroying the 3-torus. In (e) the attractor has become chaotic with Z_3 symmetry. By (f) there has been a symmetry-increasing crisis to S_3 symmetry. There is some resemblance to Figure 11b.

8   Time Series Analysis

The detection of chaos from time series is a widely studied problem, and the basic principles are now becoming fairly well understood, at least for data from well-controlled experiments. Many different techniques have been devised. The first goes back to Packard [28] and Takens [34]. More recent improvements have been obtained in particular by Broomhead and King [6] for flows, and Broomhead and King [7] for mappings.

However, the detection of symmetric chaos requires further refinements if we wish to find out what the symmetries of the attractor are. For example, if we take one of Chossat, Field and Golubitsky's D_3 attractors and plot its x- and y-coordinates as two separate time series, then the symmetry is by no means apparent (Figure 26). This is not especially surprising, since the x- and y-coordinates are not related in a symmetric fashion. If instead we plot (Figure 27) time- series for the variables x, $x/2 + \sqrt{3}y/2$, $x/2 - \sqrt{3}y/2$, which are *permuted* by D_3, we observe something rather more interesting.

First, of course, the three time series in Figure 27(a), say, are not identical. If they were, it would mean that *every individual* point on the attractor, rather than just the attractor itself, would have D_3 symmetry, so the attractor would be just the origin. What we see, however, is that all three time series have a very similar appearance.

Greg King and Ian Stewart

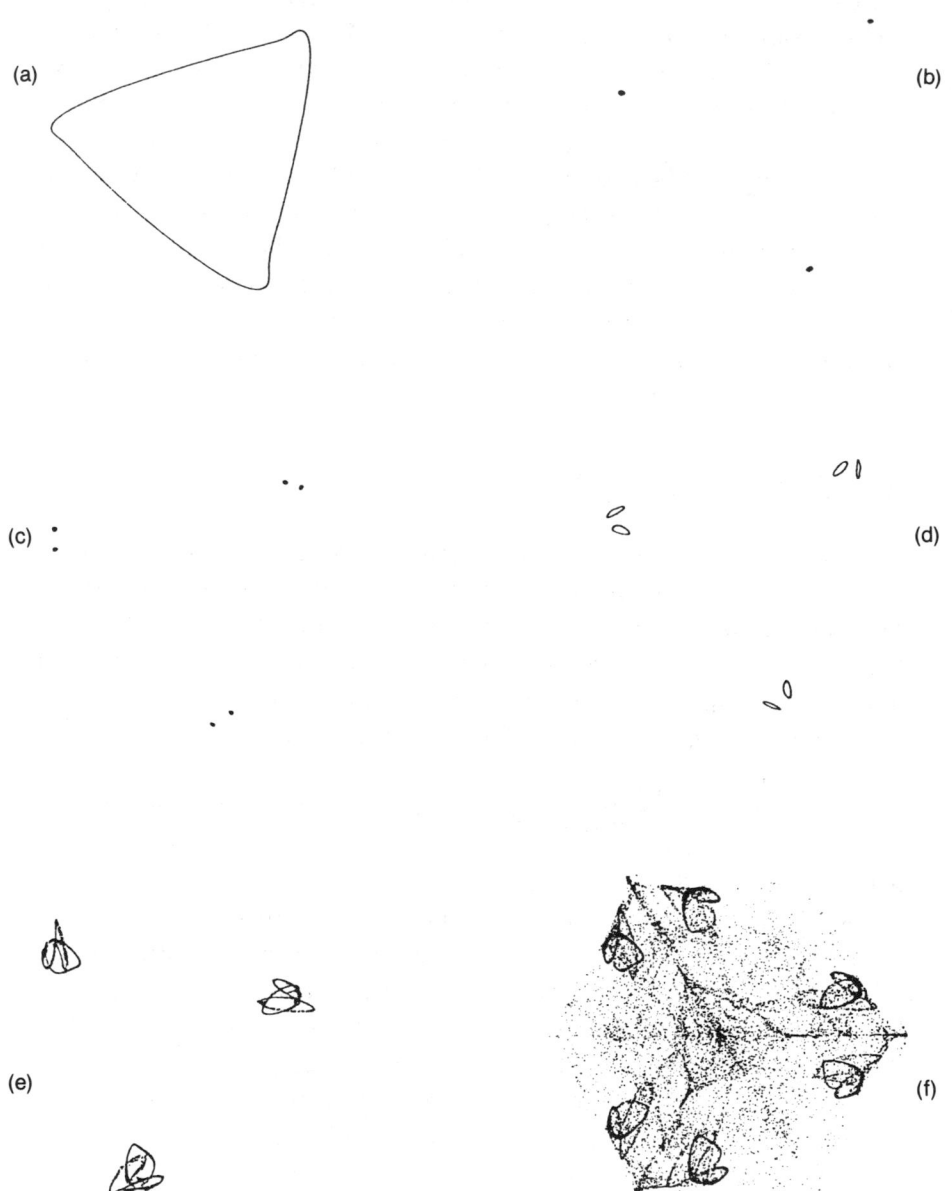

(a)

(b)

(c)

(d)

(e)

(f)

Figure 24: Bifurcation sequence in a system of three forced identical oscillators.

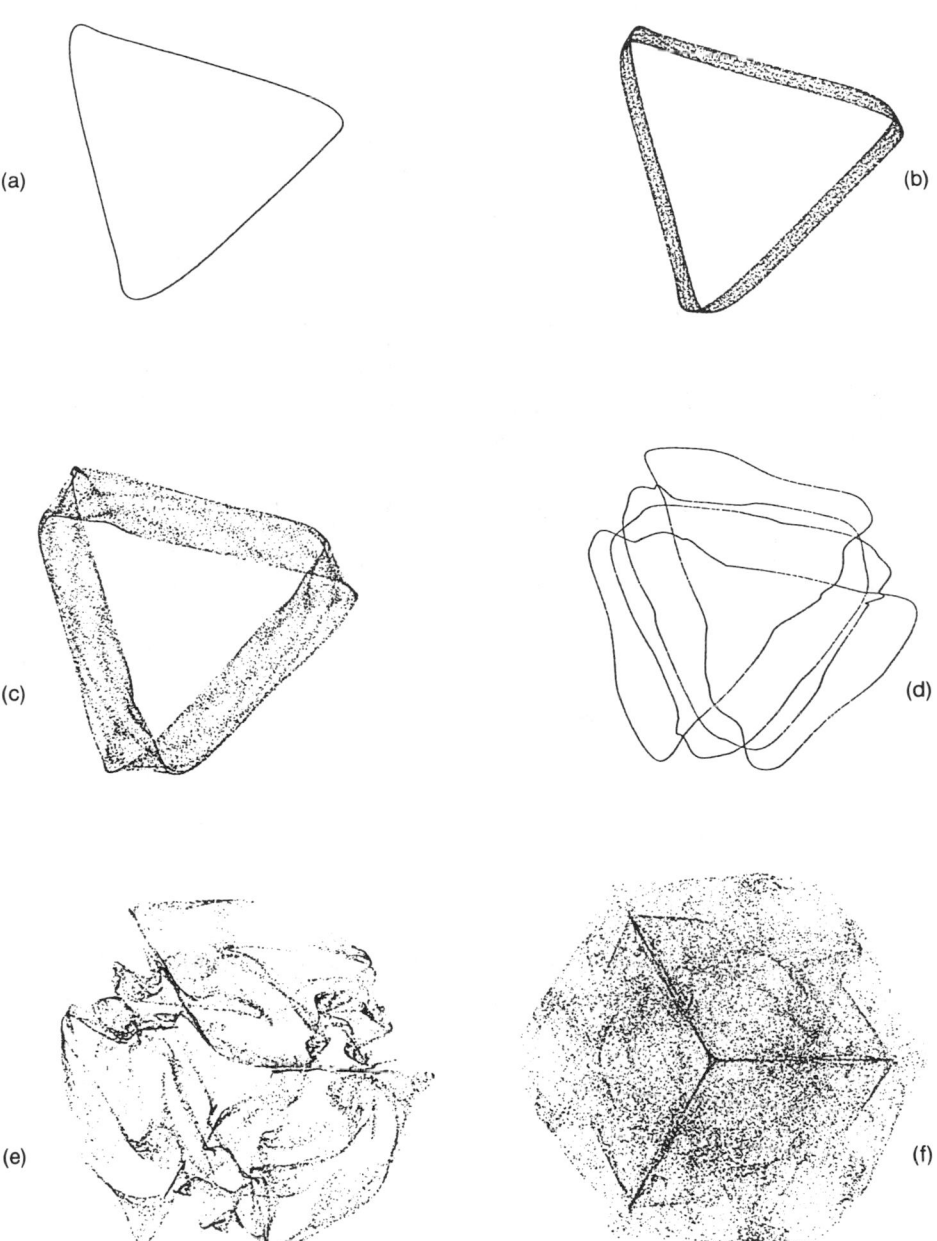

Figure 25: Alternative bifurcation sequence in a system of three forced identical oscillators.

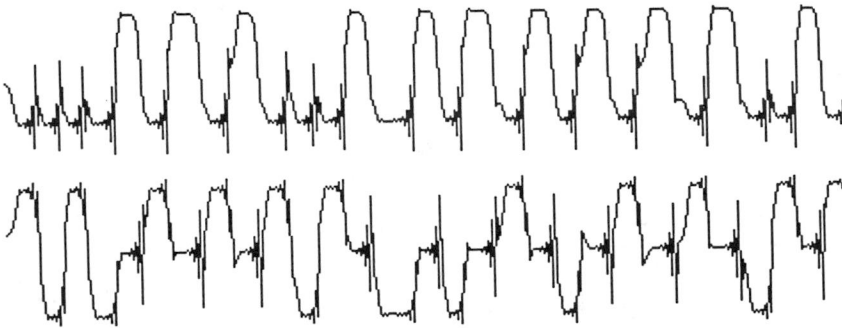

Figure 26: Two time series for the coordinates of a point in a $\mathbf{D}_3$-symmetric attractor (3): $\alpha = -.9, \beta = 0, \gamma = -.8, \lambda = 1.3$.

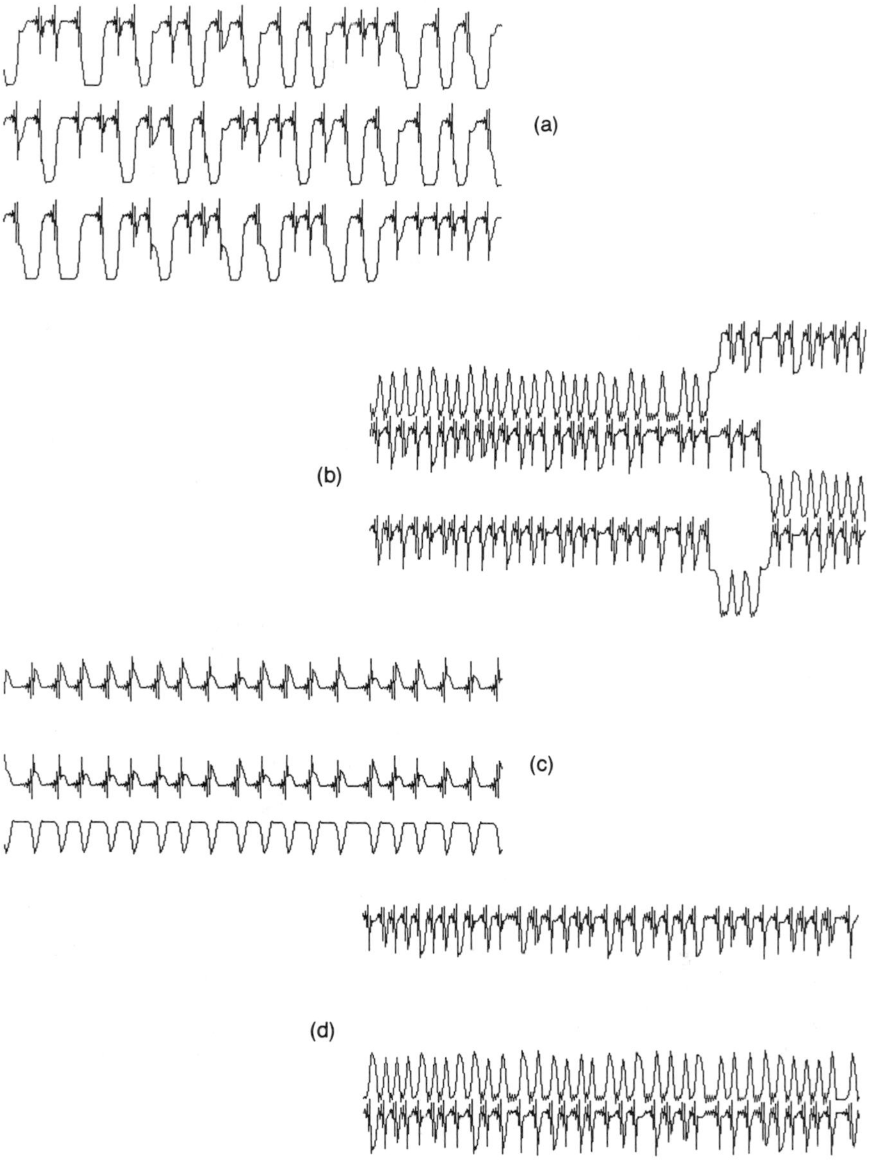

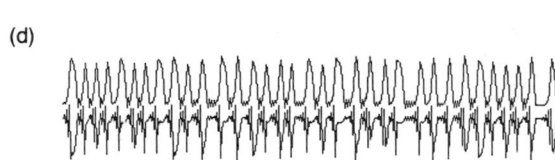

Figure 27: Four time series for symmetrically placed combinations of the coordinates of a point in an attractor for the D_3-equivariant map (3). (a) $\alpha = -.9, \beta = 0, \gamma = -.8, \lambda = 1.3$. (b) $\alpha = -1.1, \beta = .213, \gamma = 0.6, \lambda = 1.89$. (c) $\alpha = -.9, \beta = 0, \gamma = -.8, \lambda = 1.22$. (d) $\alpha = -1.1, \beta = .212, \gamma = .6, \lambda = 1.89$.

If we look at randomly selected segments from each, it is very hard to tell which is which. The same goes for Figure 27(b). However, in Figures 27(c,d) there is a clear distinction: two series look similar but the third is quite different. These correspond to $\mathbf{Z}_2$-symmetric attractors, and the symmetry is clearly 'visible' in some statistical sense in the three time series. We now explain why this happens.

8.1   Local Isomorphism

To pin down this notion of 'segmentwise resemblance' we proceed as follows. Suppose throughout this subsection that M is a manifold, $f : M \to M$ is a smooth mapping, and $A \subset M$ is an attractor for f. Let (x_t) and (y_t) be two orbits of f in A. We say that (x_t) *locally resembles* (y_t) if for all ϵ and $N_1 > N_2 > 0$ there exists T such that

$$\|(x_t) - (y_{t+T})\| < \epsilon$$

for $N_2 \leq t \leq N_1$. If (x_t) locally resembles (y_t) and (y_t) locally resembles (x_t) then we say that (x_t) and (y_t) are *locally isomorphic*.

Note that we need both time series locally to resemble each other in order to have an equivalence relation: the time series $(0, 0, 0, 0, \ldots)$ locally resembles $(1, 0, 0, , 0 \ldots)$ but not conversely. Intuitively, two time series are locally isomorphic if every finite segment of one can be matched, to within error ϵ, by a segment of the other. The importance of this notion is that any statistical properties of time series that depend upon the limiting behaviour of finite segments are the same for locally isomorphic time series. Examples are quantities such as entropy or Liapunov exponents. While local isomorphism depends upon the entire infinite length of a time series, we have already observed that the three finite time series in Figure 27 have many segments approximately in common. We now explain why, if extended to infinite length, they can be expected to be locally isomorphic.

Lemma 8.1
Suppose that (x_t) and (y_t) are dense orbits in an attractor A for a mapping f. Then (x_t) and (y_t) are locally isomorphic. Conversely if (x_t) is dense in A and (y_t) is locally isomorphic to (x_t), then (y_t) is also dense.

Proof
Choose $\epsilon > 0$ and consider an arbitrary segment of (x_t) where $N_2 \leq t \leq N_1$. Since (y_t) is dense in A, for arbitrary $\delta > 0$ we can find k such that $\|y_k - x_{N_2}\| < \delta$. By continuity of f, if we make δ sufficiently small we then have $\|(x_{N_2+t}) - (y_{k+t})\| < \epsilon$ for all t such that $0 \leq t \leq N_1 - m$, which is what we need.

Lemma 8.2
(a) If A is compact then $f(A) = A$.
(b) If A is compact and (x_t) is a dense orbit then for all k the orbit $f^k(x_t)$ is dense in A.

Proof
(a) Since A is an attractor, it has a dense orbit (x_t). If $y \in A$ then there exists a subsequence $(x_{t_n}) \longrightarrow y$. If $n > 0$ then $x_{t_n} \in f(A)$. So $y \in \overline{f(A)}$. By compactness $f(A)$ is closed, so $y \in f(A)$ and f is onto as required.
(b) This now follows since $(f^k(x_t)$ is dense in $f^k(A) = A$.

8.2   Equivariant Phase Space Reconstruction

An important technique for the experimental detection of chaos is the recognition of deterministic dynamics — especially chaos — in an experimental time series. In this section we discuss the analogous problem for equivariant dynamics: not only must the presence of deterministic dynamics be detected, but the symmetry of the corresponding attractor must be obtained.

The simplest method, which goes back to Packard *et al.* [28] and Takens [34], is to replace the time series $\{y_t\}$ by a 'moving window' of length N, given by vectors $x_i = (y_i, \ldots, y_{i+N-1})$. Then, for large enough N, generically the attractor formed by the x_i in $\mathbf{R}^N$ is topologically equivalent to that for the dynamics that gave rise to the original time series $\{y_t\}$.

There is a refinement of this method, due to Broomhead and King [6], based upon ideas from signal processing. It is formulated for signals produced by a continuous dynamical system. Broomhead and King [7] explains how to apply these ideas to Poincaré sections, that is, to mappings, by local analysis. We describe only the technique as it applies to continuous dynamical systems.

As in the Packard-Takens method, we form a 'moving window' of length N from this time series, but the difference is that we then apply principal component analysis to write the resulting vectors in $\mathbf{R}^N$ as linear combinations of eigenvectors of a correlation matrix. Intuitively, this process finds the most common patterns among the vectors in $\mathbf{R}^N$ and expresses each such vector as a linear combination of such patterns. The dominant patterns, having the largest eigenvalues, are retained: the remainder are considered to be 'noise' and are ignored.

Specifically, define an N-column matrix $X = (x_{ij})$, where $1 \leq j \leq N$ but i is arbitrary, by setting $x_{ij} = y_{i+j-1}$. Form the $N \times N$ correlation matrix $C = X^{\mathrm{T}}X$ and let its eigenvectors be $v_1, \ldots, v_N$ and corresponding eigenvalues $\sigma_1, \ldots, \sigma_N$. These eigenvalues are all real, since C is symmetric, and can be arranged in decreasing order: without loss of generality $\sigma_1 \geq \ldots \geq \sigma_N$.

Typically these eigenvalues decrease rapidly and then level off, as in Figure 28. The point at which they level off is the *noise floor* for the observations. Suppose this starts at σ_{M+1}. Let $V = \mathrm{span}(v_1, \ldots, v_M)$ and let the projection of $v \in \mathbf{R}^N$ into V be $\hat{v}$. Then the original time series is replaced by the series of M-dimensional vectors $\hat{x}_i$ where x_i is the ith row of X.

What happens if we apply this method to a time series obtained from an equivariant dynamical system? As a numerical experiment, we take the Field-Golubitsky-Chossat mapping (3) with parameters $\alpha = 1.8, \beta = 0, \gamma = 1.34164, \lambda = -1.8$, which we know produces the fully symmetric attractor of Figure 4(b). For simplicity we use the Packard-Takens approach, though similar remarks apply to the Broomhead-King refinement. We take $N = 2$ (knowing in advance that the attractor is embedded in $\mathbf{R}^2$), and let the 'experimental measurement' be the x-coordinate of the point on the attractor. Thus we plot the pairs (x_i, x_{i+1}), for an orbit (x_t, y_t) defining the attractor.

The result is shown in Figure 29. We can recognise this as a

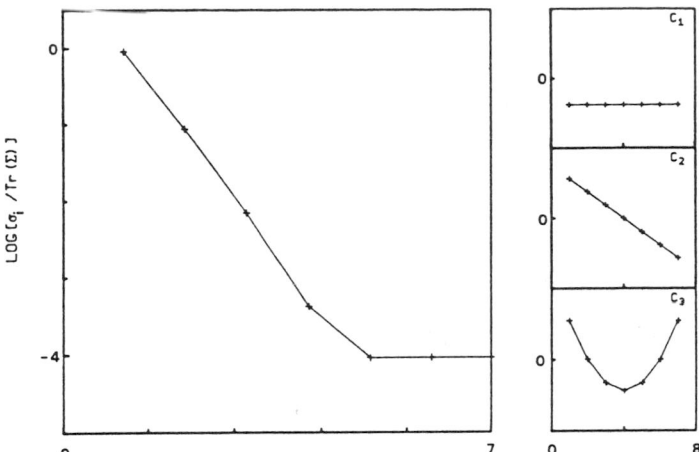

Figure 28: Typical sequence of eigenvalues for the Broomhead-King method.

twisted version of the attractor, but one that does not preserve its symmetry.

This is not really suprising. In order to preserve the attractor of the symmetry, we need firstly to make an *equivariant* observation (which the x-coordinate is *not*), and then we need to *process* it equivariantly. Let us define our terms, and then pursue this idea further.

Suppose that a phase point x follows an orbit of a dynamical system (1). Then an *observation* is just a mapping $x \mapsto \phi(x)$ into some space Y. Usually observations made in experiments are 1-dimensional, so $Y = \mathbf{R}$, but more generally Y may be multidimensional, and it is this case that is most important in equivariant systems. Suppose that f in (1) is equivariant under the action of a group G. Then we say that ϕ is *equivariant* under G if there is an action of G on Y such that for all $\gamma \in G$ we have $\phi(\gamma.x) = \gamma.\phi(x)$. One reason for requiring this conditions is that equivariant observations *preserve symmetry*, in the sense that $\Sigma_x \subset \Sigma_{\phi(x)}$ and $\Delta_x \subset \Delta_{\phi(x)}$, as can easily be verified from the definitions.

As a motivating example, suppose we seek only to detect the

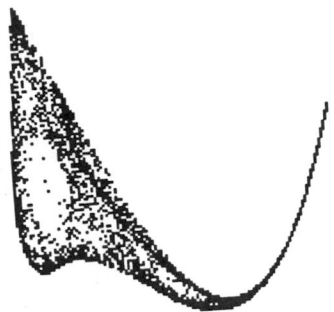

Figure 29: Reconstructed attractor using Packard-Takens method with $N = 2$. The distortion might be removed by increasing N, but the symmetry is not preserved.

presence of an attractor that is symmetric under reflection κ in the x-axis. Moreover, assume that one 'experimental measurement' at our disposal is the projection onto the line making angle $\frac{\pi}{6}$ with the x-axis. This is not equivariant, but we can supplement it by the correponding projection in its image under κ, namely, projection on to the line making angle $-\frac{\pi}{6}$. That is, we define a two-dimensional observation $g = (g_1, g_2)$ by

$$g_1(x, y) = x \cos(\frac{\pi}{6}) + y \sin(\frac{\pi}{6}) \qquad (21)$$

$$g_2(x, y) = x \cos(\frac{\pi}{6}) - y \sin(\frac{\pi}{6}). \qquad (22)$$

This observation is equivariant for κ provided we make κ act on the target $\mathbf{R}^2$ of g by $\kappa(X, Y) = (Y, X)$. That is,

$$
\begin{aligned}
g(\kappa.x, \kappa.y) &= g(x, -y) = (g_2(x, y), g_1(x, y)) \\
&= \kappa.(g_1(x, y), g_1(x, y)) = \kappa.g(x, y). \qquad (23)
\end{aligned}
$$

Now we experiment with the equivariant version of the Packard-Takens method — taking $N = 1$! Namely, we plot not (x, y) but

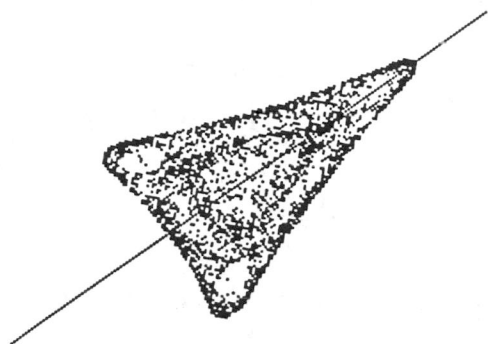

Figure 30: $\mathbf{Z}_2$-equivariant reconstruction of a $\mathbf{D}_3$-symmetric attractor preserves only the $\mathbf{Z}_2$ symmetry.

$(g_1(x, y), g_2(x, y))$. The result, shown in Figure 30, demonstrates that this reconstruction preserves the symmetry of the attractor. (The choice of $\frac{\pi}{6}$ was made to exaggerate the distortion enough to make it clear that only this symmetry survives.) Note that the appropriate symmetry is reflection in the *diagonal*: this permutes the two observations.

What about the Broomhead-King method? Here is one suggestion, appropriate to continuous systems. Instead of using a moving window X of length N we use a moving equivariant window X of length $2N$ defined by

$$x_i = (g_1(x_i, y_i), \quad \ldots, \quad g_1(x_{i+N-1}, y_{i+N-1});$$
$$g_2(x_i, y_i), \quad \ldots, \quad g_2(x_{i+N-1}, y_{i+N-1})). \quad (24)$$

We anticipate that the eigenvalue structure should be equivariant, that is, the eigenspaces should afford representations of the group $\mathbf{Z}_2$. So eigenvectors should either be *symmetric*, with the second N entries resembling the first N; or *antisymmetric*, with the second N entries resembling the negative of the first N. Moreover, if the attractor has $\mathbf{Z}_2$ isotropy, then only symmetric eigenvectors should occur above the noise floor.

Similar remarks apply to arbitrary group actions: provided an equivariant observation is taken, and used to construct an equivariant window, then the eigenvectors above the noise floor should be *fixed* by the isotropy subgroup of the attractor. In practice this will only hold approximately, but it would be easy enough to devise a measure for the amount to which the eigenvectors deviate from this type of symmetry: for example, consider the sum of squares of differences between the entries of the eigenvector and those of its image under elements of the isotropy subgroup.

9 Turbulent Taylor Vortices Revisited

What can we say about the phenomenon of turbulent Taylor vortices, which we began with, in the light of the above understanding of symmetric chaos?

There is a strong resemblance between the creation of turbulent Taylor vortices and a symmetry-increasing crisis of strange attractors. What at first sight may seem just an analogy may be a complete mathematical relationship. Although we have discussed mainly discrete dynamical systems, the same general range of phenomena should occur for continuous flows; and in any case we may discuss continuous dynamics in terms of a hypothetical Poincaré mapping. The following scenario was suggested by Golubitsky in conversation: it is summarized schematically in Figure 31. Focus only on the Z_2 'flip' symmetry whose mirror- plane is the boundary between a pair of adjacent vortices, and which for genuine Taylor vortices leads to the boundary being planar. The bifurcation from Taylor vortices to wavy vortices breaks this symmetry, but the system as a whole retains it. Thus for every Z_2-symmetry-breaking flow that can occur there is a second flow, its conjugate under the flip, which can *also* occur. Initial conditions select one from this pair. As it happens, the two flows for wavy vortices are identical up to a phase shift, because wavy vortices have a symmetry that is a flip composed with a half-period phase shift. This renders the two flows indistinguishable in the usual experiments, and indeed it is hard to think of an experiment that *would* distinguish them. However, it is clear on symmetry grounds that if one flow is a possible solution to the equations, then so must the other be.

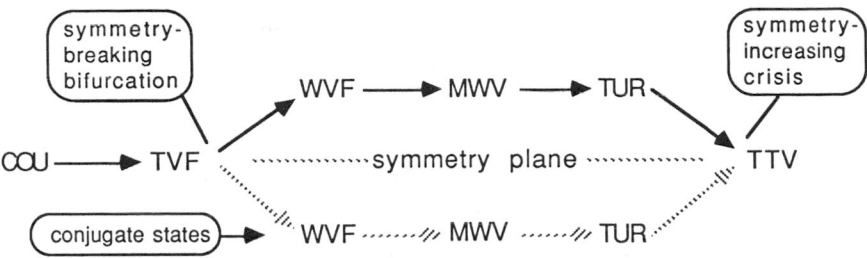

Figure 31: Schematic representation of bifurcation sequence thought to be involved in the formation of turbulent Taylor vortices. The states are labelled as follows: COU = Couette flow; TVF = Taylor vortices; WVF = wavy vortices; MWV = modulated wavy vortices; TUR = wavy turbulence; TTV = turbulent Taylor vortices.

Subsequent bifurcations to modulated wavy vortices and eventually to turbulence would not be expected to restore the Z_2 symmetry, because increased symmetry is inconsistent with the general phenomenology of those bifucations. So again the observed turbulent flow is one from a symmetrically related pair. Correspondingly we may hypothesize a pair of symmetrically related strange attractors in a suitable phase space. One attractor corresponds to the observed asymmetric flow, the other to its ghostly companion.

As the turbulence becomes stronger, it is natural for these attractors to grow in size. If so, they may well collide. What would we observe if they did?

We have seen that the Z_2 symmetry of the merged attractor is not directly observable, but that it gives rise to a *statistical* Z_2 symmetry in any suitable equivariant observation. The flow pattern itself is an equivariant observation — applying the flip to the flow is equivalent to standing on your head! So the observed flow will have a statistical Z_2 symmetry; that is, a symmetry 'on the average'. Such a flow will in particular possess a plane, the mirror-plane for the flip, on which the average axial velocity is zero.

This is precisely what Fenstermacher *et al.* [15] have observed in turbulent Taylor vortex flow.

Indeed more may be true. Experiments in Swinney's laboratory (communicated to us by Dan Lathrop) suggest that the *entire* sequence of symmetry-breaking bifurcations from Couette flow to modulated wavy vortices may be restored in a corresponding series of crises (or similar transitions) of chaotic attractors.

The symmetry-increasing crisis model shows that the most puzzling aspect of the formation of turbulent Taylor vortices — the increase in pattern as the turbulence becomes stronger — is actually natural. Only by growing in size can strange attractors collide to combine their symmetries.

There are strong analogies between this scenario, the bifurcations of the cubic logistic map, and those of the Van der Pol-Duffing oscillator reported above. The general changes in symmetry, and the changes from order to chaos, are essentially identical in all three. The differences lie in the fine detail of the dynamics.

It would be interesting to test this idea. It does not appear feasible to do so by way of numerical simulations: we are looking at a turbulent time-dependent three-dimensional flow, beyond the capabilities even of a supercomputer. However, it should be relatively easy to test it experimentally, by using a Z_2-equivariant time series. The natural way to achieve this is to use two laser probes, symmetrically placed above and below the 'boundary' between adjacent turbulent Taylor vortices, arranged to measure the axial velocity (or some velocity with a strong axial component). Indeed two beams could be produced from a single laser using a beam-splitter; however, two detectors might be necessary. The signals could be processed by an equivariant version of the method of Broomhead and King [6], as described above.

Acknowledgements
We are grateful to Peter Ashwin, Sofia Castro, Mike Field, Marty Golubitksy, Gabriela Gomes, Maciej Krupa, Dan Lathrop, Robert MacKay, and Mark Roberts for permission to include unpublished ideas of theirs and to describe work in progress, and for helpful dis-

cussions about symmetric chaos. Many of our pictures were drawn using the dynamical systems package *Kaos* for SUN workstations, written by John Guckenhimer and Swan Kim. The first author's research was partially supported by a grant from the Science and Engineering Research Council. Part of this paper derives from work done by the second author when visiting the Institute for Mathematics and its Applications at the University of Minnesota, to whom he is grateful for financial support and hospitality.

Bibliography

[1] P.Ashwin, Symmetric chaos in systems of three and four forced oscillators, *Nonlinearity*, to appear.

[2] P.Ashwin, G.P.King, and J.W.Swift, Three identical oscillators with symmetric coupling, *Nonlinearity*, to appear.

[3] J.S.Birman and R.F.Williams, Knotted periodic orbits in dynamical systems I: Lorenz equations,*Topology* **22** (1983) 47-82.

[4] R.Bowen, *On Axiom A Diffeomorphisms*, CBMS Regional Conference Series in Mathematics **35** , Amer. Math. Soc., Providence RI 1978.

[5] A.Brandstater and H.L.Swinney, Strange attractors in weakly turbulent Couette-Taylor flow, *Phys. Rev.* A **35** (1987) 2207-2220.

[6] D.S.Broomhead and G.P.King, Extracting qualitative dynamics from experimental data, *Physica* **20D** (1986) 217-236.

[7] D.S.Broomhead and G.P.King, On the qualitative analysis of experimental dynamical systems, in *Nonlinear Phenomena and Chaos*, (ed. S.Sarkar), Adam Hilger, Bristol 1986, 113.

[8] S.Castro, *Experiments in Nonlinear Dynamics: an attractor of the modified van der Pol oscillator (with quintic characteristic)*, Nonlinear Systems Laboratory report, Univ. of Warwick 1990.

[9] P.Chossat and M.Golubitsky, Iterates of maps with symmetry, *SIAM J. Math. Anal* **19** (1988) 1259-1270.

[10] P.Chossat and M.Golubitsky, Symmetry- increasing bifurcation of chaotic attractors, *Physica* D **32** (1988) 423-436.

[11] P.Collet and J.-P.Eckmann, *Iterated Maps of the Interval as Dynamical Systems*, Birkhäuser, Boston 1980.

[12] W.D.Crowe, R.Hasson, P.J.Rippon, and P.E.D. Strain-Clark, On the structure of the Mandelbar set, *Nonlinearity* 4 541-554.

[13] P.Cvitanović, *Universality in Chaos* (2nd ed.), Adam Hilger, Bristol 1989.

[14] R.L.Devaney, *An Introduction to Chaotic Dynamical Systems*, 2nd ed., Addison-Wesley, Redwood City CA 1989.

[15] P.R.Fenstermacher, H.L.Swinney, and J.P.Gollub, Dynamical instabilities and transition to chaotic Taylor vortex flow, *J. Fluid Mech.* **94** (1979) 103-128.

[16] M.J.Field and M.Golubitsky, to appear.

[17] J.G.Franjioni, C.-W-Leong, and J.M.Ottino, Symmetries within chaos: a route to effective mixing, *Phys. Fluids* A **1** (1989) 1772-1783.

[18] S.T.Gaito and G.P.King, Chaos on a catastrophe manifold, in *Quantitative Measures of Dynamical Complexity*, NATO ARW Series, Plenum Press, New York 1990.

[19] M.Golubitsky, D.G.Schaeffer, and I.N.Stewart, *Singularities and Groups in Bifurcation Theory* vol.2, Springer, New York 1988.

[20] M.G.M.Gomes and G.P.King, *Bifurcation analysis of a chaotic Van der Pol-Duffing oscillator*, preprint, University of Warwick 1990.

[21] C.Grebogi, E.Ott, F.Romeiras, and J.A.Yorke. Critical exponents for crisis induced intermittency, *Phys. Rev.* A **36** (1987) 5365-5380.

[22] J.Guckenheimer and P.J.Holmes, *Nonlinear Oscillations, Dynamical Systems, and Bifurcations of Vectro Fields*, Applied Mathematical Sciences **42** , Springer, New York 1983.

[23] P.J.Holmes, A nonlinear oscillator with a strange attractor, *Phil. Trans. R. Soc. London* A **292** (1979) 419-448.

[24] M.Kitano, T.Yabuzaki, and T.Ogawa, Symmetry- recovering crises of chaos in polarization-related optical bistability, *Phys. Rev.* A **29** (1984) 1288-1296.

[25] M.Krupa and R.M.Roberts, in preparation.

[26] R.S.MacKay and C.Tresser, Transition to topological chaos for circle maps, *Physica* **19D** (1986) 206-237.

[27] J.Milnor, *Remarks on iterated cubic maps*, preprint, SUNY Stonybrook Inst. Math. Sci. 1990.

[28] N.H.Packard, J.P.Crutchfield, J.D.Farmer, and R.S.Shaw, Geometry from a time series, *Phys. Rev. Lett.* **45** 712-.

[29] E.Piña and E.Cantoral, Symmetries of the quasicrystal mapping, *Phys. Lett.* A **135** (1989) 190-196.

[30] T.Rogers and D.C.Whitley, Chaos in the cubic mapping, *Math. Modelling* **4** (1983) 9-25.

[31] H.Sakaguchi and K.Tomita, Bifurcations of the coupled logistic map, *Progr. Theoret. Phys.* **78** (1987) 305-315.

[32] Science and Engineering Research Council, *The Remarkable World of Nonlinear Systems*, SERC , Swindon 1989.

[33] K.G.Szabó and T.Tél, On the symmetry-breaking bifurcation of chaotic attractors, *J. Stat. Phys.* **54** (1989) 925-948.

[34] F.Takens, Detecting strange attractors in turbulence, in *Dynamical Systems and Turbulence, Warwick 1980*, Lecture Notes in Math. **898** (eds. D.A.Rand and L.-S. Young), Springer, New York 1981, 366-381.

Bäcklund and Reciprocal Transformations: Gauge Connections

B.G. Konopelchenko
Institute of Nuclear Physics 630090,
Novosibirsk USSR

C. Rogers
Loughborough University of Technology, U.K.
and
University of Waterloo, Canada

1 Introduction

The rôle of **Bäcklund Transformations** (BTs) is well-established in the analysis of nonlinear evolution equations amenable to the **Inverse Scattering Transform** (IST). Thus, in particular, BTs are routinely used to construct nonlinear superposition principles whereby multi-soliton solutions may be generated. BTs can be constructed by a variety of means. Classical, jet-bundle and bilinear operator methods are described, in detail, in Rogers and Shadwick [1]. In the context of integrable systems, perhaps the most direct approach is via the so-called **Dressing Method** (DM) (Zakharov and Shabat [2]). This is linked to the **Classical Darboux Transformation** (CDT) introduced in 1882 which generates solutions to related Schrödinger equations (Darboux [3]). The connection has been described by Levi, Ragnisco and Sym [4]. Therein, the analogue of the CDT for the Zakharov-Shabat-AKNS spectral scheme was presented. The relation between the BT and DM methods for $1 + 1$

–dimensional nonlinear integrable systems has been detailed by Levi, Ragnisco and Sym in [5].

Decomposition of BTs into the product of commuting **elementary Bäcklund Transformations** (EBTs) has been treated by Konopelchenko, [6-8] and Calogero and Degasperis [9]. Generalised lattices of solutions which incorporate the Bianchi lattices associated with the standard BTs may be thereby constructed [8]. Certain classes of BTs constitute Abelian free groups, the generators of which are the EBTs [6].

Extension to higher dimensions of the established methods developed for nonlinear integrable equations in 1 + 1-dimensions is a subject of current research. BTs have been shown to play a key rôle in this area. Thus, until recently, the only known localized objects in 2 + 1-dimensions were the so-called lump solutions which decay algebraically at infinity. However, in an important new development, BTs have been exploited by M. Boiti and his co-workers to derive truly two-dimensional coherent structures which decay exponentially in all directions [10]. The associated nonlinear superposition principle was used to study the scattering properties of these localized phenomena. The 2 + 1-dimensional BTs employed were constructed by the DM based on appropriate gauge transformations (Boiti, Konopelchenko and Pempinelli [11]). The DM had been earlier used in 2 + 1-dimensions by Levi, Pilloni and Santini [12] to generate auto-BTs for the Kadomtsev-Petviashvili (KP), two-dimensional three wave and Davey-Stewartson (DS) equations, in turn.

Gauge transformations, in conjunction with reciprocal transformations, also play an important part in links between scattering schemes associated with classes of nonlinear integrable systems. Here, reciprocal transformations in 1 + 1- and 2 + 1- dimensions are introduced. The link between the 1 + 1-dimensional ZS-AKNS and WKI spectral schemes via a combination of gauge and reciprocal transformations is established. A reciprocal invariance of the $1+1$-dimensional Dym hierarchy is also exhibited. In $2+1$-dimensions, a reciprocal link between the KP and a Dym-type equation is revealed. This allows a novel invariance of the latter to be constructed.

2   The Dressing Method

The inverse spectral transform (IST) is now a well-established tool in the investigation of initial value problems for a wide class of nonlinear equations [13–17]. The key to the IST method is the correspondence between a **nonlinear system** and an associated pair of **linear** problems

$$L_1\psi = 0, \tag{2.1}$$
$$L_2\psi = 0, \tag{2.2}$$

where the operators L_1 and L_2 are parametrized by a certain set of functions $u_1(x_1,\dots,x_n),\dots\ u_k(x_1,\dots,x_n)$ of the independent variables $x_1,\dots,x_n$. The compatibility of the two linear problems (2.1), (2.2), namely, the commutativity condition

$$[L_1, L_2] = 0 \tag{2.3}$$

produces the nonlinear system for the u_i which is amenable to the IST. The standard IST method involves the application of both the direct and inverse spectral methods to the linear system (2.1) – (2.2) to solve initial value problems for the associated **integrable** nonlinear system. Dressing transformations

$$\psi \to \psi' = G\psi, \tag{2.4}$$
$$L_i \to L_i' = GL_iG^{-1} = L_i'(u_1',\dots,u_k') \tag{2.5}$$
$$i = 1, 2$$

applied to the inverse scattering formulation $(2.1) - (2.2)$ were originally introduced by Zakharov and Shabat [2] and provide, in particular, an incisive method for the construction of auto-BTs admitted by the nonlinear equations associated with the commutativity condition (2.1). Thus, it is required that the dressing transformations preserve the form of the operators L_1 and L_2 so that, since

$$[L_1', L_2'] = G[L_1, L_2]G^{-1} = 0, \tag{2.6}$$

it follows that the functions $u_1', \ldots, u_k'$ obey the same nonlinear system as the functions $u_1, \ldots, u_k$. Accordingly, the dressing transformations $(2.4) - (2.5)$ convert solutions of a given integrable nonlinear equation as generated by the compatibility condition (2.3), into solutions of the same equation.

In order to extract explicit dressing transformations, it is necessary to assume specific forms for the dressing (gauge) operator G. Thus, if one chooses the operator G as an appropriate Volterra-type integral operator then one arrives at the well-known Gelfand-Levitan-Marchenko linear integral equations and the corresponding reconstruction formulae for the potentials. The initial value problem for the underlying nonlinear integrable equations may then, in principle, be solved. The details of this approach to the IST via dressing transformations are set down in [2,14]. In what follows, it is shown how the DM may be used to construct auto-BTs and associated nonlinear superposition principles for nonlinear integrable equations amenable to the inverse scattering formulation $(2.1) - (2.2)$.

3　Bäcklund Transformations via the Dressing Method

The use of another ansätz for G allows the construction via the dressing transformations $(2.4) - (2.5)$ of associated auto-BTs. This method of construction was originally proposed in [12, 19, 20].

Here, we consider the application of the DM to the construction of auto-BTs for nonlinear evolution equations associated with the Zakharov-Shabat (ZS) and Ablowitz-Kaup-Newell-Segur (AKNS) spectral problem viz, [2, 15],

$$L_1\psi = \left[\frac{\partial}{\partial x} - \lambda\begin{pmatrix} 1 & 0 \\ 0 & -1 \end{pmatrix} - \begin{pmatrix} 0 & q \\ r & 0 \end{pmatrix}\right]\begin{pmatrix} \psi_1 \\ \psi_2 \end{pmatrix} = 0 \quad (3.1)$$

$$L_2\psi = \left[\frac{\partial}{\partial t} - V(x,t;\lambda)\right]\begin{pmatrix} \psi_1 \\ \psi_2 \end{pmatrix} = 0, \quad V = \begin{pmatrix} A & B \\ C & -A \end{pmatrix} \quad (3.2)$$

where λ is a (spectral) parameter, $\psi_i \quad i = 1, 2$ are eigenfunctions, $q(x,t)$ and $r(x,t)$ are generally complex-valued functions and $V(x,t;\lambda)$ is a 2×2 matrix with polynomial dependence on λ. The compatibility condition (2.3) with different $V(x,t;\lambda)$ gives rise to various systems of nonlinear evolution equations for $q(x,t)$ and $r(x,t)$. One important system is that corresponding to

$$V = \begin{pmatrix} 2\lambda^2 i - iqr & 2\lambda iq + iq_x \\ 2\lambda ir - ir_x & -2\lambda^2 i + iqr \end{pmatrix} \quad (3.3)$$

namely

$$\left.\begin{aligned} iq_t + q_{xx} - 2qrq = 0 \\ ir_t - r_{xx} + 2qrr = 0 \end{aligned}\right\} \quad (3.4)$$

where it is assumed that $q, r \to 0$ as $\mid x \mid \to \infty$. In the case $q = -\bar{r}$, the system (3.4) reduces to the nonlinear Schrödinger (NLS) equation

$$iq_t + q_{xx} + 2 \mid q \mid^2 q = 0. \quad (3.5)$$

Other specializations of V leading to nonlinear evolution equations of importance are given in [1].

The ansätz for G which give rise to BTs adopts the form [12,20]

$$G = \sum_{k=0}^{n} \lambda^k g_k(x,t) \qquad (3.6)$$

where $g_k(x,t)$ are 2×2 matrix-valued functions.

The simplest BT corresponds to the most elementary G of the type (3.6), namely, the 1st order matrix polynomial

$$G = \lambda g_1(x,t) + g_0(x,t). \qquad (3.7)$$

To determine the explicit form of the auto-BT corresponding to (3.7), it is noted that the ZS-AKNS system may be written as

$$\left. \begin{aligned} \psi_x &= U\psi \\ \psi_t &= V\psi \end{aligned} \right\} \qquad (3.8)$$

where

$$U = \lambda \sigma_3 + P \qquad (3.9)$$

with

$$\sigma_3 = \begin{pmatrix} 1 & 0 \\ 0 & -1 \end{pmatrix}, \ P = \begin{pmatrix} 0 & q \\ r & 0 \end{pmatrix}. \qquad (3.10), (3.11)$$

Under the dressing transformation (2.4) – (2.5) the ZS-AKNS system becomes

$$\left. \begin{aligned} \psi'_x &= U'\psi' \\ \psi'_t &= V'\psi' \end{aligned} \right\} \qquad (3.12)$$

where

$$U' = GUG^{-1} + G_x G^{-1}, \tag{3.13}$$

$$V' = GVG^{-1} + G_t G^{-1}. \tag{3.14}$$

In particular, substitution of (3.9) into (3.13) produces the following system of equations for g_0 and g_1:

$$[\sigma_3, g_1] = 0, \tag{3.15a}$$

$$\frac{\partial g_1}{\partial x} - [\sigma_3, g_0] + g_1 P - P' g_1 = 0, \tag{3.15b}$$

$$\frac{\partial g_0}{\partial x} + g_0 P - P' g_0 = 0, \tag{3.15c}$$

where

$$P' = \begin{pmatrix} 0 & q' \\ r' & 0 \end{pmatrix}. \tag{3.16}$$

The solution of the system (3.15a – 3.15c) is readily obtained if

$$g_1 = \begin{pmatrix} 0 & 0 \\ 0 & 1 \end{pmatrix}. \tag{3.17}$$

In this case, if $g_0 = (g_{0ij})$ then (3.15b) gives

$$g_{012} = -\tfrac{1}{2} q', \qquad g_{021} = -\tfrac{1}{2} r \tag{3.18}, (3.19)$$

while substitution of these expressions into the diagonal part of the matrix equation (3.15c) yields

$$g_{011,x} = 0,$$

$$g_{022} = \lambda_0 - \tfrac{1}{2} \int_{-\infty}^{x} \left(q'(x',t) r'(x',t) - q(x',t) r(x',t) \right) dx'$$

$$\tag{3.20}, (3.21)$$

where λ_0 is a complex constant. Further, with the use of (3.20) – (3.21), the off-diagonal part of (3.15c) yields

$$\frac{\partial q'}{\partial x} - q' \int_{-\infty}^{x} (q'r' - qr) \, dx' + 2\lambda_0 q' - 2g_{011}q = 0, \quad (3.22a)$$

$$\frac{\partial r}{\partial x} + r \int_{-\infty}^{x} (q'r' - qr) \, dx' - 2\lambda_0 r + 2g_{011}r' = 0, \quad (3.22b)$$

and elimination of the integral terms in (3.22) gives

$$\tfrac{1}{2}\frac{\partial}{\partial x}(q'r) = g_{011}(qr - q'r'). \quad (3.23)$$

Hence, (3.22a, b) yield, with $g_{011} = -1$

$$\left. \begin{array}{l} \dfrac{\partial q'}{\partial x} - \tfrac{1}{2}q'^2 r + 2\lambda_0 q' + 2q = 0, \\[2mm] \dfrac{\partial r}{\partial x} + \tfrac{1}{2}q'r^2 - 2\lambda_0 r - 2r' = 0. \end{array} \right\} \quad (3.24)$$

The relations (3.24) constitute the spatial part of the auto-BT associated with the dressing (gauge) transformation with

$$G := G_{\lambda_0}^{(1)}(\lambda) = \lambda \begin{pmatrix} 0 & 0 \\ 0 & 1 \end{pmatrix} + \begin{pmatrix} -1 & -\tfrac{1}{2}q' \\ -\tfrac{1}{2}r & \lambda_0 - \tfrac{1}{4}q'r \end{pmatrix}. \quad (3.25)$$

They are generic to the whole class of nonlinear evolution equations connected with the linear system (3.1).

On the other hand, the temporal part of the BT depends on the particular member of the ZS-AKNS class: thus, it is determined by the form of V. In particular, with the specialization (3.3) corresponding to the coupled NLS system (3.4), the relation (3.14) produces

$$
\left.
\begin{aligned}
i\frac{\partial q'}{\partial t} - q'\left(q'r' + qr - \tfrac{1}{2}r\frac{\partial q'}{\partial x}\right) - 2\lambda_0\frac{\partial q'}{\partial x} - 2\frac{\partial q}{\partial x} &= 0, \\
i\frac{\partial r}{\partial t} + r\left(q'r' + qr + \tfrac{1}{2}q'\frac{\partial r}{\partial x}\right) - 2\lambda_0\frac{\partial r}{\partial x} - 2\frac{\partial r'}{\partial x} &= 0
\end{aligned}
\right\}.
$$

$$(3.26)$$

The relations (3.26) represent the temporal part of the auto-BT for the coupled NLS system (3.4) corresponding to the dressing transformations $G = G^{(1)}_{\lambda_0}(\lambda)$.

Another simple BT corresponds to the choice

$$
g_1 = \begin{pmatrix} 1 & 0 \\ 0 & 0 \end{pmatrix}
\tag{3.27}
$$

in (3.15). Proceeding as in the previous case, it is readily seen that the associated dressing (gauge) transformation is

$$
G := G^{(2)}_{\mu_0}(\lambda) = \lambda\begin{pmatrix} 1 & 0 \\ 0 & 0 \end{pmatrix} + \begin{pmatrix} \mu_0 - \tfrac{1}{4}r'q & \tfrac{1}{2}q \\ \tfrac{1}{2}r' & -1 \end{pmatrix}
\tag{3.28}
$$

with corresponding generic spatial part of the auto-BT

$$
\left.
\begin{aligned}
\frac{\partial q}{\partial x} - \tfrac{1}{2}r'q^2 + 2\mu_0 q + 2q' &= 0, \\
\frac{\partial r'}{\partial x} + \tfrac{1}{2}r'^2 q - 2\mu_0 r' - 2r &= 0
\end{aligned}
\right\}.
$$

$$(3.29)$$

The temporal part of the BT in the case of the coupled NLS system (3.4) is given by (3.14) together with (3.3) as

$$
\left.
\begin{aligned}
i\frac{\partial q}{\partial t} - q\left(q'r' + qr - \tfrac{1}{2}r'\frac{\partial q'}{\partial x}\right) - 2\mu_0\frac{\partial q}{\partial x} - 2\frac{\partial q'}{\partial x} = 0 \\[2mm]
i\frac{\partial r'}{\partial t} + r'\left(q'r' + qr + \tfrac{1}{2}q\frac{\partial r}{\partial x}\right) - 2\mu_0\frac{\partial r'}{\partial x} - 2\frac{\partial r}{\partial x} = 0
\end{aligned}
\right\}.
$$

$$(3.30)$$

The BTs associated with the gauge (dressing transformations) $G^{(1)}_{\lambda_0}(\lambda)$ and $G^{(2)}_{\mu_0}(\lambda)$ play an important rôle in the analysis of the nonlinear evolution equations associated with the ZS-AKNS spectral problem $(3.1) - (3.2)$. They are termed **elementary** BTs and are here denoted by $\mathbb{B}^{(1)}_{\lambda_0}$ and $\mathbb{B}^{(2)}_{\mu_0}$ respectively. It is readily shown that any BT corresponding to a gauge operator of the form (3.6) may be obtained as an appropriate product of several $G^{(1)}_{\lambda_0}(\lambda)$ and $G^{(2)}_{\mu_0}(\lambda)$ with different λ_0 and μ_0. In particular, the general operator G linear in λ as given by (3.7) can be represented as the product of two **elementary** gauge operators $G^{(1)}_{\lambda_0}(\lambda)$ and $G^{(2)}_{\mu_0}(\lambda)$. The associated elementary BTs may likewise be composed to generate broad classes of BTs of the type

$$
\mathbb{B}^{(n_1,n_2)} := \prod_{i=1}^{n_1}\mathbb{B}^{(1)}_{\lambda_i}\ \prod_{k=1}^{n_2}\mathbb{B}^{(2)}_{\mu_k}.
$$

$$(3.31)$$

The sequential application of $G^{(1)}_{\lambda_0}(\lambda)$ and $G^{(2)}_{\mu_0}(\lambda)$ to ψ yields

$$G_{\mu_0}^{(2)}(\lambda)G_{\lambda_0}^{(1)}(\lambda)\psi$$

$$= \begin{pmatrix} \lambda + \mu_0 - \frac{1}{4}r_{12}q_1, & \frac{q_1}{2} \\ \frac{r_{12}}{2}, & -1 \end{pmatrix} \begin{pmatrix} -1, & \frac{-q_1}{2} \\ \frac{-r_0}{2}, & \lambda + \lambda_0 - \frac{r_0 q_1}{4} \end{pmatrix} \psi$$

$$(3.32)$$

while, the reversed order composition gives

$$G_{\lambda_0}^{(1)}(\lambda)G_{\mu_0}^{(2)}(\lambda)\psi$$

$$= \begin{pmatrix} -1, & -q_{21} \\ \frac{-r_2}{2}, & \lambda + \lambda_0 - \frac{r_2 q_{21}}{4} \end{pmatrix} \begin{pmatrix} \lambda + \mu_0 - \frac{1}{4}r_2 q_0, & \frac{q_0}{2} \\ \frac{-r_2}{2}, & -1 \end{pmatrix} \psi$$

$$(3.33)$$

In the above, we adopt the notation

$$q_{12} := \mathbb{B}_{\mu_0}^2 \mathbb{B}_{\lambda_0}^1 q_0 \qquad , \qquad q_{21} := \mathbb{B}_{\lambda_0}^1 \mathbb{B}_{\mu_0}^2 q_0,$$

$$r_{12} := \mathbb{B}_{\mu_0}^2 \mathbb{B}_{\lambda_0}^1 r_0 \qquad , \qquad r_{21} := \mathbb{B}_{\lambda_0}^1 \mathbb{B}_{\mu_0}^2 r_0 \qquad (3.34)$$

$$q_1 := \mathbb{B}_{\lambda_0}^1 q_0 \qquad , \qquad r_2 := \mathbb{B}_{\mu_0}^2 r_0.$$

It is seen that commutativity of the action of the elementary gauge transformations is assured by the two conditions

$$q_{21} - q_0 = \frac{2(\lambda_0 - \mu_0)}{\frac{r_2}{2} + \frac{2}{q_1}} \qquad (3.35)$$

$$r_{12} - r_0 = \frac{-2(\lambda_0 - \mu_0)}{\frac{q_1}{2} + \frac{2}{r_2}} \qquad (3.36)$$

It is noted that $(3.35) - (3.36)$ impose the simple relation

$$r_2(q_0 - q_{21}) = q_1(r_{12} - r_0) \qquad (3.37)$$

independent of any λ_0 and μ_0.

Let us now turn to the action of the BTs as set down in (3.34). The situation is represented in Figure 1

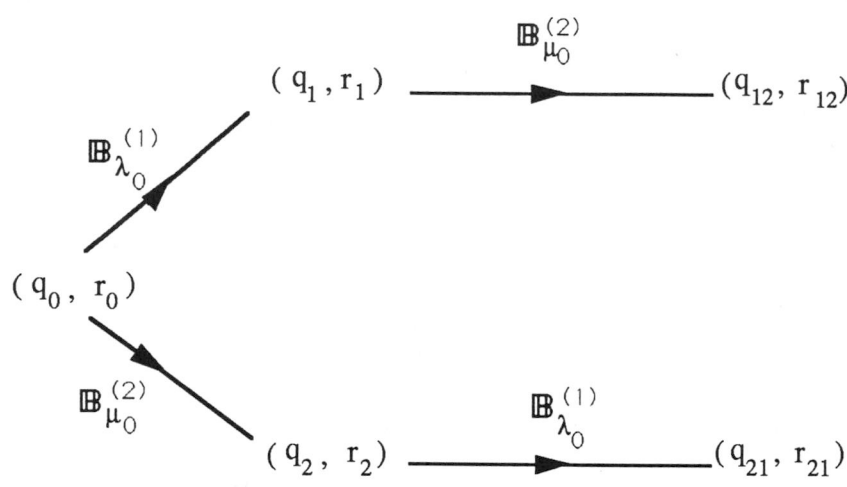

Figure 1

Action of Elementary BTs

In view of (3.24) and (3.29), the diagram is equivalent to the following system of Bäcklund relations:

$\mathbb{B}_{\lambda_0}^{(1)} : (q_0, r_0) \to (q_1, r_1)$

$$
\begin{cases}
\dfrac{\partial q_1}{\partial x} - \tfrac{1}{2} q_1^2 r_0 + 2\lambda_0 q_1 + 2q_0 = 0, & (3.38a) \\[2ex]
\dfrac{\partial r_0}{\partial x} + \tfrac{1}{2} q_1 r_0^2 - 2\lambda_0 r_0 - 2r_1 = 0, & (3.38b)
\end{cases}
$$

$\mathbb{B}^{(2)}_{\mu_0} : (q_1,r_1) \rightarrow (q_{12},r_{12})$

$$\begin{cases} \dfrac{\partial q_1}{\partial x} - \tfrac{1}{2}r_{12}q_1^2 + 2\mu_0 q_1 + 2q_{12} = 0, & (3.39a) \\[4mm] \dfrac{\partial r_{12}}{\partial x} + \tfrac{1}{2}r_{12}^2 q_1 - 2\mu_0 r_{12} - 2r_1 = 0, & (3.39b) \end{cases}$$

$\mathbb{B}^{(2)}_{\mu_0} : (q_0, r_0) \rightarrow (q_2, r_2)$

$$\begin{cases} \dfrac{\partial q_0}{\partial x} - \tfrac{1}{2}r_2 q_0^2 + 2\mu_0 q_0 + 2q_2 = 0, & (3.40a) \\[4mm] \dfrac{\partial r_2}{\partial x} + \tfrac{1}{2}r_2^2 q_0 - 2\mu_0 r_2 - 2r_0 = 0, & (3.40b) \end{cases}$$

$\mathbb{B}^{(1)}_{\lambda_0} : (q_2,r_2) \rightarrow (q_{21},r_{21})$

$$\begin{cases} \dfrac{\partial q_{21}}{\partial x} - \tfrac{1}{2}q_{21}^2 r_2 + 2\lambda_0 q_{21} + 2q_2 = 0, & (3.41a) \\[4mm] \dfrac{\partial r_2}{\partial x} + \tfrac{1}{2}q_{21}r_2^2 - 2\lambda_0 r_2 - 2r_{21} = 0. & (3.41b) \end{cases}$$

Comparison of (3.38a) and (3.39a) yields

$$-\tfrac{1}{2}q_1^2 r_0 + 2\lambda_0 q_1 + 2q_0 = -\tfrac{1}{2}r_{12}q_1^2 + 2\mu_0 q_1 + 2q_{12}, \qquad (3.42)$$

while, comparison of (3.40b) and (3.41a) gives

$$\tfrac{1}{2}r_2^2 q_0 - 2\mu_0 r_2 - 2r_0 = \tfrac{1}{2}q_{21}r_2^2 - 2\lambda_0 r_2 - 2r_{21}, \qquad (3.43)$$

whence,

$$2(q_{12} - q_0) = 2q_1(\lambda_0 - \mu_0) + \frac{q_1^2}{2}(r_{12} - r_0), \qquad (3.44)$$

and

$$2(r_{21} - r_0) = 2r_2(\mu_0 - \lambda_0) + \frac{r_2^2}{2}(q_{21} - q_0), \qquad (3.45)$$

in turn.

On use of (3.35) in (3.45) to eliminate q_{21} we obtain

$$\frac{q_1 r_2}{2}(r_{21} - r_0) + 2(r_{21} - r_0) + 2r_2(\lambda_0 - \mu_0) = 0, \qquad (3.46)$$

while (3.36) yields

$$\frac{q_1 r_2}{2}(r_{12} - r_0) + 2(r_{12} - r_0) + 2r_2(\lambda_0 - \mu_0) = 0. \qquad (3.47)$$

Subtraction of (3.46) and (3.47) gives

$$r_{12} = r_{21} := r_3 \qquad (3.48)$$

In a similar manner, (3.44) together with (3.35) − (3.36) give

$$\frac{q_1 r_2}{2}(q_{12} - q_0) + 2(q_{12} - q_0) + 2q_1(\mu_0 - \lambda_0) = 0, \qquad (3.49)$$

$$\frac{q_1 r_2}{2}(q_{21} - q_0) + 2(q_{21} - q_0) + 2q_1(\mu_0 - \lambda_0) = 0, \qquad (3.50)$$

whence, on subtraction,

$$q_{12} = q_{21} := q_3 . \qquad (3.51)$$

Hence, the composition of the elementary BTs $\mathbb{B}^{(1)}_{\lambda_0}$ and $\mathbb{B}^{(2)}_{\mu_0}$ may be represented in a **commutative Bianchi** diagram as below in Figure 2.

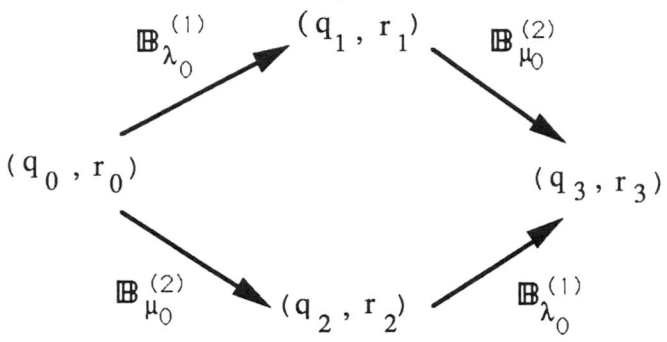

Figure 2

The Commutative Bianchi Diagram

The commutativity conditions (3.35), (3.36), namely [8]

$$q_3 = q_0 + \frac{2(\lambda_0 - \mu_0)}{\frac{r_2}{2} + \frac{2}{q_1}}, \qquad r_3 = r_0 + \frac{2(\mu_0 - \lambda_0)}{\frac{q_1}{2} + \frac{2}{r_2}} \quad (3.52), (3.53)$$

represent **nonlinear superposition principles** whereby (q_3, r_3) may be calculated purely algebraically given solution pairs $(q_0, r_0), (q_1, r_1)$ and (q_2, r_2).

Elimination of (q_1, r_1) and (q_2, r_2) from the system $(3.38) - (3.41)$ produces relations containing only (q_0, r_0) and (q_3, r_3), namely

$$\frac{\partial}{\partial x}(q_3 - q_0) = -(\lambda_0 + \mu_0)(q_3 - q_0)$$

$$+ (q_3 + q_0)\sqrt{(\lambda_0 - \mu_0)^2 + (q_3 - q_0)(r_3 - r_0)}$$
$$(3.54)$$

$$\frac{\partial}{\partial x}(r_3 - r_0) = (\lambda_0 + \mu_0)(r_3 - r_0)$$

$$+ (r_3 + r_0)\sqrt{(\lambda_0 - \mu_0)^2 + (q_3 - q_0)(r_3 - r_0)}$$

$$\tag{3.55}$$

These relations determine the simplest non-elementary BT, $\mathbb{B}^{\text{sol}}_{\lambda_0,\mu_0} = \mathbb{B}^{(1)}_{\lambda_0}\mathbb{B}^{(2)}_{\mu_0} = \mathbb{B}^{(2)}_{\mu_0}\mathbb{B}^{(1)}_{\lambda_0}$. In the special case

$$r = -\bar{q}, \qquad \mu_0 = -\lambda_0 \tag{3.56}$$

if we set $q_3 := q', q_0 := -q$ then the well known spatial BT

$$q_x + q'_x = (q - q')\sqrt{4\lambda_0^2 - |q + q'|^2} \tag{3.57}$$

for the NLS (3.5) is retrieved [1].

The existence of nonlinear superposition principles underlying auto BTs is of the greatest importance since it allows us to generate, by iteration of purely algebraic procedures, sequences of exact solutions to the nonlinear evolution equations amenable to the inverse scattering method. In particular, multi-soliton solutions may be so-generated.

Here, the procedure is illustrated for the coupled NLS system (3.4). This admits the starting trivial solution

$$P = P_{0,0} := \begin{pmatrix} 0 & 0 \\ 0 & 0 \end{pmatrix}, \tag{3.58}$$

corresponding to $(q, r) = (q_0, r_0) = (0, 0)$. Substitution in the Bäcklund relations (3.38) and (3.40) with $\lambda_0 \to \lambda_1, \mu_0 \to \mu_1$ yields, on application of $\mathbb{B}^{(1)}_{\lambda_1}$ and $\mathbb{B}^{(2)}_{\mu_1}$ to (q_0, r_0)

$$\frac{\partial q_1}{\partial x} + 2\lambda_1 q_1 = 0, \quad r_1 = 0,$$

$$q_2 = 0, \quad \frac{\partial r_2}{\partial x} - 2\mu_1 r_2 = 0,$$

whence

$$q_1 = \phi(t)e^{-2\lambda_1 x}, \quad r_1 = 0$$

$$q_2 = 0, \quad r_2 = \psi(t)e^{2\mu_1 x}.$$

The nature of $\phi(t)$ and $\psi(t)$ may be determined by the temporal part of the BT. Alternatively, on substitution back in the original NLS system (3.4) we obtain

$$q_1 = q_{(10)} = e^{4i\lambda_1^2 t - 2\lambda_1(x - x_{01})}, \quad r_1 = r_{(10)} = 0,$$

$$q_2 = q_{(01)} = 0, \quad r_2 = r_{(01)} = e^{-4i\mu_1^2 t + 2\mu_1(x - \tilde{x}_{01})}$$

where $x_{01}, \tilde{x}_{01}$ are arbitrary constants.

Application of the nonlinear superposition principle now yields.

$$q_3 = \mathbb{B}_{\lambda_1}^{(1)} \mathbb{B}_{\mu_1}^{(2)} q_0 := q_{1,1} = q_{0,0} + \frac{2(\lambda_1 - \mu_1)}{\dfrac{r_{0,1}}{2} + \dfrac{2}{q_{1,0}}},$$

$$r_3 = \mathbb{B}_{\lambda_1}^{(1)} \mathbb{B}_{\mu_1}^{(2)} r_0 := r_{1,1} = r_{0,0} + \frac{2(\mu_1 - \lambda_1)}{\dfrac{q_{1,0}}{2} + \dfrac{2}{r_{0,1}}}.$$

On repeated application of the elementary BTs to P_0 we obtain

$$\prod_{i=1}^{n_1} \mathbb{B}_{\lambda_i}^{(1)} \prod_{k=1}^{n_2} \mathbb{B}_{\mu_k}^{(2)} P_{0,0} = \begin{pmatrix} 0 & q_{n_1,n_2} \\ r_{n_1,n_2} & 0 \end{pmatrix} := P_{n_1,n_2} \qquad (3.59)$$

where the nonlinear superposition principle yields

$$q_{n_1+1,n_2+1} = q_{n_1,n_2} + \frac{2(\lambda_1 - \mu_1)}{\dfrac{r_{n_1,n_2+1}}{2} + \dfrac{2}{q_{n_1+1,n_2}}}, \qquad (3.60)$$

$$r_{n_1+1,n_2+1} = r_{n_1,n_2} + \cfrac{2(\mu_1 - \lambda_1)}{\cfrac{q_{n_1+1,n_2}}{2} + \cfrac{2}{r_{n_1,n_2+1}}}, \qquad (3.61)$$

$$n_1, n_2 = 0, 1, \ldots .$$

The solutions $P_{n_1,0}$ and P_{0,n_2} are readily derived via the Bäcklund relations and are given by

$$q_{n_1,0} = \sum_{k=1}^{n_1} e^{4i\lambda_k^2 t - 2\lambda_k(x - x_{0k})}, \qquad r_{n_1,0} = 0 \qquad (3.62)$$

$$q_{0,n_2} = 0, \qquad r_{0,n_2} = \sum_{k=1}^{n_2} e^{-4i\mu_k^2 t + 2\mu_k(x - \tilde{x}_{0k})} \qquad (3.63)$$

where x_{0k} and $\tilde{x}_{0k}$ are arbitrary constants. The remaining solutions P_{n_1,n_2} may now all be calculated by a purely algebraic procedure on use of the nonlinear superposition principles (3.60) – (3.61). Thus, in particular, on use of $P_{0,0}, P_{1,0}$ and $P_{0,1}$ it is seen that

$$q_{1,1} = 2(\lambda_1 - \mu_1) \left[\frac{e^{-4i\mu_1^2 t + 2\mu_1(x - \tilde{x}_{01})}}{2} + 2e^{-4i\lambda_1^2 t + 2\lambda_1(x - x_{01})} \right]^{-1},$$

$$(3.64)$$

$$r_{1,1} = 2(\mu_1 - \lambda_1) \left[\frac{e^{-4i\lambda_1^2 t - 2\lambda_1(x - x_{01})}}{2} + 2e^{4i\mu_1^2 t - 2\mu_1(x - \tilde{x}_{01})} \right]^{-1}.$$

$$(3.65)$$

Further, on application of (3.60) – (3.61) to $P_{0,1}, P_{1,1}, P_{0,2}$ we obtain $P_{1,2}$. Iteration of this procedure leads to any P_{n_1,n_2} [8]. This is given in terms of $q_{1,0} \ldots q_{n_1,0}$ and $r_{0,1} \ldots, r_{0,n_2}$. Thus, P_{n_1,n_2} is a **nonlinear** superposition of plane wave solutions to the decoupled **linear** Schrödinger equations

$$iq_t + q_{xx} = 0,$$

$$ir_t - r_{xx} = 0 .$$

The iterative procedure is depicted in a lattice diagram in Figure 3. Therein, q_{n_1,n_2}, r_{n_1,n_2} are denoted by q_{n_1,n_2} and r_{n_1,n_2} respectively.

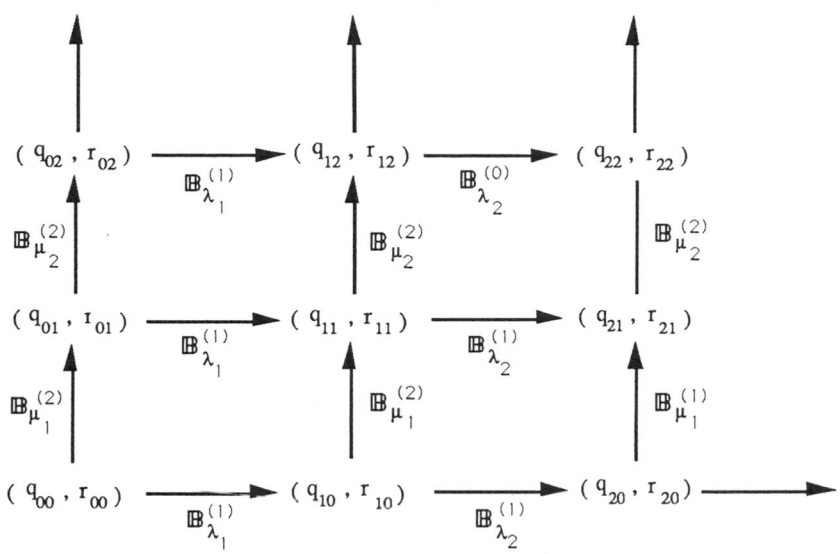

Figure 3.

The Elementary Bäcklund Transformation Lattice

It is noted that the above lattice of solutions is a consequence of the elementary BT decomposition. The soliton BTs $\mathbb{B}^{sol}_{\lambda_0,\mu_0}$ deliver the soliton ladder $P_{n,n}$ $n = 1, 2, \dots$ on the diagonal part of the lattice.

4   Elementary Bäcklund Transformations in $2 + 1$-Dimensions

Elementary BTs may also be constructed for $2+1$-dimensional integrable equations.

In the 2+1-dimensional generalization of the ZS-AKNS system we are concerned with the pair of linear equations

$$T_1\psi = 0,$$
$$T_2\psi = 0, \tag{4.1}$$

where

$$T_1 = \frac{\partial}{\partial x} + \sigma_3 \frac{\partial}{\partial y} + P(x, y, t) \tag{4.2}$$

$$T_2 = \frac{\partial}{\partial t} + V\left(x, y, t, \frac{\partial}{\partial y}\right) \tag{4.3}$$

with

$$P = \begin{pmatrix} 0 & q(x, y, t) \\ r(x, y, t) & 0 \end{pmatrix}. \tag{4.4}$$

The compatibility condition

$$[T_1, T_2] = 0 \tag{4.6}$$

with various $V\left(x, y, t, \frac{\partial}{\partial y}\right)$ generates $2+1$-dimensional integrable nonlinear systems for q, r. In particular, corresponding to

$$V = -2i\sigma_3 \frac{\partial^2}{\partial y^2} - 2iP \frac{\partial}{\partial y} + i\sigma_3 P_x - iP_y + i\sigma_3 P^2 - i\phi \tag{4.7}$$

we obtain a $2 + 1$-dimensional integrable generalisation of the system (3.4), namely the Davey-Stewartson (DS) system

$$i\frac{\partial P}{\partial t} + \sigma_3 \left(\frac{\partial^2 P}{\partial x^2} + \frac{\partial^2 P}{\partial y^2}\right) - 2\sigma_3 P^3 - [\phi, P] = 0, \left.\begin{array}{l} \\ \\ \end{array}\right\}$$

$$\left(\frac{\partial}{\partial x} + \sigma_3 \frac{\partial}{\partial y}\right)\phi - 2\frac{\partial}{\partial y}P^2 = 0 \tag{4.8}$$

where ϕ is an auxiliary 2×2 diagonal matrix [2]. The DS system was first derived in the analysis of the propagation of two-dimensional nonlinear waves in dispersive media [21].

The dressing method may be extended to integrable systems in $2 + 1$-dimensions [2]. In this case, the dressing (gauge) operator G adopts a **polynomial** form in the derivative $\dfrac{\partial}{\partial y}$, namely [11]

$$G = \sum_{k=0}^{n} G_k(x, y, t) \frac{\partial^k}{\partial y^k}. \tag{4.9}$$

The elementary dressings $G^{(1)}$ and $G^{(2)}$ may be shown to be given by

$$G^{(i)} = \alpha_i \frac{\partial}{\partial y} - \tfrac{1}{2}\sigma_3(P'\alpha_i - \alpha_i P) - \tfrac{1}{2}\sigma_3\alpha_i\partial^{-1}(P'^2 - P^2) + \beta_i$$

$$(i = 1, 2) \tag{4.10}$$

where

$$\alpha_1 = \begin{pmatrix} 0 & 0 \\ 0 & 1 \end{pmatrix}, \qquad \alpha_2 = \begin{pmatrix} 1 & 0 \\ 0 & 0 \end{pmatrix},$$

$$\beta_1 = \begin{pmatrix} 1 & 0 \\ 0 & \lambda_0 \end{pmatrix}, \qquad \beta_2 = \begin{pmatrix} \mu_0 & 0 \\ 0 & 1 \end{pmatrix}, \tag{4.11}$$

$$\partial^{-1} := \left(\frac{\partial}{\partial x} + \sigma_3 \frac{\partial}{\partial y} \right)^{-1}.$$

The spatial parts of the corresponding elementary BTs
are given by

$$\mathbb{B}^{(1)}_{\lambda_0} \begin{cases} \left(\dfrac{\partial}{\partial x} - \dfrac{\partial}{\partial y}\right) q' - \dfrac{q'}{2} \partial_1^{-1} \left(\left(\dfrac{\partial}{\partial x'} - \dfrac{\partial}{\partial y}\right) rq'\right) + 2\lambda_0 q' + 2q = 0, \\[2mm] \left(\dfrac{\partial}{\partial x} - \dfrac{\partial}{\partial y}\right) r + \dfrac{r}{2} \partial_1^{-1} \left(\left(\dfrac{\partial}{\partial x'} - \dfrac{\partial}{\partial y}\right) rq'\right) - 2\lambda_0 r - 2r' = 0, \end{cases}$$

$$(4.12)$$

$$\mathbb{B}^{(2)}_{\mu_0} \begin{cases} \left(\dfrac{\partial}{\partial x} + \dfrac{\partial}{\partial y}\right) q - \dfrac{q}{2} \partial_2^{-1} \left(\left(\dfrac{\partial}{\partial x'} + \dfrac{\partial}{\partial y}\right) qr'\right) + 2\mu_0 q + 2q' = 0, \\[2mm] \left(\dfrac{\partial}{\partial x} + \dfrac{\partial}{\partial y}\right) r' + \dfrac{r'}{2} \partial_2^{-1} \left(\left(\dfrac{\partial}{\partial x'} + \dfrac{\partial}{\partial y}\right) qr'\right) - 2\mu_0 r' - 2r = 0 \end{cases}$$

$$(4.13)$$

where

$$\partial_1^{-1} := \left(\dfrac{\partial}{\partial x} + \dfrac{\partial}{\partial y}\right)^{-1}, \quad \partial_2^{-1} := \left(\dfrac{\partial}{\partial x} - \dfrac{\partial}{\partial y}\right)^{-1}. \quad (4.14)$$

Nonlinear superposition principles which generalize
(3.52), (3.53), namely

$$q_3 = \frac{\frac{1}{4} q_1^2 r_0 - \frac{1}{2} q_1 \Gamma + (\lambda_0 - \mu_0) q_1 + q_0 + \dfrac{\partial q_1}{\partial y}}{1 + \frac{1}{4} q_1 r_2} \quad (4.15)$$

$$r_3 = \frac{\frac{1}{4} r_2^2 q_0 + \frac{1}{2} r_2 \Gamma - (\lambda_0 - \mu_0) r_2 + r_0 + \dfrac{\partial r_2}{\partial y}}{1 + \frac{1}{4} q_1 r_2} \quad (4.16)$$

with

$$\Gamma := \partial_1^{-1}(q_1 r_1 - q_0 r_0) + \partial_2^{-1}(q_2 r_2 - q_0 r_0) \quad (4.17)$$

are readily derived. The associated commutative Bianchi diagram allows the construction of a lattice of exact solutions to the DS system (4.8).

The solutions $P_{n_1,0}$ and P_{0,n_2} which generalize (3.62), (3.63) are given by

$$
\begin{cases}
q_{n_1,0} = \sum_{k=1}^{n_1} \int_{\Gamma_1} C_1^{(k)}(\mu) \cdot \\[2mm]
\quad \exp\left\{2i\left(\lambda_k^{(1)^2} - \mu^2\right)t + \mu(x+y) - \lambda_k^{(1)}(x-y)\right\} d\mu \\[4mm]
r_{n_1,0} = 0,
\end{cases}
$$

$$
\begin{cases}
q_{0,n_2} = 0, \\[4mm]
r_{0,n_2} = \sum_{k=1}^{n_2} \int_{\Gamma_2} C_2^{(k)}(\nu) \cdot \\[2mm]
\quad \exp\left\{2i\left(\lambda_k^{(2)^2} - \nu^2\right)t - \nu(x+y) + \lambda_k^{(2)}(x+y)\right\} d\nu
\end{cases}
$$
(4.19)

(4.18)

where $C_1^{(k)}(\mu), C_2^{(k)}(\nu)$ are arbitrary functions and Γ_1, Γ_2 are arbitrary contours in the complex planes of the variables μ and ν.

The simplest soliton-type solution $P_{1,1}$ of the DS equation is of the form

$$
q_{1,1} = \frac{1}{D}\int_{\Gamma_1}(\lambda_1 + \mu)C_1(\mu) \cdot
$$

$$
\exp\{2i(\lambda_1^2 - \mu^2)t + \mu(x+y) - \lambda_1(x-y)\}d\mu,
$$
(4.20)

$$
r_{1,1} = \frac{1}{D}\int_{\Gamma_2}(\lambda_2 + \nu)C_2(\nu) \cdot
$$

$$
\exp\{2i(\lambda_2^2 - \nu^2)t - \nu(x-y) + \lambda_2(x+y)\}d\nu
$$

where

$$\mathcal{D} = 1 + \tfrac{1}{4} \int_{\Gamma_1} \int_{\Gamma_2} C_1(\mu) C_2(\nu).$$

$$\exp\{2i(\lambda_1^2 + \lambda_2^2 - \mu^2 - \nu^2)t \tag{4.21}$$

$$+ (\mu + \lambda_2)(x + y) - (\nu + \lambda_2)(x - y)\} d\nu \, d\mu$$

and $C_1(\mu), C_2(\nu)$ are arbitrary functions.

The presence of arbitrary functions in the solutions $P_{n_1,0}$ and P_{0,n_2} and, hence, in the solutions P_{n_1,n_2} is the important feature of the 2+1-dimensional case. It leads to a much richer class of solutions than in the one-dimensional case. In particular, an appropriate choice of the functions $C_1^{(k)}$ and $C_2^{(k)}$ in the P_{n_1,n_2} gives rise to solutions of the DS equation which are exponentially localized in the plane [10]. The simplest solution of this type is the breather

$$q(x, y, t) = r^*(x, y, t)$$

$$= \frac{1}{\mathcal{D}} \exp\{i(\nu_1 x + \nu_2 y - \tfrac{1}{2}(\nu_1^2 + \nu_2^2)t + (\lambda^2 + \mu^2)t)\} \tag{4.22}$$

where

$$\mathcal{D} = \cosh\{(\lambda + \mu)(x - \nu_1 t) - (\lambda - \mu)(y - \nu_2 t)\}$$

$$+ \sqrt{2} \cos h\{(\lambda - \mu)(x - \nu_1 t) - (\lambda + \mu)(y - \nu_2 t) + \delta\}. \tag{4.23}$$

The recent discovery of these $2 + 1$-dimensional localized structures via BTs has injected a new interest into the study of higher-dimensional integrable systems. In particular, it has led to a study of certain boundary value problems for the DS equation and of $2 + 1$-dimensional coherent entities which have been termed **dromions** [22].

5 Reciprocal Transformations in $1+1$-Dimensions Linked Inverse Scattering Schemes.

Reciprocal transformations have been extensively employed in Continuum Mechanics not only to reveal hidden symmetries in nonlinear systems but also to solve nonlinear boundary value problems. These applications are described in detail in Rogers and Shadwick [1] and Rogers and Ames [23].

In the present context of Soliton Theory, reciprocal transformations in $1+1$-dimensions may be shown to be a key component in the link between the AKNS and WKI inverse scattering schemes [24]. Moreover, the Dym hierarchy as set forth by Calogero and Degasperis [17] is invariant under a class of reciprocal transformations. This result may be used to construct a generic auto-BT for the KdV hierarchy in a novel manner. It is this area of application of reciprocal transformations that we now describe.

In the sequel, we make use of the following result [25]:

RI

The conservation law

$$\frac{\partial}{\partial t}\left\{T\left(\frac{\partial}{\partial x};\frac{\partial}{\partial t}:u\right)\right\} + \frac{\partial}{\partial x}\left\{F\left(\frac{\partial}{\partial x};\frac{\partial}{\partial t}:u\right)\right\} = 0 \quad (5.1)$$

is transformed to the associated conservation law

$$\frac{\partial}{\partial t'}\left\{T'\left(\frac{\partial}{\partial x'};\frac{\partial}{\partial t'}:u\right)\right\} + \frac{\partial}{\partial x'}\left\{F'\left(\frac{\partial}{\partial x'};\frac{\partial}{\partial t'}:u\right)\right\} = 0$$
$$(5.2)$$

by the **reciprocal** transformation

$$dx' = T\,dx - F\,dt, \qquad t' = t$$

$$T' = \frac{1}{T(D;\partial';u)},$$

$$\left.\begin{array}{c} \\ \\ \\ \\ \\ \end{array}\right\} R \qquad (5.3)$$

$$F' = \frac{-F(\mathbb{D};\partial:u)}{T(\mathbb{D};\partial:u)}$$

where

$$\mathbb{D} := \frac{\partial}{\partial x} = \frac{1}{T'}\frac{\partial}{\partial x'},$$

$$\partial := \frac{\partial}{\partial t} = \frac{F'}{T'}\frac{\partial}{\partial x'} + \frac{\partial}{\partial t'},$$

$$0 <| T |< \infty.$$

The above result is readily demonstrated. Thus, (5.3) implies that

$$\frac{\partial x'}{\partial x} = T, \qquad \frac{\partial x'}{\partial t} = -F$$

whence compatibility yields (5.1).

Similarly, (5.3) implies that

$$\frac{\partial x}{\partial x'} = \frac{1}{T}, \qquad \frac{\partial x}{\partial t'} = \frac{F}{T}$$

and compatibility gives (5.2).

It is readily shown that $R^2 = I$ so that the transformation R is reciprocal.

Thus,

$$dx'' = T'\,dx' - F'\,dt' = T'(T\,dx - F\,dt) - F'\,dt' = dx,$$

$$dt'' = dt' = dt,$$

$$T'' = \frac{1}{T'} = T, \qquad F'' = \frac{-F'}{T'} = \frac{F}{TT'} = F.$$

The above result will now be shown to be a key component in the link between the ZS-AKNS scattering scheme and the WKI scattering scheme

$$\left.\begin{array}{c} \psi_{x^*} = U^*\psi, \\ \\ \psi_{t^*} = V^*\psi, \end{array}\right\} \tag{5.4}$$

where

$$U^* = \begin{pmatrix} -i\lambda^* & \lambda^*q^* \\ \lambda^*r^* & i\lambda^* \end{pmatrix}, \quad V^* = \begin{pmatrix} A^* & B^* \\ C^* & -A^* \end{pmatrix}. \tag{5.5}, (5.6)$$

In this connection, we recall the result of Section 3 concerning the gauge transformation G of the ZS-AKNS scheme $(3.8) - (3.11)$. In particular, we consider the gauge transformation $G = g^{-1}$ given by

$$\left.\begin{array}{c} \tilde{\psi} = g^{-1}\psi, \\ \\ \tilde{U} = g^{-1}Ug - g^{-1}g_x, \\ \\ \tilde{V} = g^{-1}Vg - g^{-1}g_t \end{array}\right\} g^{-1} \tag{5.7}$$

where, as in Zakharov and Takhtadzhyan [26], g is taken as the solution of the system $(3.8) - (3.11)$ with $\lambda = 0$ so that

$$g_x = \begin{pmatrix} 0 & q \\ r & 0 \end{pmatrix} g, \tag{5.8}$$

$$g_t = \begin{pmatrix} A(x,t;0) & B(x,t;0) \\ C(x,t;0) & -A(x,t;0) \end{pmatrix}. \tag{5.9}$$

It is observed that the gauge transformation g^{-1} has inverse g with

$$\left.\begin{array}{l} \psi = g\tilde{\psi}, \\[2mm] U = g\tilde{U}g^{-1} + g_x g^{-1}, \\[2mm] V = g\tilde{V}g^{-1} + g_t g^{-1}, \end{array}\right\} \; g \qquad (5.10)$$

whence, with the choice of g given by (5.8)–(5.9), the inverse scattering scheme

$$\left.\begin{array}{l} \tilde{\psi}_x = \tilde{U}\,\tilde{\psi}, \\[2mm] \tilde{\psi}_t = \tilde{V}\,\tilde{\psi} \end{array}\right\} \qquad (5.11)$$

with

$$\tilde{U} = \lambda S, \qquad (5.12)$$

$$\tilde{V} = g^{-1}\begin{pmatrix} A - A\,|_{\lambda=0} & B - B\,|_{\lambda=0} \\[2mm] C - C\,|_{\lambda=0} & -A + A\,|_{\lambda=0} \end{pmatrix} g \qquad (5.13)$$

where

$$S = g^{-1}\begin{pmatrix} 1 & 0 \\ 0 & -1 \end{pmatrix} g = \begin{pmatrix} a & b \\ c & -a \end{pmatrix} \qquad (5.14)$$

with

$$a^2 + bc = 1 \qquad (5.15)$$

is obtained. The inverse scattering scheme (5.11) gauge equivalent to the ZS-AKNS scheme (3.8)-(3.11) has compatibility condition

$$S_t - \lambda^{-1}\tilde{V}_x + S\tilde{V} - \tilde{V}S = 0, \qquad (5.16)$$

where $\tilde{V}$ is given by (5.13).

We next apply a reciprocal transformation R to the inverse scattering scheme (5.11).

Thus, if (5.16) delivers the conservation law

$$a_t + \epsilon_x = 0 \qquad (5.17)$$

then introduction of x^*, t^* according to

$$dx^* = a\,dx - \epsilon\,dt, \qquad t^* = t \ \} R \qquad (5.18)$$

leads to the reciprocally associated inverse scattering scheme

$$\left. \begin{array}{l} \tilde{\psi}_{x^*} = U^* \tilde{\psi} \\[2mm] \tilde{\psi}_{t^*} = V^* \tilde{\psi} \end{array} \right\} \qquad (5.19)$$

where

$$U^* = \frac{\tilde{U}}{a}, \qquad V^* = \tilde{V} + \frac{\epsilon\tilde{U}}{a} . \qquad (5.20),(5.21)$$

If, following the work of Ishimori [27], Wadati and Sogo [28] we impose the further transformation

$$\left. \begin{array}{l} q^* = \dfrac{-ib}{\sqrt{(1-bc)}} \\[5mm] r^* = \dfrac{-ic}{\sqrt{(1-bc)}} \end{array} \right\} T \qquad (5.22)$$

with inverse

$$\left. \begin{array}{l} b = \dfrac{iq^*}{\sqrt{(1-q^*r^*)}} \\[5mm] c = \dfrac{ir^*}{\sqrt{(1-q^*r^*)}} \end{array} \right\} T^{-1} \qquad (5.23)$$

then the inverse scattering scheme (5.19) is taken to the
WKI system (5.4) with $\psi \to \tilde{\psi}$ and $\lambda = -\lambda^* i$.

Hence, the Ishimori-Wadati-Sogo procedure whereby
the ZS-AKNS inverse scattering scheme is linked to the
WKI scheme may be decomposed according the the following diagram:

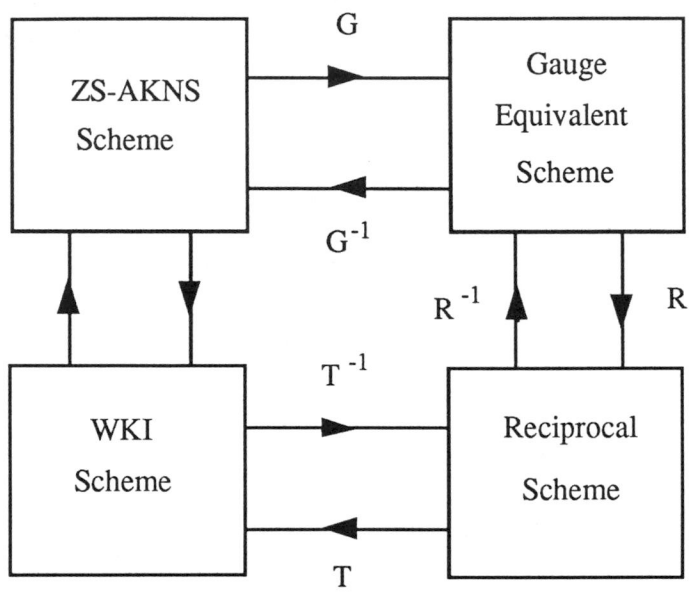

Figure 4

The link between the ZS-AKNS and

WKI Inverse Scattering Schemes

The above procedure may be used to link a generalization
of the pair of coupled nonlinear evolution equations (3.4),
namely

$$
\left.\begin{array}{l}
iq_t + \alpha i(q_{xxx} - 6qrq_x) + \beta(q_{xx} - 2qrq) = 0 \\
ir_t + \alpha i(r_{xxx} - 6qrr_x) - \beta(r_{xx} - 2qrr) = 0
\end{array}\right\}
\qquad (5.24)
$$

of the ZS-AKNS system with the coupled nonlinear pair

$$
\left.\begin{array}{l}
q_{t^*}^* + \alpha \left[\dfrac{q_{x^*}^*}{(1 - r^*q^*)^{\frac{3}{2}}} \right]_{x^*x^*} - \beta i \left[\dfrac{q^*}{(1 - r^*q^*)^{\frac{1}{2}}} \right]_{x^*x^*} = 0 \\[4mm]
r_{t^*}^* + \alpha \left[\dfrac{r_{x^*}^*}{(1 - r^*q^*)^{\frac{3}{2}}} \right]_{x^*x^*} + \beta i \left[\dfrac{r^*}{(1 - r^*q^*)^{\frac{1}{2}}} \right]_{x^*x^*} = 0
\end{array}\right\}
$$
$$(5.25)$$

of the WKI system (see Ishimori [27]). In particular, the specializations $r = r^* = -1, \alpha = -1, \beta = 0$ together with the substitution $\rho^2 = 1 + q^*$ in (5.25) produces the link between the KdV equation

$$
q_t - q_{xxx} + 6qq_x = 0 \tag{5.26}
$$

and the Dym equation

$$
(\rho^2)_t + 2(\rho^{-1})_{xxx} = 0. \tag{5.27}
$$

6 Reciprocal Invariance of the Dym Hierarchy

Here, we consider the Dym hierarchy [9]

$$
\rho_t = \rho^{-1}\{-\mathbf{D}^3\mathbf{r}\mathbf{I}r\}^n\rho\rho_x, \qquad n = 1, 2, \dots \tag{5.28}
$$

where the operators $\mathbf{D}$ and $\mathbf{I}$ are defined by

$$
\mathbf{D}\phi(x,t) := \phi_x, \qquad \mathbf{I}\phi(x,t) := \int_x^\infty dy\,\phi(y,t) \qquad (5.29),(5.30)
$$

while $\mathbf{r}$ denotes the multiplicative operator given by

$$
\mathbf{r}\phi(x,t) := r\phi(x,t) \tag{5.31}
$$

where $r = \rho^{-1} \neq 0$. It is assumed that $1 - \rho$ and $1 - r$ are **bona fide** potentials [9]. In particular, it is required that $\lim\limits_{x \to \pm\infty} \rho = 1$.

The hierarchy (5.28) may be re-written as

$$\rho_t + \varepsilon_{n,x} = 0, \qquad n = 1, 2, \ldots \qquad (5.32)$$

where

$$\varepsilon_n = \int_x^\infty \rho_t(y, t) \, dy \qquad (5.33)$$

and the flux terms ε_n are generated iteratively by the relations

$$\varepsilon_n = -\int_x^\infty \rho^{-1}[\rho^{-1}\varepsilon_{n-1}]_{xxx} \, dx, \qquad n = 1, 2, \ldots, \qquad \varepsilon_0 = 0.$$
$$(5.34)$$

Thus, (5.32) together with (5.34) yields

$$(\rho^2)_t + 2(\rho^{-1}\varepsilon_{n-1})_{xxx} = 0, \qquad n = 1, 2, \ldots \ . \qquad (5.35)$$

It is noted that the base equation of the nonlinear hierarchy (5.35) corresponding to $n = 1$ is the Dym equation (5.27).

Now, the reciprocal theorem RI shown that

$$\frac{\partial u}{\partial t} + \frac{\partial}{\partial x}\Lambda(\mathbf{D}^{(0)}u, \mathbf{D}^{(1)}u, \ldots, \mathbf{D}^{(n)}u) = 0 \qquad (5.36)$$

is taken to

$$\frac{\partial u'}{\partial t'} + \frac{\partial}{\partial x'}\left\{-u'\Lambda\left[\mathbb{D}^{(0)}\left(\frac{1}{u'}\right), \mathbb{D}^{(1)}\left(\frac{1}{u'}\right), \ldots, \mathbb{D}^{(n)}\left(\frac{1}{u'}\right)\right]\right\} = 0$$

$$\left(\mathbb{D} := \frac{1}{u'}\frac{\partial}{\partial x'}\right)$$

$$(5.37)$$

under the reciprocal transformation

$$dx' = u\, dx - \Lambda \left(\mathbf{D}^{(0)}u, \mathbf{D}^{(1)}u, \ldots, \mathbf{D}^{(n)}u \right) dt, \ dt' = dt, \left.\begin{matrix} \\ \\ \\ \\ \\ \\ \\ \\ \end{matrix}\right\} R$$

$$u' = \frac{1}{u},$$

$$0 < |\, u\, | < \infty \ .$$

$$(5.38)$$

Accordingly, under the reciprocal transformation

$$dx' = \rho^2 dx - 2 \left(\rho^{-1} \varepsilon_{n-1} \right)_{xx} dt, \ dt' = dt, \left.\begin{matrix} \\ \\ \\ \\ \\ \end{matrix}\right\} R' \quad (5.39)$$

$$\rho' = \frac{1}{\rho},$$

$$0 < |\, \rho\, | < \infty \ .$$

the hierarchy (5.35) becomes

$$(\rho'^2)_{t'} - 2 \left[\left(\rho^{-1} \varepsilon_{n-1} \right)_x \right]_{x'x'} = 0, \quad n = 1, 2, \ldots \ . \quad (5.40)$$

The invariance of the Dym equation under the reciprocal transformation (5.39) in the case $n = 1$ was established in [24]. The general result was established in [29]. In order to prove the result for the hierarchy, it is required to show that if we introduce ε'_n $n = 1, 2, \ldots$, accordingly to

$$(\rho'^{-1} \varepsilon'_n)_x + (\rho^{-1} \varepsilon_n)_x = 0, \quad n = 1, 2, \ldots, \quad (5.41)$$

then

$$\varepsilon'_n = -\int_x^\infty \rho'^{-1} [\rho'^{-1} \varepsilon'_{n-1}]_{x'x'x'} dx', \quad n = 1, 2, \ldots \ . \ (5.42)$$

Now, the relation (5.41) yields

$$\varepsilon'_n = \rho^{-1} \int_x^\infty \rho^2 (\rho^{-1}\varepsilon_n)_x \, dx, \qquad , n = 1, 2, \ldots . \qquad (5.43)$$

whence

$$\varepsilon'_{n,x'} - \rho'^{-1}[\rho'^{-1}\varepsilon'_{n-1}]_{x'x'x'} = (\rho^{-1})_{x'}\rho\varepsilon'_n - \rho^{-1}(\rho^{-1}\varepsilon_n)_x$$

$$+ \rho'[\rho'^2(\rho^{-1}\varepsilon_{n-1})_{xx}]_x$$

$$= \rho^{-1}(\rho^{-1})_x[\varepsilon'_n - \varepsilon_n + 2\rho^{-1}(\rho^{-1}\varepsilon_{n-1})_{xx}]$$

$$(5.44)$$

on use of (5.34). However, the relation (5.43) also shows that

$$\varepsilon'_n = -\rho^{-1}[\rho\varepsilon_n + 2\int_x^\infty \rho_x\varepsilon_n \, dx]$$

$$= \varepsilon_n + 2\rho^{-1}\int_x^\infty \rho\varepsilon_{n,x} \, dx$$

$$= \varepsilon_n + 2\rho^{-1}\int_x^\infty [\rho^{-1}\varepsilon_{n-1}]_{xxx} \, dx = \varepsilon_n - 2\rho^{-1}(\rho^{-1}\varepsilon_{n-1})_{xx}$$

whence,

$$\varepsilon'_n - \varepsilon_n + 2\rho^{-1}(\rho^{-1}\varepsilon_{n-1})_{xx} = 0. \qquad (5.46)$$

Accordingly, (5.44) shows that

$$\varepsilon'_{n,x'} - \rho'^{-1}[\rho'^{-1}\varepsilon'_{n-1}]_{x'x'x'} = 0, \qquad n = 1, 2, \ldots \qquad (5.47)$$

so that the relation (5.42) is established and the invariance of the Dym hierarchy (5.28) under the reciprocal transformation (5.39) follows. Use of the link between the WKI and ZS-AKNS scattering schemes established in the previous section may be used to show that this invariance of the Dym hierarchy induces the parameter-independent part of the auto-BT for the KdV hierarchy

$$q_t = \mathbf{K}^n q_x, \qquad n = 1, 2, \ldots \tag{5.48}$$

where $\mathbf{K}$ is the integro-differential operator defined by

$$\mathbf{K} := \frac{\partial^2}{\partial x^2} - 4q + 2q_x \int_x^\infty dy . \tag{5.49}$$

The Miura transformation

$$q = u_x + u^2 \tag{5.50}$$

between the KdV hierarchy given by (5.42), (5.43) and the mKdV hierarchy

$$u_t = \mathbf{M}^n u_x, \qquad n = 1, 2, \ldots \tag{5.51}$$

where

$$\mathbf{M} := \frac{\partial^2}{\partial x^2} - 4u^2 + 4u_x \int_x^\infty dy\, u \tag{5.52}$$

also emerges naturally via the linking process. The details are given in [29].

Recently, systems analogous to the Dym hierarchy have been introduced which are related to the Caudrey-Dodd-Gibbon (CDG) and Kaup - Kupershmidt (KK) hierarchies in the same way that the Dym hierarchy is related to the mKdV and KdV hierarchies. Explicit auto-BTs for the CDG and KK hierarchies may be generated via reciprocal invariance. The details are given in [30].

7 Reciprocal Transformations in 2 + 1-Dimensions. Application to the Kadomtsev-Petviashvili Equation.

The notion of reciprocal transformations in 2 + 1-dimensions was introduced in [31]. Here it is shown that the integrable generalization of the Dym equation to 2 + 1-dimensions as given by Konopelchenko and Dubrovsky in [32] is linked by such a reciprocal transformation to the singularity manifold equation associated with application of the Painlevé test to the Kadomtsev-Petviashvili (KP) equation [33]. This result is used to generate a novel involutory invariance of the 2 + 1-dimensional Dym equation. The following extensions to 2 + 1-dimensions of the reciprocal result RI is used in the sequel [31].

RII

The integro-differential equation

$$
\partial_t T(\partial_x; \partial_y; \partial_t; \partial_x^{-1} : u(x, y, t))
$$
$$
+ \partial_x F(\partial_x; \partial_y; \partial_t; \partial_x^{-1} : u(x, y, t)) = 0 \tag{7.1}
$$

is taken under the reciprocal transformation

$$
\left. dx' = T\,dx - F\,dt + \partial_x^{-1} T_y\,dy, \qquad y' = y, \quad t' = t \atop 0 <|\, T \,|< \infty \right\} \; RII \tag{7.2}
$$

to the associated equation

$$
\partial_{t'} T' + \partial_{x'} F' = 0 \tag{7.3}
$$

where

$$T' = \frac{1}{T(\mathbf{D}_1'; \mathbf{D}_2'; \mathbf{D}_3'; I_1' : u)} \tag{7.4}$$

$$F' = \frac{-F(\mathbf{D}_1'; \mathbf{D}_2'; \mathbf{D}_3'; I_1' : u)}{T(\mathbf{D}_1'; \mathbf{D}_2'; \mathbf{D}_3'; I_1' : u)} \tag{7.5}$$

and

$$\mathbf{D}_1' := \partial_x = (1/T')\partial_{x'},$$

$$\mathbf{D}_2' := \partial_y = [-(\partial_{x'}^{-1}T_{y'}')/T']\partial_{x'} + \partial_{y'}$$

$$\mathbf{D}_3' := \partial_t = (F'/T')\partial_{x'} + \partial_{t'} \tag{7.6}$$

$$I_1' := \partial_x^{-1} = \partial_{x'}^{-1}\mathbf{T}'$$

In the above, ∂_x^{-1} is the integral operator defined by

$$\partial_x^{-1}\phi := \int_a^x \phi(\sigma, y, t)\, d\sigma \tag{7.7}$$

In addition, $\mathbf{T}'$ is the multiplicative operator defined by

$$\mathbf{T}'\phi := T'\phi \tag{7.8}$$

The above result is readily established along the lines of RI [31].

We next turn to the Lax system with operators [32]

$$L = \partial_x^2 + \alpha u(x, y, t)\partial_x + \beta u(x, y, t) + \partial_y,$$

$$T = -4\partial_x^3 - 6\alpha u \partial_x^2$$

$$- [3\alpha u_x + \frac{3}{2}\alpha^2 u^2 + 6\beta u - 3\alpha(\partial_x^{-1}u_y)]\partial_x \tag{7.9}$$

$$- 3\beta u_x - \frac{3}{2}\alpha\beta u^2 + 3\beta\partial_x^{-1}u_y + \partial_t.$$

The commutativity condition

$$[L, T] = 0 \qquad (7.10)$$

produces the class of nonlinear evolution equations in $2 + 1$-dimensions

$$u_t = u_{xxx} + 6\beta u u_x - \frac{3}{2}\alpha^2 u^2 u_x + 3\partial_x^{-1} u_{yy} - 3\alpha u_x \partial_x^{-1} u_y \ . \ (7.11)$$

The specialisation $\alpha = 0$ produces a Kadomtsev-Petviash-vili equation, while $\beta = 0$ yields a modified Kadomtsev-Petviashvili equation [32].

If we set $\alpha = 0, \beta = 1/6$ together with $x^* = x, y^* = y/\sqrt{3}$, $t^* = -t$ then the KP equation is obtained in the form considered by Weiss [34], namely

$$u_{y^* y^*} + (u_{t^*} + u u_{x^*} + u_{x^* x^* x^*})_{x^*} = 0 \ . \qquad (7.12)$$

On the other hand, if we take [32]

$$L = r^2(x, y, t)\partial_x^2 + \partial_y,$$
$$T = 4r^3\partial_x^3 + [6r^2 r_x - 6r^2\partial_x^{-1}(r_y/r^2)]\partial_x^2 - \partial_t, \qquad (7.13)$$

then the commutativity condition (7.10) produces a generalization of the Dym equation to $2 + 1$-dimensions, viz

$$r_t = r^3 r_{xxx} + 3r^{-1}[r^2\partial_x^{-1}(r_y/r^2)]_y. \qquad (7.14)$$

It will now be shown that the latter equation is linked by a $2 + 1$-dimensional reciprocal transformation to the singularity manifold equation associated with the KP equation.

Thus, the $2+1$-dimensional Dym equation (7.14) may be re-written in the form

$$\rho_t + \varepsilon_x = 0, \tag{7.15}$$

where $\rho = r^{-1}$ and

$$\varepsilon = \partial_x^{-1}[\rho^{-1}(\rho^{-1})_{xxx} - 3\rho^3(\rho^{-2}\partial_x^{-1}\rho_y)_y]. \tag{7.16}$$

Under the reciprocal transformation

$$\left. \begin{array}{l} d\bar{x} = \rho\, dx - \varepsilon\, dt + \partial_x^{-1}\rho_y\, dy, \qquad \bar{y} = y, \quad \bar{t} = t \\[2mm] \bar{\rho} = \dfrac{1}{\rho} \\[4mm] 0 < |\rho| < \infty \end{array} \right\} \;\; \bar{R} \quad (7.17)$$

we obtain

$$\bar{\rho}_{\bar{t}} + \bar{\varepsilon}_{\bar{x}} = 0, \tag{7.18}$$

where

$$\bar{\varepsilon} = \bar{\varepsilon}_1 + \bar{\varepsilon}_2 \tag{7.19}$$

with

$$\bar{\varepsilon}_1 = -(1/\rho)\partial_x^{-1}[\rho^{-1}(\rho^{-1})_{xxx}] = -\bar{\rho}\partial_x^{-1}\{\partial_{\bar{x}}[\bar{\rho}^{-1}\partial_{\bar{x}}(\bar{\rho}_{\bar{x}}/\bar{\rho})]\}$$

$$= -\bar{\rho}\partial_{\bar{x}}^{-1}\{\bar{\rho}\partial_{\bar{x}}[\bar{\rho}^{-1}\partial_{\bar{x}}(\bar{\rho}_{\bar{x}}/\bar{\rho})]\} = -\bar{\rho}[\partial_x(\bar{\rho}_{\bar{x}}/\bar{\rho}) - \tfrac{1}{2}(\bar{\rho}_{\bar{x}}/\bar{\rho})^2]$$

$$= \bar{\rho}\left[\frac{3}{2}(\bar{\rho}_{\bar{x}}/\bar{\rho})^2 - \bar{\rho}_{\bar{x}\bar{x}}/\bar{\rho}\right],$$

$$\bar{\varepsilon}_2 = (3/\rho)\partial_x^{-1}[\rho^3(\rho^{-2}\partial_x^{-1}\rho_y)_y] = \frac{3}{\rho}\partial_x^{-1}\left\{\rho^3\left[\frac{1}{\rho}\left(\frac{1}{\rho}\partial_x^{-1}\rho_y\right)\right]_y\right\}$$

$$= \frac{3}{\rho}\partial_x^{-1}\left\{\rho^3\left\{-\frac{1}{\rho^2}\rho_y\left(\frac{1}{\rho}\partial_x^{-1}\rho_y\right)\right\} + \frac{1}{\rho}\left[-\bar{\rho}\frac{\partial}{\partial\bar{y}}\left(\frac{1}{\bar{\rho}}\partial_{\bar{x}}^{-1}\bar{\rho}_{\bar{y}}\right)\right]\right\}$$

since

$$\partial_y\left(\frac{1}{\rho}\partial_x^{-1}\rho_y\right) = -\bar{\rho}\partial_{\bar{y}}\left(\frac{1}{\rho}\partial_{\bar{x}}^{-1}\bar{\rho}_{\bar{y}}\right).$$

But,

$$\partial_{\bar{x}}^{-1}\bar{\rho}_{\bar{y}} = -\frac{\partial_x^{-1}\rho_y}{\rho}$$

so that

$$\bar{\varepsilon}_2 = 3\bar{\rho}\partial_{\bar{x}}^{-1}[-(1/\bar{\rho})(\partial_{\bar{x}}^{-1}\bar{\rho}_{\bar{y}})^2(1/\bar{\rho})_{\bar{x}} - (1/\bar{\rho})\partial_{\bar{x}}^{-1}\bar{\rho}_{\bar{y}\bar{y}}].$$

Thus, the reciprocal associate under $\bar{R}$ of the 2+1-dimensional Dym equation is

$$\bar{\rho}_{\bar{t}} - [\bar{\rho}\{\bar{\rho}_{\bar{x}\bar{x}}/\bar{\rho} - 3(\bar{\rho}_{\bar{x}}/\bar{\rho})^2/2$$

$$+ 3\partial_{\bar{x}}^{-1}[(1/\bar{\rho})(\partial_{\bar{x}}^{-1}\bar{\rho}_{\bar{y}})^2(1/\bar{\rho})_{\bar{x}} \qquad (7.20)$$

$$+ (1/\bar{\rho})\partial_{\bar{x}}^{-1}\bar{\rho}_{\bar{y}\bar{y}}]\}]_{\bar{x}} = 0.$$

If we now introduce the potential $\bar{\phi}$ according to

$$\bar{\rho} = \bar{\phi}_{\bar{x}} = x_{\bar{x}} \qquad (7.21)$$

then (7.20) yields

$$-3\frac{\partial}{\partial\bar{y}}\left(\frac{\bar{\phi}_{\bar{y}}}{\bar{\phi}_{\bar{x}}}\right) + \frac{\partial}{\partial\bar{x}}\left(\frac{\bar{\phi}_{\bar{t}}}{\bar{\phi}_{\bar{x}}} - \{\bar{\phi};\bar{x}\} - \frac{3}{2}\left(\frac{\bar{\phi}_{\bar{y}}}{\bar{\phi}_{\bar{x}}}\right)^2\right) = 0 \quad (7.22)$$

where

$$\{\bar{\phi};\bar{x}\} = \frac{\partial}{\partial\bar{x}}\left(\frac{\bar{\phi}_{\bar{x}\bar{x}}}{\bar{\phi}_{\bar{x}}}\right) - \frac{1}{2}\left(\frac{\bar{\phi}_{\bar{x}\bar{x}}}{\bar{\phi}_{\bar{x}}}\right)^2 \qquad (7.23)$$

denotes the Schwarzian derivative of $\bar{\phi}$.

Under the change of variables

$$x^* = \bar{x}, \ y^* = \bar{y}/\sqrt{3}, \ t^* = -\bar{t} \tag{7.24}$$

(7.22) produces the singularity manifold equation

$$\frac{\partial}{\partial y^*}\left(\frac{\bar{\phi}_{y^*}}{\bar{\phi}_{x^*}}\right) + \frac{\partial}{\partial x^*}\left(\frac{\bar{\phi}_{t^*}}{\bar{\phi}_{x^*}} + \{\bar{\phi}; x^*\} + \frac{1}{2}\left(\frac{\bar{\phi}_{y^*}}{\bar{\phi}_{x^*}}\right)^2\right) = 0 \tag{7.25}$$

associated with the KP equation [34]. The link to the latter is obtained via the relation

$$u = -\frac{\bar{\phi}_{t^*}}{\bar{\phi}_{x^*}} - \frac{4\bar{\phi}_{x^*x^*x^*}}{\bar{\phi}_{x^*}} + 3\left(\frac{\bar{\phi}_{x^*x^*}}{\bar{\phi}_{x^*}}\right)^2 - \left(\frac{\bar{\phi}_{y^*}}{\bar{\phi}_{x^*}}\right)^2. \tag{7.26}$$

Combination of the invariance of the singularity manifold equation (7.25) under the Möbius transformation M:

$$\phi^k = \frac{a\bar{\phi} + b}{c\bar{\phi} + d}, \qquad ad - bc \neq 0 \tag{7.27}$$

and the association of the $2 + 1$-dimension of the $2 + 1$-dimensional Dym equation (7.15) with (7.22) under the reciprocal transformation

$$x = \partial_{\bar{x}}^{-1}\bar{\rho} = \bar{\phi}, \qquad \bar{x} = \partial_x^{-1}\rho = \phi,$$
$$y = \bar{y}, \quad t = \bar{t}, \quad \rho = 1/\bar{\rho}, \tag{7.28}$$

induces a novel invariance of the $2 + 1$-dimensional Dym equation as indicated in Figure 5.

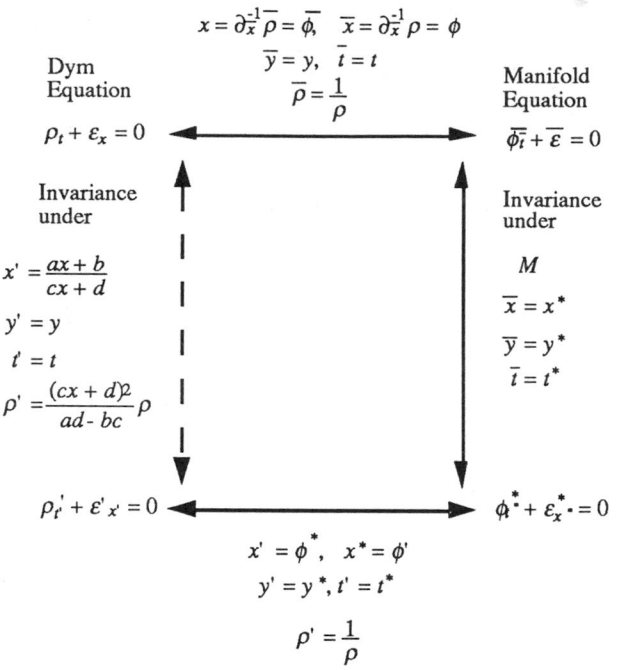

Figure 5

The   Induced   Invariance   of   the   2 + 1-Dimensional
Dym Equation

Thus, the invariance is given by

$$x' = \frac{ax + b}{cx + d},$$

$$t' = t, \qquad ad - bc \neq 0 \tag{7.29}$$

$$\rho' = \frac{(cx + d)^2 \rho}{ad - bc} \ .$$

The combination of 2 + 1-dimensional reciprocal transformations with gauge transformations to produce new

integrable systems in a manner analogous to that of Section 5 for $1 + 1$-dimensional systems is suggested. Indeed, gauge transformations have been recently used in $2 + 1$-dimensions by Konopelchenko and Matkarimov [35] to link two nonlinear integrable systems of importance, namely, the Ishimori system [36]

$$\mathbf{S}_t(x, y, t) + \mathbf{S} \times (\mathbf{S}_{xx} + \alpha^2 \mathbf{S}_{yy}) + \phi_x \mathbf{S}_y + \phi_y \mathbf{S}_x = 0,$$

$$\phi_{xx} - \alpha^2 \phi_{yy} + 2\alpha^2 \mathbf{S}.(\mathbf{S}_x \times \mathbf{S}_y) = 0$$

$$\mathbf{S}(x, y, t).\mathbf{S}(x, y, t) = 1, \tag{7.30}$$

$$\alpha^2 = \pm 1$$

with the Davey-Stewartson system. The Ishimori system represents the $2 + 1$-dimensional integrable extension of the continuous Heisenberg ferromagnet model in $1 + 1$-dimensions, namely [37,38]

$$\mathbf{S}_t + \mathbf{S} \times \mathbf{S}_{xx} = 0,$$

$$\mathbf{S}.\mathbf{S} = 1 \tag{7.31}$$

In [27, 28], (7.31) has been linked by a combination of gauge and reciprocal transformations to the Shimuzi-Wadati equation [39]

$$q_t \pm i(q/\sqrt{1+ \mid q \mid^2})_{xx} = 0 \tag{7.32}$$

An analogous procedure is available in $2 + 1$-dimensions to construct an integrable $2+1$-dimensional version of the Shimuzi-Wadati equation.

To conclude, it is noted that reciprocal transformations may also be employed in soliton theory to analyse the distinctive symmetry structure of interconnected integrable nonlinear hierarchies [40].

Acknowledgements

Support under - Natural Sciences of Canada Grant. No. A0879 is gratefully acknowledged (C.R.)

References

[1] C. Rogers, and W.F. Shadwick, *Bäcklund Transformations and Their Applications*, Academic Press, New York, 1982.

[2] V.E. Zakharov and A.B. Shabat, Funct. Anal. Pril. **8**, 43, 1974.

[3] G. Darboux, C.R. Acad. Sci. Paris **94**, 1456, 1882.

[4] D. Levi, O. Ragnisco and A. Sym, Il Nuovo. Cimento, **83B**, 34, 1984.

[5] D. Levi, O. Ragnisco and A. Sym, Lett. Nuovo. Cimento, **33**, 401, 1982.

[6] B.G. Konopelchenko, Phys. Lett. **74A**, 189, 1979.

[7] B.G. Konopelchenko, Phys. Lett. **100B**, 254, 1981.

[8] B.G. Konopelchenko, Phys. Lett. **87A**, 445, 1982.

[9] F. Calogero and A. Degasperis, Physica **14D**, 103, 1984.

[10] M. Boiti, J. Leon, L. Martina and F. Pempinelli, Phys. Lett. **132A**, 432, 1988.

[11] M. Boiti, B.G. Konopelchenko and F. Pempinelli, Inverse Problems **1**, 35, 1985.

[12] D. Levi, L. Pilloni and P.M. Santini, Phys. Lett. **81A**, 419, 1981.

[13] G.L. Lamb Jr., *Elements of Soliton Theory*, Wiley, New York, 1980.

[14] V.E. Zakharov, S.V. Manakov, S.P. Novikov and L.P. Pitaevski, *Theory of Solitons. The Inverse Problem Method*, Nauka, Moscow, 1980, Plenum, 1984.

[15] M.J. Ablowitz and H. Segur, *Solitons and the Inverse Scattering Transform*, Philadelphia, SIAM, 1982.

[16] R.K. Dodd, J.C Eilbeck, J.D. Gibbon and H.C. Morris, *Solitons and Nonlinear Waves*, Academic Press, New York, 1982.

[17] F. Calogero and A. Degasperis, *Spectral Transform and Solitons : Tools to Solve and Investigate Nonlinear Evolution Equations I*, North Holland, Amsterdam, 1982.

[18] B.G. Konopelchenko, *Nonlinear Integrable Equations*, Lecture Notes in Physics, Vol 270, Springer Verlag, Heidelberg, 1987.

[19] D.V. Chudnovsky and G.V. Chudnovsky, J. Math. Phys. **22**, 2518, 1981.

[20] M. Boiti and G.Z. Tu, Il Nuovo. Cimento **71B**, 253, 1982.

[21] A. Davey and K. Stewartson, Proc. Roy. Soc. London Ser A **338**, 101, 1974.

[22] A.S. Fokas and P.M. Santini, Preprint INS#121, Institute for Nonlinear Studies, Clarkson University, 1989.

[23] C. Rogers and W.F. Ames, *Nonlinear Boundary Value Problems in Science and Engineering*, Academic Press, New York, 1989.

[24] C. Rogers and P. Wong, Physica Scripta **30**, 10, 1984.

[25] J.G. Kingston and C. Rogers, Phys. Lett. **92A**, 261, 1982.

[26] V.E. Zakharov and L.A. Takhtadzyan, Theor, Math. Phys. **38**, 17, 1979.

[27] Y. Ishimori, J. Phys. Soc. Japan **51**, 3036, 1982.

[28] M. Wadati and K. Sogo, J. Phys. Soc. Japan **52**, 394, 1983.

[29] C. Rogers and M.C. Nucci, Physica Scripta **33**, 289, 1986.

[30] C. Rogers and S. Carillo, Physica Scripta **36**, 865, 1987.

[31] C. Rogers, J. Phys. A. Math. Gen. **49**, L 491, 1986.

[32] B.G. Konopelchencko and V.G. Dubrovsky, Phys. Lett. **102A**, 15, 1984.

[33] C. Rogers, Phys. Lett. **120A**, 15, 1987.

[34] J. Weiss, J. Math. Phys. **26**, 2174, 1985.

[35] B.G. Konopelchenko and B.T. Matkarimov, Preprint Università Degli Studi, Dipartimento do Fisica, Lecce, 1988.

[36] Y. Ishimori, Progr, Theor. Phys. **72**, 33, 1984.

[37] M. Lakshmanan, Phys. Lett. A **61**, 53, 1977.

[38] L.A. Takhatadzyan, Phys. Lett. A **64**, 253, 1977.

[39] T. Shimizu and M. Wadati, Prog. Theor. Phys. **63**, 808, 1980.

[40] S. Carillo and B. Fuchssteiner, J. Math. Phys, **30**, 1606, 1989.

Nonlinear Reaction-Diffusion Systems

R. H. Martin, Jr.
North Carolina State University

M. Pierre
Université de Nancy I

1.1 Introduction

Suppose m is a positive integer, $\mathbf{R}_+ = [0, \infty)$, and $\mathbf{F} = (F_i)_1^m$ is a locally Lipschitz function from $\mathbf{R}_+^m$ into $\mathbf{R}^m$ that is quasi-positive:

$$(1.1) \qquad \xi = (\xi_i)_1^m \in \mathbf{R}_+^m \quad \text{and} \quad \xi_k = 0 \Rightarrow F_k(\xi) \geq 0.$$

Under these circumstances the ordinary differential equation

$$(ODE) \qquad y' = F(y), \qquad y(0) = \zeta \in \mathbf{R}_+^m, \qquad t \geq 0$$

has a unique, noncontinuable, nonnegative solution $y(\cdot\,;\zeta)$ on an interval of the form $[0, b_\zeta)$ where $0 < b_\zeta \leq \infty$ for each $\zeta \in \mathbf{R}_+^m$. Furthermore, if $b_\zeta < \infty$ then $|y(t;\zeta)| \to \infty$ as $t \uparrow b_\zeta$.

Suppose now that Ω is a bounded domain in $\mathbf{R}^N$ with smooth boundary $\partial\Omega$. Let ∂_ν denote the outward normal derivative on $\partial\Omega$ and let Δ denote the Laplacian on Ω. In conjunction with (ODE) consider the reaction-diffusion system

$$(RDS) \quad \begin{cases} u_t = D\Delta u + F(u) & \text{on} \quad \Omega \times (0, \infty) \\ B[u] = 0 & \text{on} \quad \partial\Omega \times (0, \infty) \\ u = z & \text{on} \quad \Omega \times \{0\} \end{cases}$$

where $D = \text{diag}\,(d_1, \ldots, d_m)$ is a diagonal matrix with $d_i > 0$ for each i, $u = (u_i)_1^m$, $\Delta u = (\Delta u_i)_1^m$, and $B[u] = (B_i[u_i])_1^m$ is a (possibly nonhomogeneous) boundary operator of the form

$$(1.2) \quad \begin{cases} B_i[u_i] = \lambda_i \partial_\nu u_i + (1 - \lambda_i)(u_i - \alpha_i). \\ \text{with} \quad 0 \leq \lambda_i \leq 1 \quad \text{and} \quad \alpha_i \geq 0 \quad \text{for} \quad i = 1, \ldots, m. \end{cases}$$

The initial function $z = (z_i)_1^m$ is assumed nonnegative and continuous on $\overline{\Omega}$.

The local Lipschitz assumption on F may be used to show that (RDS) has a unique noncontinuable solution $u(\cdot, z)$ on $\overline{\Omega} \times [0, T(z))$, where $0 < T(z) \leq \infty$. Furthermore, if $T(z) < \infty$ then it must be the case that <u>both</u>

$$(1.3) \quad \begin{cases} \lim_{t \to T(z)-} \max\{\|u_i(\cdot, t)\|_\infty;\ i = 1, \ldots, m\} = \infty \\ \text{where} \quad \|\cdot\|_\infty \quad \text{denotes the essential supremum over} \quad \Omega, \end{cases}$$

and

$$(1.4) \quad \begin{cases} \lim_{t \to T(z)-} \sup \max\left\{\int_0^t \int_\Omega |F_i(u(x,t))|^r \, dx\, dt : i = 1, \ldots, m\right\} = \infty \\ \text{for some} \quad r > 1. \end{cases}$$

Since (RDS) can be viewed as a semilinear differential equation involving an analytic semigroup and a relatively continuous perturbation in any of the spaces $L^p(\Omega)^m$, $1 \leq p < \infty$, the assertions in (1.3) and (1.4) may be deduced from Ball [2, Theorem 3.1], for example. Note further that the quasi-positive assumption (1.1) on F and the maximum principle imply that each solution to (RDS) remains nonnegative so long as it exists.

The fundamental consideration in this paper is the following:

$$(1.5) \quad \begin{cases} \text{under what circumstances does global existence} \\ \text{of each solution to (ODE) [i.e., } b_\zeta = \infty \text{ for all } \zeta] \\ \text{imply global existence of each solution to (RDS)} \\ \text{[i.e., } T(z) = \infty \text{ for all } z]? \end{cases}$$

For $m = 1$ global existence for (ODE) always implies the same for (RDS). This can easily be established using the maximum principle and differential inequalities. It also carries over to the system case $m \geq 2$ whenever F is quasimonotone:

$$\forall k \in \{1, \ldots, m\}, \quad F_k(\xi_1, \ldots, \xi_m) \quad \text{is nondecreasing in} \quad \xi_i, \quad i \neq k.$$

These techniques are indicated in [16], for example.

A general answer to (1.5) is unknown even when restricted to some basic classes of systems. Our purpose is to give some results in that direction.

1.2   A Negative Result

It is possible to destroy global existence for solutions to an ordinary differential equation system by the addition of diffusion in one component. Following Morgan [18], the nonnegative solutions to

$$(2.1) \quad \begin{cases} y_1' = y_1, & y_1(0) = \zeta_1 > 0 \\ y_2' = y_2^2 - |3 - y_1|y_2^3, & y_2(0) = \zeta_2 > 0 \end{cases} \quad t \geq 0$$

exist for all $t \geq 0$ (see [18]). However, the mixed diffusion and ordinary system

$$(2.2) \quad \begin{cases} u_t = \pi^{-2}u_{xx} + u & \text{on} \quad (0,1) \times (0,\infty) \\ v_t = v^2 - |3 - u|v^3 & \\ u_x = 0 & \text{on} \quad \{0,1\} \times (0,\infty) \\ u(x,0) = 6\cos(\pi x), \quad v(x,0) = v_o(x) \quad 0 < x < 1 \end{cases}$$

where v_o is smooth and $v_o(1/3) > 0$, has finite blow-up time. For note that $u(x,t) \equiv 6\cos\pi x$, and hence $u(1/3, t) \equiv 3$ and

$$v_t(1/3, t) = [v(1/3, t)]^2$$

Thus $v(1/3, t)$ blows up as $t \uparrow 1/v_o(1/3)$. A similar example involving an important scientific model is given in the next section.

Remark 2.1   It is interesting to note that if the associated vector field in (2.1) is perturbed by

$$\begin{pmatrix} y_1 \\ y_2 \end{pmatrix} \to \begin{pmatrix} -y_1 \\ -\delta y_2 \end{pmatrix}$$

then the resulting system has finite time blow-up for all $\delta \geq 0$ and all u_o, v_o with $u_o = 3$ and $v_o > 0$.

We now give an example of a system where both of the diffusion coefficients are positive and a solution blows up in finite time even though the associated ordinary equation has the global existence property for all nonnegative initial data.

Consider the reaction-diffusion system

(2.3)
$$\begin{cases} u_t = a\Delta u + 2aN \\[2mm] v_t = \Delta v + 2(N-t)^+ v^2 - 8|1-u|v^3 & (x,t) \in \Omega \times (-1,\infty) \\[2mm] u = 0, \quad v = (1+t^2)^{-1} & (x,t) \in \partial\Omega \times (-1,\infty) \end{cases}$$

where Ω is the unit ball in $\mathbf{R}^N$ and $(r)^+ = r$ if $r \geq 0$, $(r)^+ = 0$ if $r < 0$. Also, consider the associated ordinary differential system

(2.4)
$$\begin{cases} y_1' = 2aN, & y_1(-1) = \zeta_1 \\[2mm] y_2' = 2(N-t)^+ y_2^2 - 8|1-y_1|y_2^3, & y_2(-1) = \zeta_2 \end{cases} \qquad t \geq 0$$

For convenience we use the time interval $(-1,\infty)$ instead of $(0,\infty)$.

Proposition 2.1   *Assume that*

(2.5)
$$a > (N+1)^2/4N.$$

Then the solution to (2.4) exists on $[-1,\infty)$ for all initial data ζ_1, $\zeta_2 \geq 0$. However,

(2.6)
$$u(r,t) = 1 - r^2, \quad v(r,t) = (t^2 + r^2)^{-1}$$

is a nonnegative solution to (2.3) on $\overline{\Omega} \times (-1, 0)$ which blows up in $L^\infty(\Omega)$ as $t \uparrow 0$. More precisely

  (i)   if $N > 4$, then (u, v) is a weak solution to (2.3) on the whole set $\overline{\Omega} \times (-1, N)$ but blows up in $L^{N/2}(\Omega)$ at $t = 0$.

  (ii)   if $1 \leq N < 4$, then (u, v) cannot be extended into a weak solution to (2.3) around $t = 0$.

Remarks.   By weak solution, we mean that (2.3) holds in the sense of distributions.

Above example shows at the same time that

(a)   global existence for all nonnegative initial data in O.D.E. (2.4) does not imply global existence in associated R.D.S. (2.3).

(b)   weak solutions of R.D.S. can be bounded (and even C^∞) at two different times t_1, t_2, but not in (t_1, t_2).

(c)   solutions of R.D.S. might exist on $\overline{\Omega} \times (t_1, t_2) \setminus \{(x_o, t_o)\}$ and blow up at $(x_o, t_o) \in \Omega \times (t_1, t_2)$ only.

The latter pathological situations (b) and (c) can also happen for equations ($m = 1$). Indeed v as defined in (2.6) satisfies

$$\begin{cases} v_t = \Delta v + 2(N - t)v^2 - 8r^2 v^3 & (x, t) \in \Omega \times (-1, N) \\ v = (1 + t^2)^{-1} & (x, t) \in \partial\Omega \times (-1, N). \end{cases}$$

Proof of Proposition 2.1   To check that (2.6) defines a solution of (2.3), note that

$$u_t = 0, \qquad \Delta u = -2N$$

$$v_t = \frac{-2t}{(t^2 + r^2)^2}, \quad v_{x_i} = \frac{-2x_i}{(t^2 + r^2)^2}, \quad v_{x_i x_i} = \frac{-2}{(t^2 + r^2)^2} + \frac{8x_i^2}{(t^2 + r^2)^3}$$

Hence

$$(2.7) \quad v_t - \Delta v = \frac{-2t}{(t^2 + r^2)^2} + \frac{2N}{(t^2 + r^2)^2} - \frac{8r^2}{(t^2 + r^2)^3} = 2(N-t)v^2 - 8(1-u)v^3.$$

This computation is valid for $t \in (-1, 0)$. It is also valid in the sense of distributions on $\Omega \times (-1, \infty)$ if $N > 4$, since all quantities in (2.7) are then locally integrable on $\Omega \times (-1, \infty)$. However, note that the right-hand side of (2.7) is not locally integrable if $1 \le N \le 3$ which proves (ii).

Now, suppose for contradiction that a solution (y_1, y_2) to (2.4) blows up at a finite time $t^* > -1$. Since

$$(2.8) \qquad y_1(t) = 2aN(t + 1) + \zeta_1 \quad \text{on} \quad (-1, t^*),$$

it is the case that $y_2(t) \to \infty$ as $t \uparrow t^*$. Furthermore it is clear from the form of the vector field in (2.4) that

$$-1 < t^* \le N \quad \text{and} \quad 1 - y_1(t) \to 0 \quad \text{as} \quad t \uparrow t^*.$$

Hence, using (2.8)

$$(2.9) \qquad\qquad 2aN(t^* + 1) + \zeta_1 = 1.$$

It follows from (2.4), (2.8), (2.9) that

(2.10) $y_2' = 2(N - t)y_2^2 - 16aN(t^* - t)y_2^3.$

In particular $y_2' \leq 2(N + 1)y_2^2$ so that for all $t_o < t < t^*$

$$y_2(t_o)^{-1} - y_2(t)^{-1} \leq 2(N + 1)(t - t_o)$$

and by letting $t \uparrow t^*$

(2.11) $y_2(t_o) \geq 1/\{2(N + 1)(t^* - t_o)\}.$

On the other hand, $y'(t_o) > 0$ for t_o close enough to t^* so that (2.10) implies

(2.12) $2(N - t_o) \geq 16aN(t^* - t_o)y_2(t_o).$

Combining (2.11) and (2.12) leads to

$$2(N - t_o) \geq 8aN/(N + 1) \implies (N + 1)^2 \geq 4aN.$$

This contradicts (2.5) and completes the proof.

Remark. Note that equation $y' = 2y^2 - (t^* - t)y^3$ which is similar to (2.10) has the solution $y(t) = (t^* - t)^{-1}$. This justifies the need of above proof.

1.3 A Typical Example

We now consider a very simple 2×2 system of P.D.E. for which the associated O.D.E. has the global existence and positivity properties but where usual arguments like maximum principle or invariant regions techniques do not apply:

(3.1)
$$\begin{cases} u_t = d_1 \Delta u - uh(v) \\ \\ v_t = d_2 \Delta v + uh(v) \end{cases} \quad \text{on} \quad \Omega \times (0, \infty)$$

subject to the boundary conditions of the form

(3.2)
$$\begin{cases} \lambda_1 \partial_\nu u + (1 - \lambda_1)(u - \alpha) = 0 \\ \\ \lambda_2 \partial_\nu v + (1 - \lambda_2)(v - \beta) = 0 \end{cases} \quad \text{on} \quad \partial\Omega \times (0, \infty)$$

where α, $\beta \geq 0$, $0 \leq \lambda_1, \lambda_2 \leq 1$ and $h : \mathbf{R}^+ \to \mathbf{R}^+$ is C^1 with $h(0) = 0$. To minimize subscripts, we use (u, v) instead of (u_1, u_2).

Here global nonnegative solutions exist for the corresponding O.D.E.

$$(3.3) \qquad \begin{cases} y_1' = -y_1 h(y_2), & y_1(0) = \zeta_1 \geq 0 \\ y_2' = y_1 h(y_2), & y_2(0) = \zeta_2 \geq 0. \end{cases}$$

However, global existence in (3.1), (3.2) is not yet completely understood. Let us first indicate some easy facts.

Note that condition (1.1) holds for the nonlinear terms so that local existence of nonnegative solutions to (3.1), (3.2) is assured as explained in the introduction.

If $d_1 = d_2$ and $\lambda_1 = \lambda_2$, then $u + v$ satisfies the heat equation with standard boundary conditions. Therefore global existence is obvious.

In all cases, since $-uh(v) \leq 0$, the maximum principle implies u is uniformly bounded by $\|u(\cdot, 0)\|_\infty$. As a consequence if

$$(3.4) \qquad h(v) \leq Lv + M \quad \forall v \geq 0$$

for some L, $M > 0$, then from Gronwall's inequality we get a uniform bound on $v(x, t)$ for $(x, t) \in \Omega \times (0, T)$ and for all T. Therefore nonnegative solutions exist globally.

Outside these situations, the question becomes more complicated even when we have obvious uniform *a priori* $L^1(\Omega)$-estimates. For example when $\lambda_1 = \lambda_2 = 1$

$$\int_\Omega u(t) + v(t) = \int_\Omega u_o + v_o \quad \forall t > 0.$$

Let us summarize some recent results.

R 1.   Assume polynomial growth for h:

$$(3.5) \qquad \exists p > 1, \quad \exists L, M > 0 \qquad h(v) \leq Lv^p + M \quad \forall v > 0$$

and

$$(3.6) \qquad 0 < \lambda_1, \lambda_2 \leq 1 \quad \text{or} \quad 0 = \lambda_1 = \lambda_2.$$

Then (3.1), (3.2) has global nonnegative classical solutions.

When p is small enough this can be proved using "bootstrap" arguments deducing L^∞-estimates from L^1-estimates (see [1]). A proof for all p was first given in [17]. A similar result was obtained in [12]. A different and more general proof is given in next section. Note that global existence is unresolved if $\lambda_1 = 1$, $\lambda_2 = 0$.

R 2. This restriction of polynomial growth has been slightly improved to allow $h(v) \sim e^{v^\gamma}$, $\gamma < 1$ in [7]. The method is different and seems more specific to the particular structure of (3.1) while the next section carries over to more general systems.

R 3. A rather particular approach indicated in Section 7 provides global existence of classical solutions of (3.1) for *any* growth of h but with

$$\Omega = \mathbf{R}^N, \qquad d_1 \le d_2.$$

A completely different and very general method based on the use of *a priori* estimates in $L^1(\Omega)$ allows us to handle any kind of growth for h and works for general diffusion. However, it only provides weak solutions. In particular, we do not know whether or not they remain in $L^\infty(\Omega)$ for all time. On the other hand, L^1-initial data are possible:

R 4. Assume (3.6) holds and $u_o, v_o \in L^1(\Omega)$, $u_o, v_o \ge 0$. Then (3.1), (3.2) has a global "weak" solution with $u(\cdot, 0) = u_o(\cdot)$, $v(\cdot, 0) = v_o(\cdot)$. (Note that we do not assume (3.5) here.) The idea of the proof will be indicated in Section 8.

1.4   The Basic Method for Polynomial Growth

The basic method will be presented here for a 2×2 system with general and *different* diffusions. Extensions and limits of the method will be described in Sections 5 and 6. For $r = 1, 2$, we set

$$(4.1) \qquad A^r u = \partial_i(a_{ij}^r \partial_j u) + b_i^r \partial_i u + c^r u$$

where we use the usual summation convention on indices and

$$(4.2) \qquad a_{ij}^r, b_i^r \in C^1(\overline{\Omega}), \quad c^r \in C^0(\overline{\Omega}) \qquad r = 1, 2$$

$$(4.3) \qquad a_{ij}^r \xi_i \xi_j \ge \alpha |\xi|^2 \quad \text{for some} \quad \alpha > 0, \quad r = 1, 2.$$

We consider the following system

$$(4.4) \qquad \begin{cases} u_t = A^1 u + f(u, v) \\ \\ v_t = A^2 v + g(u, v) \end{cases} \quad \text{on} \quad \Omega \times (0, \infty)$$

subject to the boundary conditions

$$(4.5) \quad \begin{cases} \lambda_1 \partial_\nu^1 u + (1 - \lambda_1)(u - \alpha_1) = 0 \\ \\ \lambda_2 \partial_\nu^2 v + (1 - \lambda_2)(v - \alpha_2) = 0 \end{cases} \quad \text{on} \quad \partial\Omega \times (0, \infty)$$

where ∂_ν^r is the conormal derivative on $\partial\Omega$ corresponding to A^r, that is

$$\partial_\nu^r u = a_{ij}^r \partial_j u \, n_i$$

where $\vec{n} = (n_1, \ldots, n_N)$ is the unit outward normal derivative to $\partial\Omega$. We assume

$$(4.6) \qquad \alpha_1, \alpha_2 \geq 0, \quad 0 < \lambda_1, \lambda_2 \leq 1 \quad \text{or} \quad 0 = \lambda_1 = \lambda_2$$

$$(4.7) \quad u = u_o, \quad v = v_o \quad \text{on} \quad \Omega \times \{0\}, \quad u_o, v_o \in L^\infty(\Omega), \quad u_o, v_o \geq 0.$$

In addition suppose that

$$f(u, v) = f(x, t, u(x, t), v(x, t)), \quad g(u, v) = g(x, t, u(x, t), v(x, t))$$

are measurable on $\overline{\Omega} \times [0, \infty)^3$ and satisfy the usual locally Lipschitz continuity, namely

$$|f(x, t, u, v) - f(x, t, \hat{u}, \hat{v})| + |g(x, t, u, v) - g(x, t, \hat{u}, \hat{v})| \leq K(r)(|u - \hat{u}| + |v - \hat{v}|)$$

a.e. x, t and for all $0 \leq u, v, \hat{u}, \hat{v} \leq r$.

Also,

$$(4.8) \qquad f(x, t, 0, v) \geq 0, \quad g(x, t, u, 0) \geq 0 \quad \forall u, v \geq 0, \quad \text{a.e. } x, t$$

and

$$(4.9) \quad f(x, t, u, v) + g(x, t, u, v) \leq L(u + v) + M \quad \forall u, v \geq 0, \quad \text{a.e. } x, t$$

where $L, M \geq 0$.

According to the discussions in the introduction, this system has a unique noncontinuable solution in $\overline{\Omega} \times [0, \hat{T})$. If $\hat{T} < \infty$, then

$$(4.10) \qquad \lim_{t \uparrow \hat{T}} |u(\cdot, t)|_\infty + |v(\cdot, t)|_\infty = \infty$$

(see for instance [14] for this more general situation).

We now state a proposition which contains the main idea of the basic method. Essentially, it exploits the "balance condition" (4.9). For this we introduce the following notation.

Notation: $Q_T = \Omega \times (0, T)$, $\quad \Sigma_T = \partial \Omega \times (0, T)$, $\quad \|u\|_p = \left[\int_\Omega |u(x)|^p dx \right]^{1/p}$, $\|u\|_{p,T} = \left[\int_{Q_T} |u(x, t)|^p dx \, dt \right]^{1/p}$, $\|u\|_{p,\Sigma_T} = \left[\int_{\Sigma_T} |u(\sigma, t)|^p d\sigma \, dt \right]^{1/p}$.

Proposition 4.1 *(L^p-estimates). Assume (4.1)–(4.9). For all $p \in (1, \infty)$ and all $T > 0$, there exists C_1, C_2 such that*

(4.11) $$\|v\|_{p,T} \leq C_1 + C_2 \left(\|u_0\|_p + \|v_0\|_p + \|u\|_{p,\Sigma_T} + \|u\|_{p,T} \right)$$

where C_1, C_2 depend only on the data, p and T.

Remark. By symmetry, we also have

$$\|u\|_{p,T} \leq C_1 + C_2 \left(\|u_0\|_p + \|v_0\|_p + \|v\|_{p,\Sigma_T} + \|v\|_{p,T} \right).$$

In other words, the L^p-norms of u and v on Q_T can only blow up at the same time T. As a consequence, the next result naturally follows from Proposition 4.1.

Theorem 4.2 *Assume that the noncontinuable solution (u, v) to (4.4)–(4.7) on $\Omega \times (0, \widehat{T})$ has the property that there exists a continuous, increasing function $N_1 : [0, \infty) \to [0, \infty)$ such that*

(4.12) $$\|u(\cdot, t)\|_\infty \leq N_1(t) \quad for \quad 0 \leq t \leq \widehat{T}$$

and that there are positive numbers $L_1(r)$, $M_1(r)$ and σ such that

(4.13)
$$|g(u, v)| \leq L_1(r)|v|^\sigma + M_1(r)$$
$$for \ all \quad (u, v) \in [0, \infty)^2 \quad with \quad |u| \leq r.$$

Then $\widehat{T} = \infty$ and hence (u, v) exists globally.

Remark. The condition (4.12) is satisfied when for instance

$$f(u, v) \leq Lu + M \quad for \quad u \geq 0$$

which is the case in example of Section 1.3. This is a direct consequence of the maximum principle. But more general systems can be treated in this way (see the discussion in Section 5).

Proof of Proposition 4.1 By adding up the equations in (4.4) and using (4.9) we obtain

$$(4.14) \qquad u_t - A^1 u + v_t - A^2 v \leq L(u + v) + M.$$

In order to use a duality argument, we introduce $\chi \in C^\infty(Q_T)$, $\chi \geq 0$ and ϕ a nonnegative solution of

$$(4.15) \quad \begin{cases} -\phi_t - (A^2)^*\phi = \chi & \text{on} \quad Q_T \\ \lambda_2 (\partial_\nu^2)^*\phi + (1 - \lambda_2)\phi \; - \lambda_2 \vec{b}^2 \cdot \vec{n}\phi = 0 & \text{on} \quad \Sigma_T = \partial\Omega \times (0, T) \\ \phi(\cdot, T) = 0 & \text{on} \quad \Omega \end{cases}$$

where $(A^r)^*$ denotes the adjoint of A^r and $(\partial_\nu^r)^*$ its associated conormal derivative

$$(A^r)^*\phi = \partial_j^r(a_{ij}^r \partial_i \phi) - \partial_i(b_i^r \phi) + c^r \phi, \quad (\partial_\nu^r)^*\phi = a_{ij}^r \partial_i \phi n_j$$

and $\vec{b}^r = (b_1^r, \ldots, b_N^r)$. Recall (see e.g. [24]) that ϕ exists and is nonnegative. Moreover, for all $q \in (1, \infty)$ and $T > 0$ there exists C such that

$$(4.16) \qquad \|\phi\|_{W^{2,1,q}(Q_T)} + \|\phi(\cdot, 0)\|_q \leq C\|\chi\|_{q,T}$$

where $W^{2,1,q}(Q_T) = \{\phi \in L^q(Q_T); \quad \phi_t, \; \phi_{x_i}, \; \phi_{x_i x_j} \in L^q(Q_T)\}$ with its natural norm (see [24, p. 133]).

We multiply (4.14) by $e^{-tL}\phi$ to obtain

$$(4.17) \qquad \frac{\partial}{\partial t}\left(e^{-tL}(u + v)\right)\phi - e^{-tL}(A^1 u + A^2 v)\phi \leq M e^{-tL}\phi.$$

Now we use integration by parts

$$\int_\Omega \phi A^2 v = \int_\Omega v(A^2)^*\phi + \int_{\partial\Omega} \phi \partial_\nu^2 v - v(\partial_\nu^2)^*\phi + v\phi\vec{b}^2 \cdot \vec{n}$$

and the same for u to obtain

$$-\int_\Omega (u_o + v_o)\phi(\cdot, 0) - \int_{Q_T} e^{-tL}(u + v)\phi_t - \int_{Q_T} e^{-tL}\left[u(A^1)^*\phi + v(A^2)^*\phi\right] \cdots$$

$$\cdots - \int_{\Sigma_T} e^{-tL}\left\{\phi(\partial_\nu^1 u + \partial_\nu^2 v) - u(\partial_\nu^1)^*\phi - v(\partial_\nu^2)^*\phi + u\phi\vec{b}^1 \cdot \vec{n} + v\phi\vec{b}^2 \cdot \vec{n}\right\}$$

$$\leq \int_{Q_T} M e^{-tL}\phi.$$

Using the definition of ϕ in (4.15), we rewrite this as

$$(4.18) \quad \begin{cases} \displaystyle\int_{Q_T} e^{-tL}(u+v)\chi \\ \displaystyle\leq \int_\Omega (u_o+v_o)\phi(\cdot,0) + \int_{Q_T} e^{-tL}u\left((A^1)^*\phi - (A^2)^*\phi\right) + Me^{-tL}\phi \\ \displaystyle+ \int_{\Sigma_T} e^{-tL}\left\{\phi(\partial_\nu^1 u + \partial_\nu^2 v + u\vec{b}^1\cdot\vec{n} + v\vec{b}^2\cdot\vec{n}) - u(\partial_\nu^1)^*\phi - v(\partial_\nu^2)^*\phi\right\}. \end{cases}$$

Denote by I_1, I_2, I_3 the integrals on the right hand side in (4.18) and let $p \in (1, \infty)$ and $q = p/(p-1)$. Using (4.16), we have

$$(4.19) \quad I_1 + I_2 \leq C\left[\|u_o + v_o\|_p + \|u\|_{p,T} + M(meas\, Q_T)^{1/p}\right]\|\chi\|_{q,T}.$$

To bound the boundary term I_3, we use (4.5) and (4.15). Assume first $0 < \lambda_1, \lambda_2 \leq 1$.

$$(4.20) \quad \begin{aligned} I_3 &= \int_{\Sigma_T} e^{-tL}\left\{-u\left[\lambda_1^{-1}(1-\lambda_1)\phi - \vec{b}^1\cdot\vec{n}\phi + (\partial_\nu^1)^*\phi\right]\right. \\ &\quad\left. + \left(\alpha_1\lambda_1^{-1}(1-\lambda_1) + \alpha_2\lambda_2^{-1}(1-\lambda_2)\right)\phi\right\} \\ &\leq \left(\alpha_1\lambda_1^{-1} + \alpha_2\lambda_2^{-1}\right)\int_{\Sigma_T}\phi + C\|u\|_{p,\Sigma_T}\cdot\|\phi\|_{L^q(0,T;W^{1,q}(\partial\Omega))} \end{aligned}$$

where C depends only on a_{ij}^1, $\vec{b}^1$, λ^1. By (4.16) and the continuity of the trace operator from $W^{2,q}(\Omega)$ into $W^{1,q}(\partial\Omega)$, (see [14]), we obtain

$$(4.21) \quad I_3 \leq \left[\left(\alpha_1\lambda_1^{-1} + \alpha_2\lambda_2^{-1}\right)(meas\,\Sigma_T)^{1/p} + C\|u\|_{p,\Sigma_T}\right]C\|\chi\|_{q,T}.$$

If now $\lambda_1 = \lambda_2 = 0$, then $\phi = 0$, $u = \alpha_1$, $v = \alpha_2$ on $\partial\Omega$ so that I_3 can be bounded by

$$I_3 \leq \int_{\Sigma_T} \alpha_1|(\partial_\nu^1)^*\phi| + \alpha_2|(\partial_\nu^2)^*\phi| \leq C(\alpha_1+\alpha_2)(meas\,\Sigma_T)^{1/p}\|\phi\|_{L^q(0,T;W^{1,q}(\partial\Omega))},$$

or, by (4.16) and the continuity of the trace

$$(4.22) \quad I_3 \leq C(\alpha_1 + \alpha_2)(meas\,\Sigma_T)^{1/p}C\|\chi\|_{q,T}.$$

Finally, from (4.18), (4.19), (4.21), (4.22) we obtain the existence of C_1, C_2 depending only on the data and T such that

$$\int_{Q_T}(u+v)\chi \leq [C_1 + C_2(\|u_o + v_o\|_p + \|u\|_{p,T} + \|u\|_{p,\Sigma_T})]\|\chi\|_{q,T}.$$

Since χ is arbitrary nonnegative in $C_o^\infty(Q_T)$ and u, $v \geq 0$, this implies

$$\|u + v\|_{q,T} \leq C_1 + C_2 \left(\|u_o + v_o\|_p + \|u\|_{p,T} + \|u\|_{p,\Sigma_T}\right).$$

The result of Proposition 4.1 follows.

Remark. According to the proof above, it appears that we do not need the term $\|u\|_{p,\Sigma_T}$ in (4.11) if $0 = \lambda_1 = \lambda_2$ (see (4.22)) or if there exists $d > 0$ such that

$$\lambda_1^{-1}(1 - \lambda_1)\phi + (\partial_\nu^1)^*\phi - \vec{b}^1 \cdot \vec{n}\phi \geq d\left[\lambda_2^{-1}(1 - \lambda_2)\phi + (\partial_\nu^2)^*\phi - \vec{b}^2 \cdot \vec{n}\phi\right] = 0.$$

This can be seen from (4.20). This is the case if $A^1 = dA^2$ and $\lambda_1 \leq \lambda_2/(d + \lambda_2 - d\lambda_2)$ which essentially means that the boundary conditions are of the same type for u and v.

Proof of Theorem 4.2. If (4.12) holds, then (4.11) in Proposition 4.1 applied with $T = \widehat{T}$ shows

$$\|v\|_{p,\widehat{T}} \leq C_1 + 3C_2 N_1(\widehat{T})$$

where C_1, C_2 depend on the data and on $\widehat{T}$. Therefore, for all $p < \infty$, the L^p-norms of u and v are finite on $Q_{\widehat{T}}$. From the polynomial growth assumption (4.13), this implies that $g(u, v)$ is also in $L^p(Q_{\widehat{T}})$ for all $p \in (1, \infty)$. Going back to the second equation in (4.4) and taking $p > (N + 1)/2$, we deduce that v is $L^\infty(Q_{\widehat{T}})$ (see [14]):

$$\|v\|_{L^\infty(Q_{\widehat{T}})} \leq C\|g(u, v)\|_{L^p(Q_{\widehat{T}})} < +\infty.$$

This implies $\widehat{T} = \infty$.

1.5 Extensions of the Basic Method

The method described in Section 4 applies to a large class of sytems. One of them is the widely analyzed so-called "Brusselator" which appears as an important model in chemical reactions and has the form

(5.1)
$$\begin{cases} u_t = d_1\Delta u - uv^2 + Bv \\ v_t = d_2\Delta v + uv^2 - (B + 1)v + A \end{cases}$$

where A and B are positive constants and the boundary conditions are as in (3.2). The usual boundary conditions for this model are

$$u = B/A, \quad v = A \quad \text{or} \quad \partial_\nu u = \partial_\nu v = 0 \quad \text{on} \quad \partial\Omega \times (0, \infty).$$

Note that $u = B/A$ and $v = A$ is an equilibrium solution in each of these cases.

Theorem 4.2 applies in this situation and the only nonobvious hypothesis to verify is the *a priori* L^∞-estimate (4.12) on u. One way is to argue as follows (a second way is described later). Choose any $t_o > 0$ and use the inequality $v_t \geq d_2 \Delta v - (B + 1)v + A$ to show that

$$m(t) = \min\{v(x, s); \quad x \in \overline{\Omega}, \quad t_o \leq s \leq t\} > 0$$

for all $t > t_o$ where the solution exists. Now this estimate can be used in the first equation in (5.1) to obtain the L^∞-estimate on u [observe that $-uv^2 + Bv = v(B - uv) \leq v(B - m(t)u)$]. Details can be found in Hollis, Martin, Pierre [9]. Thus Theorem 4.2 applies to the Brusselator. A more careful use of these techniques shows further that each solution to (5.1) is uniformly bounded, see [9]:

$$\sup_{t \geq 0} \|u(\cdot, t)\|_\infty + \|v(\cdot, t)\|_\infty < \infty.$$

In fact the work of Bénilan and Labani [5] establishes the existence of a constant C, independent of the initial conditions u_o and v_o, such that

$$\lim_{t \to \infty} \sup \|u(\cdot, t)\|_\infty + \|v(\cdot, t)\|_\infty \leq C.$$

In particular, this shows there is a compact attractor for the solutions of (5.1).

Instead of assuming L^∞-estimate (4.12) in Theorem 4.2, we could have more generally assumed that the nonlinear term in the first equation is bounded above by a *linear* function of u and v as in the Brusselator. Indeed, by Proposition 4.1, we know that L^p-norms of u and v on Q_T are essentially equivalent (the term $\|u\|_{p,\Sigma_T}$ can be dispensed with if the boundary conditions are of the same type for u and v). Next, we can use a Gronwall's type inequality to obtain an L^p-bound on u. For instance, for the Brusselator with $\partial_\nu u = \partial_\nu v = 0$, we have

$$\|u(t)\|_p \leq \|u_o\|_p + B \int_o^t \|v(s)\|_p ds \leq \|u_o\|_p + Bt^{1/q}\|v\|_{p,t}$$

which implies by (4.11) that

$$\|u(t)\|_p \leq C_1(t) + C_2(t)\|u(t)\|_{p,t}$$

Setting $g(t) = \|u(t)\|_p^p$, this can be written

$$g(t) \leq \left\{ C_1(t) + C_2(t) \left[\int_o^t g(s)ds \right]^{1/p} \right\}^p \leq 2^p C_1(t) + 2^p C_2(t) \int_o^t g(s)ds,$$

which is a linear Gronwall inequality. Solving this inequality bounds $\|u\|_{p,t}$ for $t < \infty$. In particular this method yields global existence in *any dimension* for systems like the following studied in [23]:

$$u_t - d_1 \Delta u = w - uv$$

$$v_t - d_2 \Delta v = w - uv \quad \text{on} \quad \Omega \times (0, \infty)$$

$$w_t - d_3 \Delta w = uv - w$$

$$\frac{\partial u}{\partial n} = \frac{\partial v}{\partial n} = \frac{\partial w}{\partial n} = 0 \quad \text{on} \quad \partial\Omega \times (0, \infty).$$

All these "L^p-techniques" can also be extended to $m \times m$ systems. Let us describe the kind of assumptions that are needed for these methods to apply. Consider

(5.2)
$$\begin{cases} u_t^i - d_i \Delta u^i = f_i(u) & \text{on} \quad \Omega \times (0, T) \\ \frac{\partial u^i}{\partial n} = 0 & \text{on} \quad \partial\Omega \times (0, T) \end{cases}$$

where $u = (u^1, \ldots, u^m)$ and $d_i > 0$. Assume $f = (f_i)$ is locally Lipschitz continuous and quasi-positive (see (1.1)). Assume moreover that

(5.3)
$$\begin{cases} \text{there exists a } m \times m \text{ *triangular and invertible* matrix } P \text{ with} \\ \text{nonnegative entries, } C > 0 \text{ and } b \in \mathbf{R}^m \text{ such that} \\ \forall r \in \mathbf{R}^m, \quad r \geq 0, \qquad Pf(r) \leq Cr + b \end{cases}$$

where the usual order in $\mathbf{R}^m$ is used, and

(5.4)   the growth of f is at most polynomial.

Then (5.2) has a global classical solution for all $u_o, v_o \in L^\infty(\Omega)$ with $u_o, v_o \geq 0$ (see [19] for a proof). The linear upper bound in (5.3) can even be weakened to some polynomial upper bound with small degree depending on the dimension (see again [19]).

This method also covers the case when reaction terms satisfy some "Liapunov-type" condition such as

(5.5) $$\nabla H(r) \cdot f(r) \leq CH(r) + b$$

where $H(r) = \sum_{i=1}^{m} H_i(r_i)$ with $H_i : [0, \infty) \to [0, \infty)$ convex, regular and $H_i(r_i) \to \infty$ as $r_i \to \infty$. Indeed, it is easy to check that if u is a nonnegative solution of (5.2), then $v^i = H_i(u^i)$ is a subsolution of

$$v_t^i - d_i \Delta v^i \leq H_i'(u^i) f_i(u), \qquad i = 1, \ldots, m.$$

Then (5.5) appears for system (5.6) as one of the *linear* relations (5.3). Since the method carries over to nonnegative subsolutions (and not only solutions), global existence will also be obtained for $m \times m$ systems satisfying m conditions like (5.5) with a *triangular structure*. All of these ideas are combined in [19] where details and several examples are given. Note that uniform L^∞-bounds in time can also be obtained in that framework (see [20]).

Finally we mention some similar results in [12] with $\Omega = \mathbb{R}^N$. It is also proved that if *only one* balance relation holds in an $m \times m$ system like

(5.7) $$\sum_{i=1}^{m} f_i(u) = 0$$

and if each f_i has less than quadratic growth, then global existence follows from direct L^∞ estimates on ∇u.

1.6   Limits of the Method and Open Problems

Although many systems can be treated by the above method, it is not sufficient to understand all the systems for which the two main properties of positivity and preservation or dissipativity of mass hold. We will list here a few difficulties that cannot be overcome by above approach and we will illustrate them by significant open problems. It would be interesting to develop new techniques to treat them or to provide good and significant negative results like actual blow up in finite time. We will successively consider the following questions

  − "Non-triangular" systems
  − "Critical" boundary conditions
  − Nonlinear diffusions
  − Growth faster than polynomial

1.6.1 "Non-triangular" Systems

We first indicate an example which does not generally fall within the scope of the above theory:

(6.1)
$$\begin{cases} u_t - d_1 \Delta u = \lambda u^p v^q - u^r v^s & \text{on } \Omega \times (0, \infty) \\ v_t - d_2 \Delta v = u^r v^s - u^p v^q \\ \dfrac{\partial u}{\partial n} = \dfrac{\partial v}{\partial n} = 0 & \text{on } \partial\Omega \times (0, \infty) \end{cases}$$

where $0 < \lambda \leq 1$, $p, q, r, s > 1$. If we denote by $f(u, v)$ (resp. $g(u, v)$) the nonlinearity in the first (resp. second) equation, we have

(6.2)
$$\begin{cases} f(u, v) + g(u, v) = (\lambda - 1)u^p v^q \leq 0 \\ f(u, v) + \lambda g(u, v) = (\lambda - 1)u^r v^s \leq 0. \end{cases}$$

Therefore, the nonlinear terms satisfy a relation like (5.3) but with a *nontriangular* matrix (except if $\lambda = 0$). As a consequence, none of the components u and v can be a priori controlled. According to Proposition 4.1, for all $p \in (1, \infty)$, the norms of u and v in $L^p(Q_T)$ are equivalent. But they might blow up together for the same T. Except for some specific choices of p, q, r, s (for instance $p = q$, $r = s = p + 1$ or another similar favorable combination), the problem of global existence in (6.1) seems to be unresolved. Note that the limit case $\lambda = 1$ is obviously harder since relations in (6.2) reduce to only one. Moreover, one loses an easy L^1-estimate on the nonlinear terms, namely,

(6.3) $(1 - \lambda) \displaystyle\int_{Q_T} u^p v^q \leq \int_\Omega u(0) + v(0),\ (1 - \lambda) \int_{Q_T} u^r v^s \leq \int_\Omega u(0) + \lambda v(0)$

which follows from (6.2). We will see in Section 8 that these estimates will prove to be useful.

A similar difficulty is encountered in next example (see [11])

(6.4)
$$\begin{cases} u_t - d_1 \Delta u = -c(x)u^p v^q & \text{on } \Omega \times (0, \infty) \\ v_t - d_2 \Delta v = c(x)u^p v^q \\ \dfrac{\partial u}{\partial n} = \dfrac{\partial v}{\partial n} = 0 & \text{on } \partial\Omega \times (0, \infty) \end{cases}$$

where $p, q \geq 1$ and $c : \Omega \to \mathbf{R}$ is a regular function *with variable sign on* Ω. Again uniform L^1-estimate and equivalence of L^p-norms hold, but global existence does not follow. Partial "local" estimates have been obtained for (6.4) in [11] where this kind of systems are introduced. Note that numerical experiments show that if d_1 and d_2 are very different $(d_2 \approx 10^{-2}d_1)$, $c(x)u^p v^q$ can become very large at some time. However, in the situations considered, the solutions seem to become "smoother" after a while.

There are actually elliptic versions of the same questions which could perhaps be looked at first to provide some hints and which are also of interest by themselves. One such example is

$$(6.5) \quad \begin{cases} u - \Delta u = \lambda u^p v^q - u^r v^s + F(x) & \\ & \text{on } \Omega \\ v - \Delta v - \mu v_{x_1 x_1} = u^r v^s - u^p v^q + G(x) & \\ \dfrac{\partial u}{\partial n} = \dfrac{\partial v}{\partial n} = 0 & \text{on } \partial\Omega \end{cases}$$

where (again) $0 < \lambda \leq 1$, $p, q, r, s > 1$, F, G are nonnegative and smooth given functions on Ω, $\mu > 0$ and where we denote $x = (x_1, \ldots, x_N)$. Again, we easily prove (as in Proposition 4.1) that

$$\forall p \in (1, \infty) \qquad \|u\|_{L^p(\Omega)} \sim \|v\|_{L^p(\Omega)}.$$

If μ is small enough, this allows us to conclude existence of a solution in (6.5). For large μ, the question of existence seems to be open in general.

1.6.2   "Critical" Boundary Conditions

Although general boundary conditions of type (4.5) can be handled by our basic method, some "critical" cases are not (see the extra hypothesis (4.6)). For instance, in the case of the Brusselator (5.1), it is not known if solutions subject to the boundary conditions

$$u = B/A \quad \text{and} \quad \frac{\partial v}{\partial n} = 0 \quad \text{on } \partial\Omega \times (0, \infty)$$

exist or not for all time. The same question holds for the simpler system (even if $d_1 = d_2$)

$$\begin{cases} u_t - d_1 \Delta u = -uv^2 \\ \\ v_t - d_2 \Delta v = uv^2 \end{cases} \quad \text{on } \Omega \times (0, \infty)$$

$$u = 1, \quad \frac{\partial v}{\partial n} = 0 \qquad \text{on } \partial\Omega \times (0, \infty).$$

It would probably be worthwhile to first understand existence in the elliptic version

$$\begin{cases} u - d_1 \Delta u = -uv^2 + F \\ \\ v - d_2 \Delta v = uv^2 + G \end{cases} \quad \text{on } \Omega$$

$$u = 1, \quad \frac{\partial v}{\partial n} = 0 \qquad \text{on } \partial\Omega$$

when F, G are given nonnegative smooth functions on Ω.

One result in that direction is obtained by Hollis [8] for the hybrid system of "Brusselator type"

(6.6)
$$\begin{cases} u_t = d_1 \Delta u - uv^2 + Bv \\ \\ v_t = uv^2 - (B+1)v + A \end{cases} \quad \text{on } \quad \Omega \times (0, \infty)$$

$$u = B/A \qquad \text{on } \quad \partial\Omega \times (0, \infty).$$

It is easy to see that solutions to the associated O.D.E. exist and are uniformly bounded on $[0, \infty)$. However, Hollis establishes the following:

(6.7)
$$\begin{cases} \text{if } \max\{v_o(x) : x \in \partial\Omega\} > \max\{A, A/B\} \text{ then there} \\ \\ \text{is a } T^* < \infty \text{ such that } \quad \lim_{t \to T^*-} \|v(\cdot, t)\|_\infty = \infty. \end{cases}$$

To see that (6.7) is valid set

$$\Lambda(\alpha) \equiv \frac{B + 1 + [(B+1)^2 - 4\alpha B]^{1/2}}{2\alpha B/A} \qquad \text{for} \quad 0 < \alpha \le 1$$

and consider the ordinary differential equation

$$w' = (\alpha B/A)w^2 - (B+1)w + \Lambda, \quad t > 0, \quad w(0) = w_o$$

and observe that $w_o > \Lambda(\alpha)$ implies $w(t) \uparrow \infty$ as $t \uparrow T_\alpha$ for some $T_\alpha < \infty$. If it is assumed that u, v exists on $(0, \infty)$ then there is an $x_1 \in \partial\Omega$ and number $\delta > 0$ and $0 < \alpha < 1$ such that $v_o(x) > \Lambda(\alpha)$ for all $x \in \Omega$ with $|x - x_1| < \delta$. Also, since u is continuous on $\overline{\Omega} \times [0, T_\alpha]$ there is an $x_o \in \Omega$ such that $|x_o - x_1| < \delta$ and $u(x_o, t) \geq \alpha B/A$ for all $0 \leq t \leq T_\alpha$. But this implies $v(x_o, t) \geq w(t)$ for $t \in [0, T_\alpha)$ where $w_o = v_o(x_o) > \Lambda(\alpha)$. Since $w(t) \uparrow \infty$ as $t \uparrow T_\alpha$ we have a contradiction and this shows that the solution to (6.6) must blow-up in finite time.

1.6.3   Nonlinear Diffusions

Our basic method heavily relies on L^p-regularity theory for *linear* parabolic operators. Obviously, this approach fails when the diffusions are nonlinear. Let us indicate some of the corresponding systems. Considering 2×2 systems with the same nonlinearity as in Section 3 leads to the equations

(6.8)
$$
\begin{cases}
u_t - \Delta\phi(u) = -uh(v) \\
\qquad\qquad\qquad\qquad\quad \text{on} \quad \Omega \times (0, \infty) \\
v_t - \Delta\psi(v) = uh(v) \\
\dfrac{\partial\phi(u)}{\partial n} = \dfrac{\partial\psi(v)}{\partial n} = 0 \quad \text{on} \quad \partial\Omega \times (0, \infty)
\end{cases}
$$

where ϕ, $\psi : [0, \infty)$ are regular increasing functions from $[0, \infty)$ into $[0, \infty)$ with $\phi(0) = \psi(0) = 0$. Local existence is easy for nonnegative L^∞-initial data. Global existence is open in general.

However, the "L^1-technique" introduced in [21] that we will describe in Section 8 turns out to be very helpful for these nonlinear situations. For instance, partial results can be obtained for specific choices of ϕ and ψ (see [13]). More interesting is the possibility of proving global existence of weak solutions in the following

(6.9)
$$
\begin{cases}
u_t - \Delta_p u = -uh(v) + F \\
v_t - \Delta_p v = uh(v) + G \\
u = v = 0 \quad \text{on} \quad \partial\Omega \times (0, \infty)
\end{cases}
$$

where $\Delta_p u = \operatorname{div}(|\nabla u|^{p-2}\nabla u)$ and $p > 1$ (see [13]). Here, the basic method would fail.

The problem becomes even more complicated when the diffusions are of different types, such as (Δ_p, Δ_q) with $p \neq q$, or $(\Delta\phi(\cdot), \Delta\psi(\cdot))$, or more general nonisotropic diffusions. Diffusions with first order terms of the following form are studied in [10]:

(6.10)
$$\begin{cases} u_t - d_1 \Delta u + \text{div}\,(u^\ell) = -uv^\beta \\ \qquad\qquad\qquad\qquad\qquad\qquad \text{on}\quad \Omega \times (0,\,\infty) \\ v_t - d_2 \Delta v + \text{div}\,(v^m) = uv^\beta \\ u = v = 0 \quad \text{on}\quad \partial\Omega \times (0,\,\infty). \end{cases}$$

For dimension $N > 1$, global existence is obtained in [10] for

$$(m > \beta \geq 1, \quad \ell \geq 1) \quad \text{or} \quad \left(m \leq \frac{N+2}{2}, \quad \beta, \ell \geq 1 \right).$$

The method is an extension of the one described in Section 4 for the non obvious cases (small values of m). A direct method of supersolutions takes care of $m > \beta$. However a gap still exists for the values of m between $(N+2)/2$ and β.

Note that elliptic version of the same problems could be considered first. Some could be treated using the "L^p-technique" of Section 4 such as

(6.11)
$$\begin{cases} u - \Delta\phi(u) = -uv^\beta + F \\ v - \Delta\psi(v) = uv^\beta + G \\ + \text{boundary conditions} \end{cases}$$

and

(6.12)
$$\begin{cases} u - \Delta_p u = -uv^\beta + F \\ v - \Delta_p v = uv^\beta + G \\ + \text{boundary conditions}. \end{cases}$$

However, many existence questions are left open for similar elliptic systems.

1.6.4   About Polynomial growth

It is well-known that L^p-regularity theory fails for $p = 1$. This is why L^∞-estimates cannot be directly obtained by the duality method used in Proposition 4.1. Therefore, the assumption of polynomial growth is necessary for these conclusions. Note that L^p-theory can be extended to some Orlicz spaces. Although it has not been explicitly written yet, the same approach should allow nonlinear growth up to less than exponential growth.

This is confirmed in a specific case with a different technique in [7] where, as mentioned before, growth of the form e^{v^γ} with $\gamma < 1$ is handled.

One of the main advantages of the L^1-technique described in Section 8 is that it does not require any growth assumption to provide global existence of weak solutions. However the question of existence of *classical* solutions remains open even in simple systems like (3.1). We give next a very particular result in that direction.

1.7   A Particular Result Without Polynomial Growth Assumption

We consider the following system on the whole space $\mathbf{R}^N$:

$$(7.1) \quad \begin{cases} u_t - d_1 \Delta u = -f(u, v) \\[2mm] v_t - d_2 \Delta v = f(u, v) \\[2mm] u(\cdot, 0) = u_o(\cdot), \quad v(\cdot, 0) = v_o(\cdot) \end{cases} \quad \begin{aligned} & \text{on} \quad \mathbf{R}^N \times (0, \infty) \\[6mm] & \text{on} \quad \mathbf{R}^N \end{aligned}$$

where $f : [0, \infty)^2 \to [0, \infty)$ is locally Lipschitz continuous and satisfies

$$(7.2) \qquad\qquad \forall u, v \geq 0 \qquad f(0, v) = 0 \leq f(u, 0).$$

(Note that $f(u, v) \geq 0$ for all $u, v \geq 0$).

Theorem 7.1 *Assume*

$$(7.3) \qquad\qquad\qquad d_2 \leq d_1$$

and

$$(7.4) \qquad u_o, v_o \in L^\infty(\mathbf{R}^N), \qquad u_o, v_o \geq 0 \text{ a.e. on } \mathbf{R}^N.$$

Then (7.1) has a classical solution on $\Omega \times (0, \infty)$.

Lemma 7.2 *Let g be a nonnegative regular function on $\mathbf{R}^N \times (0, \infty)$. For all $\lambda > 0$, let w_λ be the solution of*

$$(7.5) \quad \begin{cases} w_{\lambda t} - \lambda \Delta w_\lambda = g & \text{on} \quad \mathbf{R}^N \times (0, \infty), \qquad g \not\equiv 0 \\[2mm] w_\lambda(\cdot, 0) = 0 & \text{on} \quad \mathbf{R}^N \end{cases}$$

Then

$$(7.6) \qquad \lambda \longmapsto \lambda^{N/2} w_\lambda(t, x) \quad \text{is increasing for all } (t, x).$$

Proof. We use (for instance) the explicit representation of w_λ in terms of g, namely

$$w_\lambda(t,\, x) = \int_o^t \int_{\mathbf{R}^N} (4\pi\lambda s)^{-N/2} \exp(-|y|^2/4\lambda s) g(t - s,\, x - y) ds\, dy.$$

Since $g \geq 0$, (7.6) follows.

Proof of Theorem 7.1   Local existence of a classical nonnegative solution $(u,\, v)$ to (7.1) on a maximal interval of existence $[0,\, T_m)$ is obtained as usual.

We denote by $S_\lambda(t)$ the semigroup generated by $-\lambda\Delta$ on $\mathbf{R}^N$ and by z_λ the solution of

(7.7)
$$\begin{cases} z_{\lambda t} - \lambda\Delta z_\lambda = f(u,\, v) \quad \text{on} \quad \mathbf{R}^N \times (0,\, T_m) \\ z_\lambda(\cdot,\, 0) = 0 \quad \text{on} \quad \mathbf{R}^N. \end{cases}$$

Then, by linearity, we have for all $t \in (0,\, T_m)$

(7.8)
$$u(t) = S_{d_1}(t)u_o - z_{d_1}(t)$$

(7.9)
$$v(t) = S_{d_2}(t)v_o + z_{d_2}(t).$$

By Lemma 7.2, since $d_2 \leq d_1$ we have

(7.10)
$$z_{d_2}(t) \leq (d_1/d_2)^{N/2} z_{d_1}(t).$$

From (7.8)–(7.10) and $u \geq 0$ we deduce

(7.11)
$$v(t) \leq S_{d_2}(t)v_o + (d_1/d_2)^{N/2} S_{d_1}(t)u_o.$$

This provides a uniform L^∞-estimate for v. On the other hand we already know that

$$u(t) \leq S_{d_1}(t)u_o.$$

This proves global existence of classical solutions.

1.8   The L^1-technique

Again we will describe this technique for the 2×2 system introduced in (4.1)–(4.9). For simplicity, we assume that the diffusions are linear and dissipative in $L^p(\Omega)$ for all $p \in [1,\, \infty]$. More precisely, we assume (4.1), (4.2), (4.3) and

(8.1) $\alpha_1 = \alpha_2 = 0, \quad 0 \le \lambda_1, \lambda_2 \le 1$ (instead of (4.6))

(8.2)   For $r = 1, 2, \; c^r \le 0$ on Ω, $-\sum b_{ix_i}^r + c^r \le 0$ on Ω, $\sum_i b_i^r n_i \le 0$ on $\partial\Omega$.

We consider A^r as an unbounded operator in $L^1(\Omega)$ with the domain

(8.3) $D(A^r) = \{u \in H^2(\Omega); \; \lambda_r \partial_\nu^r u + (1 - \lambda_r)u = 0 \text{ on } \partial\Omega\}$

Under these assumptions, the closure of $-A^r$ generates a linear semigroup of contractions $S^r(t)$ in $L^1(\Omega)$ which satisfies

(8.4) $\|S^r(t)u_o\|_p \le \|u_o\|_p$ for all $p \in [1, \infty]$

(8.5) $S^r(t)$ is compact from $L^1(\Omega)$ into $L^1(\Omega)$ for all $t > 0$.

(see Brézis-Strauss [6] for a proof and for more information about the domain of the closure of A^r). The functions f and g are taken as in Section 4.

By a solution of (4.4)–(4.5) on $\Omega \times (0, \infty)$ we mean that

(8.6)
$$\begin{cases} u, v \in C\left([0, \infty); \; L^1(\Omega)\right) \\[2mm] f(u, v), \; g(u, v) \in L^1(Q_T) \quad \text{for all} \quad T > 0 \\[2mm] u(t) = S^1(t)u_o + \displaystyle\int_o^t S^1(t - s)f(\cdot, \; s, \; u(s), \; v(s))ds \\[2mm] v(t) = S^2(t)v_o + \displaystyle\int_o^t S^2(t - s)g(\cdot, \; s, \; u(s), \; v(s))ds \end{cases} \qquad \forall t \ge 0$$

where here

(8.7) $u_o, \; v_o \in L^1(\Omega), \qquad u_o, \; v_o \ge 0.$

To approximate (8.6)–(8.7) we truncate the nonlinearities f and g and the initial data as well.

(8.8) $u_o^n(x) = \min(u_o(x), n), \qquad v_o^n(x) = \min(v_o(x), n)$

(8.9) $f_n(x, t, u, v) = \psi_n\left(|u| + |v|\right)\left(f(x, t, u, v) - f(x, t, 0, 0)\right) + f(x, t, 0, 0)$

(8.10) $g_n(x, t, u, v) = \psi_n\left(|u| + |v|\right)\left(g(x, t, u, v) - g(x, t, 0, 0)\right) + g(x, t, 0, 0)$

where $\psi_n \in C_o^\infty(\mathbf{R})$ is such that $0 \le \psi_n \le 1$ and

$$\psi_n(r) = \begin{cases} 1 & \text{if } |r| \le n \\[2mm] 0 & \text{if } |r| \ge n + 1 \end{cases}$$

Note that $f(\cdot, \cdot, 0, 0) \in L^\infty(\Omega \times [0, \infty))$ and that f_n, g_n satisfy the same hypotheses as f and g with the same constants independent of n. In addition there exists $M(n)$ such that

(8.11)   a.e. x, t, $\forall u, v \geq 0$, $|f_n(x, t, u, v)| + |g_n(x, t, u, v)| \leq M(n)$.

Proposition 8.1   *We assume (4.1), (4.2), (4.3), (4.8), (4.9), (8.1), (8.2), (8.7)–(8.10). Then, for all n, there exists a classical nonnegative solution (u_n, v_n) of*

(8.12)
$$
\begin{cases}
\partial_t u_n = A^1 u_n + f_n(u_n, v_n) \\
\qquad\qquad\qquad\qquad\qquad \text{on} \quad \Omega \times (0, \infty) \\
\partial_t v_n = A^2 u_n + g_n(u_n, v_n) \\
u_n(\cdot, 0) = u_o^n, \quad v_n(\cdot, 0) = v_o^n, \quad \text{on} \quad \Omega \\
\lambda_1 \partial_\nu^1 u_n + (1-\lambda_1) u_n = 0, \\
\qquad\qquad\qquad\qquad\qquad \text{on} \quad \partial\Omega \times (0, \infty) \\
\lambda_2 \partial_\nu^2 v_n + (1-\lambda_2) v_n = 0.
\end{cases}
$$

Proof   Local existence is classical. Global existence follows from (8.11) which ensures that, for all t for which (u_n, v_n) exists,

$$
\|u_n(t)\|_\infty + \|v_n(t)\|_\infty \leq M(n, t)
$$

with $M(n, t) < \infty$ for all n, $t > 0$.

Our goal is now to prove that, under suitable additional assumptions the solution (u_n, v_n) of (8.12) converges to a solution of (8.6) as n tends to ∞. This will be realized by obtaining appropriate estimates on u_n, v_n.

Thanks to the two main structure assumptions (4.8), (4.9) on the non-linearity, we immediately have

Lemma 8.2   *Under assumptions of Proposition 8.1*

(8.13) $\displaystyle \int_\Omega u_n(t) + v_n(t) \leq e^{tL} \left[\int_\Omega u_o + v_o + ML^{-1} \left(1 - e^{-tL}\right) \right].$

Proof   From (8.12) and (4.9) we have

(8.14) $\partial_t(u_n + v_n) \leq A^1 u_n + A^2 v_n + L(u_n + v_n) + M.$

We integrate this over Ω. Since $u_n, v_n \geq 0$ and A^1, A^2 are dissipative in $L^1(\Omega)$

(8.15) $\int_\Omega A^1 u_n \leq 0, \qquad \int_\Omega A^2 v_n \leq 0.$

As a consequence, $\alpha_n(t) = u_n(t) + v_n(t)$ satisfies the differential inequality

$$\alpha_n'(t) \leq L\alpha_n(t) + M.$$

The relation (8.13) follows.

In order to get better estimates on u_n, v_n, we need to put more structure on f, g. According to examples we assume that

(8.16) $\begin{cases} \exists \alpha, \beta \geq 0, \quad \alpha \neq \beta \quad \text{and} \quad L_1, M_1 \geq 0 \quad \text{such that} \\ \forall u, v \geq 0 \quad \alpha f(u, v) + \beta g(u, v) \leq L_1(u, v) + M_1 \end{cases}$

Remark   The condition $\alpha \neq \beta$ ensures that relation (8.16) be independent of the previous assumption (4.9) concerning $f + g$. Note that both are satisfied for the model example of Section 3 with $\alpha = 1$, $\beta = 0 = L_1 = M_1 = L = M$.

Lemma 8.3 *If the assumptions of Proposition 8.1 hold and if (8.16) holds, then for all $T > 0$, there exists $R(T)$ depending only on T, $\|u_0\|_1$, $\|v_0\|_1$, L, M L_1, M_1, α, β such that*

(8.17) $\int_{Q_T} |f(u_n, v_n)| + |g(u_n, v_n)| \leq R(T).$

Proof.   Integrating each of the two first equations in (8.12) and using (8.15) lead to

(8.18) $-\int_\Omega u_o^n \leq \int_{Q_T} f(u_n, v_n), \qquad -\int_\Omega v_o^n \leq \int_{Q_T} g(u_n, v_n).$

We set

(8.19) $\begin{cases} R_n = M + L(u_n + v_n) - [f_n(u_n, v_n) + g_n(u_n, v_n)] \geq 0 \\ S_n = M_1 + L_1(u_n + v_n) - [\alpha f_n(u_n, v_n) + \beta g_n(u_n, v_n)] \geq 0 \end{cases}$

where the nonnegativeness of R_n and S_n follows by (4.9) and (8.16), respectively. Therefore, by (8.18),

(8.20) $\int_{Q_T} |R_n| = \int_{Q_T} R_n \leq \int_{Q_T} M + L(u_n + v_n) + \int_\Omega u_o^n + v_o^n$

$$(8.21) \qquad \int_{Q_T} |S_n| = \int_{Q_T} S_n \le \int_{Q_T} M_1 + L_1(u_n + v_n) + \int_{\Omega} \alpha u_o^n + \beta v_o^n.$$

The relations (8.19) can be rewritten as

$$(8.22) \quad \begin{cases} S_n - \beta R_n = M_1 - \beta M + (L_1 - \beta L)(u_n + v_n) + (\beta - \alpha)f_n(u_n, v_n) \\ S_n - \alpha R_n = M_1 - \alpha M + (L_1 - \alpha L)(u_n + v_n) + (\alpha - \beta)g_n(u_n, v_n). \end{cases}$$

From (8.20)–(8.22), we obtain

$$|\beta - \alpha| \int_{Q_T} |f(u_n, v_n)| + |g(u_n, v_n)|$$
$$\le \int_{Q_T} 4M_1 + 2(\beta + \alpha)M + (4L_1 + 2(\beta + \alpha)L)(u_n + v_n)$$
$$+ \int_{\Omega} (3(\alpha + \beta)u_o^n + (\alpha + 3\beta)(\alpha - \beta)v_o^n.$$

The estimate (8.17) follows using also (8.13) and (8.8).

Estimate (8.17) of Lemma 8.3 provides compactness for the sequence (u_n, v_n) thanks to the following result.

Lemma 8.4 *For $r = 1, 2$ and for all $T > 0$, the mapping*

$$(8.23) \qquad (w_o, g) \in L^1(\Omega) \times L^1(Q_T) \longmapsto w \in L^1(Q_T)$$

where

$$(8.24) \qquad w(t) = S^r(t)w_o + \int_o^t S^r(t - s)g(s)\, ds$$

is compact. Moreover, it is continuous from $L^1(\Omega) \times L^1(Q_T)$ into $C([0, T]; L^1(\Omega))$.

Proof Continuity into $C([0, T]); L^1(\Omega))$ is classical. Compactness into $L^1(Q_T)$ comes from the fact that $S^r(t)$ is compact for $t > 0$ and the general abstract result in Baras, Hassan and Véron [4].

Proposition 8.5 *Under assumptions of Lemma 8.3, there exists a subsequence (u_{n_k}, v_{n_k}) and $(u, v) \in L^1_{loc}([0, \infty]; L^1(\Omega))^2$ such that*

$$\lim_{k \to \infty} (u_{n_k}, v_{n_k}) = (u, v) \quad \text{in} \quad L^1(Q_T)^2 \quad \text{for all} \quad T > 0.$$

Proof. We have

$$(8.26) \qquad u_n(t) = S^1(t)u_o^n + \int_o^t S^1(t-s)f_n\left(u_n(s),\, v_n(s)\right) ds$$

$$(8.27) \qquad v_n(t) = S^2(t)v_o^n + \int_o^t S^2(t-s)g_n\left(u_n(s),\, v_n(s)\right) ds$$

where $(u_o^n,\, v_o^n)$ converges to $(u_o,\, v_o)$ in $L^1(\Omega)$ and where $f_n(u_n,\, v_n)$, $g_n(u_n,\, v_n)$ are bounded in $L^1(Q_T)$ for all T independently of n by Lemma 8.3. Therefore, by Lemma 8.4 $(u_n,\, v_n)$ is precompact in $L^1(Q_T)$ for all T. Using a sequence T_k with $\lim_{k\to\infty} T_k = \infty$ and a diagonal process, we obtain (8.25).

Comments   It is important to notice here that the arguments given above are *very general*. They are based only on the facts that

$$(8.28) \qquad \text{the semigroups } S^r(t) \text{ are nonexpansive and compact in } L^1(\Omega)$$

$$(8.29) \qquad \begin{pmatrix} 1 & 1 \\ \alpha & \beta \end{pmatrix} \begin{pmatrix} f(u,\, v) \\ g(u,\, v) \end{pmatrix} \leq \begin{pmatrix} L & M \\ L_1 & M_1 \end{pmatrix} \begin{pmatrix} u+v \\ 1 \end{pmatrix} \quad \alpha,\, \beta \geq 0,\; \alpha \neq \beta.$$

As a consequence the conclusion (8.25) (and therefore the existence of the limit $(u,\, v)$) is valid under assumptions (8.28), (8.29) even if the semigroups are nonlinear. Indeed, Lemma 8.4 is valid for compact nonlinear semigroups of contractions if w is defined as the "weak solution" (in the sense of the nonlinear semigroup theory) of

$$(8.30) \qquad \begin{cases} \dfrac{dw}{dt} - Aw \ni g & \text{on} \quad (0,\, \infty) \\[2mm] w(0) = w_o \end{cases}$$

where A is the corresponding nonlinear dissipative operator (see Baras [3] for a proof of this compactness result).

All the details of this generalization can be found in [13].

At this point we have reached the main difficulty of the problem. We need to prove that $(u,\, v)$ is a solution to our problem (8.6). Since $(u_{n_k},\, v_{n_k})$ converges in $L^1(Q_T)$ to $(u,\, v)$, up to extracting a new subsequence, we can assume that the convergence also holds almost everywhere. Since f and g are continuous with respect to u and v and by construction of $f_n,\, g_n$ (see (8.9), (8.10)), we can assume

$$(8.31) \qquad \begin{cases} \displaystyle\lim_{k\to\infty} f_{n_k}(u_{n_k},\, v_{n_k}) = f(u,\, v) & \text{a.e.}\,(x,\, t) \\[2mm] \displaystyle\lim_{k\to\infty} g_{n_k}(u_{n_k},\, v_{n_k}) = g(u,\, v) & \text{a.e.}\,(x,\, t). \end{cases}$$

However this is not sufficient to pass to the limit in (8.26), (8.27). We need that the convergence in (8.31) holds in $L^1(Q_T)$ for all T (recall the continuity result in Lemma 8.4).

The gap can be filled in by the next lemma.

Lemma 8.6 *Let σ_n be a sequence in $L^1(Q_T)$ and σ in $L^1(Q_T)$ such that*

$$(8.32) \qquad\qquad \sigma_n \to \sigma \qquad a.e. \ on \quad Q_T$$

$$(8.33) \qquad\qquad \sigma_n \ is \ uniformly \ integrable \ in \ L^1(Q_T).$$

Then σ_n converges to σ in $L^1(Q_T)$.

Proof This is a classical result (see a proof in [21]). Recall that (8.33) means that for all $\epsilon > 0$, there exists $\delta > 0$ such that

$$(K \subset Q_T \text{ measurable, meas } K < \delta) \Rightarrow \int_K |\sigma_n| < \epsilon \qquad \forall n.$$

Our goal is now to show that $f(u_n, v_n)$, $g(u_n, v_n)$ are not only bounded in $L^1(Q_T)$ but uniformly integrable. We cannot accomplish this without an extra structure assumption on f, g, namely

$$(8.34) \qquad\qquad \exists L_2, M_2 \geq 0 \qquad f(u, v) \leq L_2(u + v) + M_2.$$

This is a nothing but assuming a "triangular" structure for the nonlinear part as in Sections 4–5 (in other words we ask that $\alpha = 1$, $\beta = 0$ in (8.16)).

Theorem 8.7 *Let us make the same assumptions as in Proposition 8.1. Assume moreover (8.34). Then, for all u_o, v_o as in (8.7), there exists a solution (u, v) to (8.6).*

The proof of this theorem requires also the following result

Lemma 8.8 *Let U be a bounded open set in $\mathbf{R}^d$ and let σ_n be a sequence in $L^1(U)$. Then the following statements are equivalent:*

$$(8.35) \qquad\qquad \sigma_n \ is \ uniformly \ integrable \ in \ L^1(U)$$

$$(8.36) \quad
\begin{cases}
\text{There exists } J : (0,\infty) \to (0,\infty) \text{ with } J(0^+) = 0 \text{ and} \\[2mm]
\text{(a)} \quad J \text{ is convex, } J' \text{ is concave, } J' \geq 0 \\[2mm]
\text{(b)} \quad \lim_{r \to +\infty} \dfrac{J(r)}{r} = +\infty \\[2mm]
\text{(c)} \quad \sup_n \int_U J(|\sigma_n|) < \infty.
\end{cases}$$

Proof   This lemma is classical except perhaps for the possibility of choosing J' concave (see [15] for a proof).

Proof of Theorem 8.7   By Proposition 8.1, we have existence of the solution (u_n, v_n) to the approximate problem (8.12) (or (8.26), (8.27)). By Proposition 8.5, up to extracting a subsequence, we can assume that

(8.37) $\lim_{n\to\infty}(u_n,v_n) = (u,v)$ a.e. $x, t \in \Omega \times (0, \infty)$ and in $L^1(Q_T)$ for all T

(8.38) $\lim_{n\to\infty} f_n(u_n,v_n) = f(u,v)$ a.e. $x, t \in \Omega \times (0, \infty)$

(8.39) $\lim_{n\to\infty} g_n(u_n,v_n) = g(u,v)$ a.e. $x, t \in \Omega \times (0, \infty)$.

We now fix $T > 0$ arbitrary and J as in (8.36) with (8.36c) replaced by

(8.40) $\sup_n \int_{Q_T} J(u_n + v_n) < \infty,$ $\sup_n \int_\Omega J(2u_o^n + v_o^n) < \infty.$

This is possible by Lemma 8.8 since $u_n + v_n$ converges in $L^1(Q_T)$ and $2u_o^n + v_o^n$ converges in $L^1(\Omega)$. Next we set

(8.41) $j(r) = \int_o^r \min\left(J'(s), (J^*)^{-1}(s)\right)\, ds$

where J^* is the conjugate function of J. We easily check that j satisfies (8.36a), (8.36b) and

(8.42) $\forall r \geq 0$ $j(r) \leq J(r)$ $J^*(j'(r)) \leq r.$

Our goal is to show that

(8.43) $\sup_n \int_{Q_T} j'(2u_n + v_n)\left(|f_n(u_n, v_n)| + |g_n(u_n, v_n)|\right) < \infty.$

Let us first indicate how the proof of Theorem 8.7 can then be finished.
    Estimate (8.43) implies the uniform integrability of $f_n(u_n, v_n)$, $g_n(u_n, v_n)$ on Q_T. Indeed let $\epsilon > 0$ and Q_T be measurable. Then

$$\int_K |f_n(u_n, v_n)| \leq \int_{K\cap[2u_n+v_n<\mu]} \sup_{0\leq r,s\leq\mu} |f_n(x,t,r,s)|\, dx\, dt$$
$$+ \int_{K\cap[2u_n+v_n>\mu]} |f_n(u_n, v_n)| = I_1 + I_2$$

The second integral I_2 is estimated by

$$I_2 \leq \frac{1}{j'(\mu)} \int_{Q_T} j'(2u_n + v_n)|f_n(u_n, v_n)|.$$

Thanks to (8.43) this can be made less than $\frac{\epsilon}{2}$ by choosing $\mu = \mu(\epsilon)$ large enough depending only on ϵ. For I_1, we notice that by the assumptions on f, g (see (4.8), (4.9)) and the choice of f_n, g_n (see (8.9), (8.10)),

$$\sup_{0 \leq r, s \leq \mu} |f_n(x, t, r, s)| \leq M + 2\mu K(\mu)$$

As a consequence

$$I_1 \leq [M + 2\mu(\epsilon)K(\mu(\epsilon))] \operatorname{meas} K < \frac{\epsilon}{2} \quad \text{if meas } K < \epsilon/2[M + 2\mu(\epsilon)K(\mu(\epsilon))].$$

This proves uniform integrability of $f_n(u_n, v_n)$ and similarly of $g_n(u_n, v_n)$. Using now (8.38), (8.39) and Lemma 8.6, we obtain that $f_n(u_n, v_n)$, $g_n(u_n, v_n)$ converge in $L^1(Q_T)$ to $f_n(u, v)$, $g_n(u, v)$. Then the proof of Theorem 8.7 is finished by passing to the limit in the approximate equations (8.26), (8.27).

Let us now prove (8.43). We set

$$R_n = M_2 + L_2(u_n + v_n) - f_n(u_n, v_n) \geq 0 \quad \text{by (8.34)}$$

$$S_n = M + L(u_n + v_n) - f_n(u_n, v_n) - g_n(u_n, v_n) \geq 0 \quad \text{by (4.9)}.$$

Combining the two first equations in (8.12), we obtain

$$(8.45) \quad (2u_n + v_n)_t - (2A^1 u_n + A^2 v_n) + R_n + S_n = M + M_2 + (L_1 + L_2)(u_n + v_n).$$

We multiply by $j'(2u_n + v_n)$ and integrate over Q_T to obtain

$$(8.46) \quad \begin{aligned} \int_\Omega j(2u_n + v_n)(T) - J_1 + \int_{Q_T} j'(2u_n + v_n)(R_n + S_n) \\ = J_2 + \int_\Omega j(2u_o^n + v_o^n) \end{aligned}$$

where

$$J_1 = \int_{Q_T} j'(2u_n + v_n)(2A^1 u_n + A^2 v_n)$$
$$J_2 = \int_{Q_T} j'(2u_n + v_n)[M + M_2 + (L + L_2)(u_n + v_n)].$$

To bound J_2 we use that (see (8.41), (8.42))

$$(8.47) \quad j'(r) \cdot s \leq J(s) + J^*(j'(r)) \leq J(s) + r.$$

Therefore

$$J_2 \leq \int_{Q_T} J(M + M_2) + 2u_n + v_n + (L_1 + L_2)[J(u_n + v_n) + 2u_n + v_n]$$

which is bounded independently of n by the choice of J (see (8.40)) and by Lemma 8.2.

If we can control J_1, then from (8.46), the above estimate of J_2 and (8.40) again, we obtain our estimate (8.43) since, by definition of R_n, S_n

$$|f_n| + |g_n| \leq R_n + 2S_n + M + 2M_2 + (L + 2L_2)(u_n + v_n)$$

and by (8.47)

(8.48) $$j'(2u_n + v_n)(u_n + v_n) \leq J(u_n + v_n) + 2u_n + v_n.$$

Let us estimate J_1. We have, after integration by parts

$$J_1 = \int_{\Sigma_T} j'(2u_n + v_n)\left[2\partial_\nu^1 u_n + \partial_\nu^2 v_n + 2\vec{b}^1 \cdot \vec{n} u_n + \vec{b}^2 \cdot \vec{n} v_n\right]$$
$$- \int_{Q_T} j''(2u_n+v_n)\left[a_{ij}^1 \partial_j(2u_n)\partial_i(2u_n+v_n) + a_{ij}^2 \partial_j(v_n)\partial_i(2u_n+v_n)\right]$$
$$- \int_{Q_T} 2u_n\left[j''(2u_n+v_n)b_i^1\partial_i(2u_n+v_n) + j'(2u_n+v_n)\partial_i b_i^1\right]$$
$$- \int_{Q_T} v_n\left[j''(2u_n+v_n) + b_i^2\partial_i(2u_n+v_n) + j'(2u_n+v_n)\partial_i b_i^2\right]$$
$$+ \int_{Q_T} j'(2u_n+v_n)(2c^1 u_n + c^2 v_n).$$

According to assumptions (8.2) and the boundary conditions in (8.12), the first integral is nonnegative. So are the integrals involving $\partial_i b_i^\tau$ combined with the last integral. Now, we use ellipticity of a_{ij}^2 in the second integral to get

$$J_1 \leq \int_{Q_T} j''(2u_n+v_n)\left[-\alpha|\nabla(2u_n+v_n)|^2+\left(a_{ij}^1 - a_{ij}^2\right)\partial_j(2u_n)\partial_i(2u_n+v_n)\right]$$
$$- \int_{Q_T} j''(2u_n+v_n)(2u_n b_i^1 + b_i^2 v_n)\partial_i(2u_n+v_n).$$

Now use Young's inequality and absorb the terms in $j''(\ )|\nabla(2u_n+v_n)|^2$ with $-\alpha$ to obtain a constant C depending on α, a_{ij}^τ, b_i^τ such that

(8.49) $$J_1 \leq C\int_{Q_T} j''(2u_n+v_n)\left[|\nabla u_n|^2 + u_n^2 + v_n^2\right].$$

Since j' is concave, $j''(r) \leq j'(r)/r$ and we have

$$\int_{Q_T} j''(2u_n + v_n)(u_n^2 + v_n^2) \leq \int_{Q_T} j'(2u_n + v_n)(u_n + v_n)$$

which is treated as in (8.48) to get a uniform bound. The last term in (8.49) is estimated by using again the equation in u and the assumption (8.34) on f. We have (using ellipticity of a_{ij}^1)

$$\int_\Omega j(u_n)(T) + \int_{Q_T} j''(u_n)\,\alpha|\nabla u_n|^2 \leq \int_{\sum_T} j'(u_n)\partial_\nu^1 u_n + \vec{b}^1 \cdot \vec{n}j(u_n)$$

$$+ \int_{Q_T} -\partial_i b_i^1 j(u_n) + c^1 j'(u_n) + j'(u_n)\,(M_2 + L_2(u_n + v_n)) + \int_\Omega j(u_n^o).$$

Adding the information that $j''(u_n) \geq j''(2u_n + v_n)$, we now easily deduce that $\int_{Q_T} j''(2u_n + v_n)|\nabla u_n|^2$ is uniformly bounded. By (8.49), so is J_1, and the proof is complete.

Remarks The method described is a straightforward extension to general linear diffusions of the method introduced in [21]. As we mentioned before, applications to nonlinear diffusions like Δ_p have been made in [13].

Obviously, the most difficult part in above proof is the last one where uniform integrability of the nonlinear terms $f(u_n, v_n)$, $g(u_n, v_n)$ is established. It is in that proof that the stronger assumption (8.34) has to be made. Moreover, we also use the specific structure of the diffusions. For instance, this proof does not seem to generalize as such to nonlinear diffusions of the form $(-\Delta\phi(\cdot), -\Delta\psi(\cdot))$ where ϕ, ψ are different increasing functions (see [13] for some positive results in that direction). Similarly, we do not know how to extend it to systems like (6.1) where (8.36) does not hold.

Let us point out that there are "good" reasons for which some important work has to be made in the last part. Indeed, all the estimates up to that point and, in particular, Lemmas 8.2, 8.3 and Proposition 8.5 remain valid if u_o^n, v_o^n converge to finite measures u_o, v_o rather than in L^1. But, one can find cases for which uniform integrability fails and for which solutions (at least in usual sense) do not exist. For example consider

$$u_t - \Delta u = -uv^\gamma$$
$$\qquad\qquad\qquad \text{on} \quad B(0, 1) \times (0, T)$$
$$v_t - \Delta v = uv^\gamma$$

$$u = v = 0 \qquad\qquad \text{on} \quad \partial B(0, 1) \times (0, T)$$

$$u(\cdot, 0) = v(\cdot, 0) = \delta \quad \text{Dirac mass at the origin}$$

where $\gamma > N/(N-2)$. In this case $f(u_n, v_n) \to 0$ a.e. but not in $L^1(Q_T)$. Note that here the diffusions are the same. This emphasizes the fact that even when the diffusions are the same, the case of L^1-initial data is not completely obvious.

Finally, an interesting question is to know whether the solutions obtained are in L^∞ for $t > 0$ (and therefore classical) although the initial data are only L^1. In most cases one can show that this is indeed the case if again f and g have polynomial growth. The proof strongly relies on the "L^p-techniques" of Section 4 coupled with well-known regularizing effects of semigroups $S^r(t)$ having the property

$$\|S^r(t)u_o\|_\infty \leq Ct^{-\alpha}\|u_o\|_1.$$

This will be studied in detail elsewhere.

Bibliography

1. N. D. Alikakos, L^p bounds of solutions of reaction-diffusion equations, Comm. in P.D.E., 4 (1979), 827–868.

2. J. Ball, Remarks on blow-up and nonexistence theorems for nonlinear evolution equations, Quart. J. Math., Oxford Ser. (2) 28 (1977), 473–486.

3. P. Baras, Compacité de l'opérateur $f \to u$ solution d'une équation non linéaire $du/dt + Au \ni f$, C.R.A.S. Paris, Série A, 286 (1978).

4. P. Baras, J. C. Hassan and L. Véron, Compacité de l'opérateur définissant la solution d'une équation non homogine, C.R.A.S. Paris, Série A, 284 (1977), 799–802.

5. Ph. Bénilan and H. Labani, Existence of attractors for the Brusselator, in Recent Advances in Nonlinear Elliptic and Parabolic Problems, Ph. Bénilan, M. Chipot, L. C. Evans and M. Pierre, Eds., Pitman Res. Notes in Math., 208 (1989), 139–151.

6. H. Brézis and W. A. Strauss, Semilinear second-order elliptic equations in L^1, J. Math. Soc. Japan, 25 (1973), 565–590

7. A. Haraux and A. Youkana, On a result of K. Masuda concerning reaction-diffusion equations, Tohoku Math. J., 40 (1988), 159–163.

8. S. Hollis, Globally Bounded Solutions of Reaction-Diffusion Systems, PhD. Thesis, North Carolina State University, 1986.

9. S. Hollis, R. Martin, and M. Pierre, Global existence and boundedness in reaction diffusion systems, SIAM J. Math. Anal., 18 (1987), 744–761.

10. S. Hollis and J. Morgan, Global existence and asymptotic decay for systems of convective reaction-diffusion equations, to appear.

11. S. Hollis and J. Morgan, Interior estimates for a class of reaction-diffusion system from L^1-a priori estimates, to appear.

12. Ya. I. Kanel, Cauchy's problem for semilinear parabolic equations with balance condition, Diff. Uravneniya, 20, No. 10 (1984), 1753–1760.

13. E. Laamri, Existence globale pour des systèmes de réaction-diffusion dans L^1, Thèse de 3e cycle, Un. de Nancy I (1988).

14. O. A. Ladyzenskaja, V. A. Solonnikov and N. N. Uralceva, *Linear and Quasilinear Equations of Parabolic Type*, Transl. Math. Monographs, 23, A.M.S., Providence, R.I. (1968).

15. Lê-Châu-Hoàn, *Etude de la classe des opérateurs m-accrétofs de $L^1(\Omega)$ et accrétifs dans $L^\infty(\Omega)$*, Thèse de 3e cycle, Un. de Paris VI (1977).

16. R. Martin, *Global existence questions for reaction-diffusion systems*, in Semigroups, Theory and Applications, Brézis, Crandall and Kappel (eds.) Pitman Res. Notes in Math., No. 141, Longman, Essex, 1986.

17. K. Masuda, *On the global existence and asymptotic behavior of reaction-diffusion equations*, Hokkaido Math. J., 12 (1983), 360–370.

18. J. Morgan, *On a question of blow-up for semilinear parabolic systems*, Diff. and Int. Equs., to appear.

19. J. Morgan, *Global existence for semilinear parabolic systems*, SIAM J. Math. Anal., 20, No. 5 (1989), 1128–1144.

20. J. Morgan, *Boundedness and decay results for reaction-diffusion systems*, SIAM J. Math. Anal., to appear.

21. M. Pierre, *An L^1-method to prove global existence in some reaction-diffusion systems*, in Contributions to Nonlinear Partial Differential Equations, J. I. Diaz and P. L. Lions, Eds., Pitman Res. Notes in Math. (1987), 220–231.

22. M. Pierre, *Global existence in some reaction-diffusion systems*, Delft Progress Rep., 10 (1985), 283–289.

23. F. Rothe, *Global solutions of reaction-diffusion systems*, Lecture Notes in Math., 1072, Springer Verlag, Berlin (1984).

24. V. Solonnikov, *Boundary value problems of mathematical physics, III*, in Proceedings of the Steklov Institute of Mathematics, O. A. Ladyzenskaja, Ed., American Math. Soc., Providence, R.I. (1967).

Riccati-type Pseudopotentials and Their Applications

M.C. Nucci

Dipartimento di Matematica, Università di Perugia,
06100 Perugia, Italy
and
School of Mathematics, Georgia Institute of Technology,
Atlanta, GA 30332, U.S.A.

1 Introduction

Innumerable papers are dedicated to the study of nonlinear evolution (NLE) equations. These equations can be divided into 3 groups [6]:

- S-integrable equations which are solvable by a linear spectral problem i.e. the inverse scattering transform [7], [2].

- C-integrable equations which are linearizable by an appropriate change of variables.

- Non-integrable equations.

S-integrable equations possess a rich structure and have wide applicability to many different physical phenomena. The most famous is the Korteweg-de Vries equation [22]. Other examples of S-integrable equations are the sine-Gordon, Harry Dym, nonlinear Schrödinger, Boussinesq, and Kadomtsev-Petviashvili equations [2]. If a NLE equation is S-integrable then the two coupled equations of its spectral problem are called Lax equations [21]. Another approach to the spectral problem is given in [1]. It gives rise to the so-called AKNS system. A NLE equation is generally believed to be S-integrable if an auto-Bäcklund transformation [35] is found. Another test for S-integrability is the Painlevé property [44], which is widely used

[38]-[43] but is not always reliable [11]. For example, the Harry Dym equation does not possess the Painlevé property [38], although it is S-integrable [2].

In a fundamental and outstanding paper [37], Wahlquist and Estabrook introduced the idea of pseudopotential to the study of NLE equations in 1+1 dimensions. They used the differential forms approach to produce pseudopotentials. In particular they obtained the Lax equations and the auto-Bäcklund transformation of the Korteweg-de Vries equation. Since then many papers have been dedicated to pseudopotentials. In [25] some references are given. Several authors have found Lax equations and auto-Bäcklund transformations of NLE equations by using different types of pseudopotentials. It is known that a non-Abelian pseudopotential gives rise to an auto-Bäcklund transformation, and if it is linear with an arbitrary parameter it generates the Lax equations [2].

This monograph is devoted to those pseudopotentials which have defining equations of the Riccati-type [25]-[30].

This type of pseudopotential generates the Lax equations, auto-Bäcklund transformation, and singularity manifold equation of the corresponding NLE equation in 1+1 dimensions [25]-[28] by using the properties of the Riccati ordinary differential equation.

This technique can be generalized to NLE equations in 2+1 dimensions [27], [28] by imposing the defining equations of the pseudopotential to be of a Riccati-type in one-space variable.

Lax equations and auto-Bäcklund transformations for an equation with higher-order (≥ 3) scattering can also be derived [30] if a pseudopotential exists such that its defining equations are of a type given by a member of the Riccati-chain [3].

A topic related to Riccati-type pseudopotentials is the derivation of novel S-integrable equations and their auto-Bäcklund transformation from the singularity manifold equations and their invariance under the Möbius group, respectively [29]. A well-known example is the link between the Harry Dym and the singularity manifold equation of the Korteweg-de Vries equation [38].

Finally, we consider equations which are not S-integrable. These equations possess Riccati-type pseudopotentials which derive from local conservation laws. The trivial origin of the corresponding Lax equations [8] annhilates the importance of these types of pseudopotentials.

In the following, the direct approach [13] is used to search for pseudopotentials. The extensive calculations involved with this technique may be greatly simplified by means of a Computer Algebra system. We use ad-hoc interactive REDUCE (version 3.3 [17]) programs.

2 Riccati-type pseudopotentials for NLE equations in 1+1 dimensions

2.1 The Riccati equation

The Riccati equation [18]:

$$r'(x) = f_0(x) + f_1(x)r(x) + f_2(x)r^2(x) \tag{2.1}$$

has some interesting properties [18]. It is related to the linear second order equation:

$$V''(x) = \left[\frac{f_2'(x)}{f_2(x)} - f_1(x)\right] V'(x) - f_0(x)f_2(x)V(x) \tag{2.2}$$

if we set:

$$r(x) = -\frac{1}{f_2(x)}[\log V(x)]' \tag{2.3}$$

It is formally invariant under a Möbius transformation operating on the dependent variable r, so that if we assume:

$$r = \frac{a + br^*}{c + dr^*} \tag{2.4}$$

then r^* satisfies the Riccati equation:

$$\begin{aligned}
(bc - ad)r^{*'} &= c^2 f_0 + acf_1 + a^2 f_2 \\
&\quad + [2cdf_0 + (ad + bc)f_1 + 2abf_2]r^* \\
&\quad + (d^2 f_0 + bdf_1 + b^2 f_2)r^{*2}
\end{aligned} \tag{2.5}$$

where a, b, c, d, are arbitary constants such that $ad - bc \neq 0$. Another property of (2.1) is that the cross ratio of four solutions r_i ($i = 1, 2, 3, 4$) of (2.1) is a constant, i.e.:

$$\frac{(r_1 - r_3)(r_2 - r_4)}{(r_1 - r_4)(r_2 - r_3)} = constant \tag{2.6}$$

If V_1 and V_2 are two linearly independent solutions of the equation:

$$V''(x) = P_1(x)V' + P_0(x)V(x) \tag{2.7}$$

then:

$$w(x) = \frac{V_1}{V_2} \tag{2.8}$$

satisfies the equation:

$$\{w; x\} = P_1'(x) - \frac{P_1^2(x)}{2} - 2P_0(x) \tag{2.9}$$

where $\{w; x\}$ denotes the Schwarzian derivative of w, i.e.:

$$\{w; x\} = \left(\frac{w''}{w'}\right)' - \frac{1}{2}\left(\frac{w''}{w'}\right)^2 \tag{2.10}$$

A remarkable property of (2.9) is its invariance under the Möbius group. It is easy to show that the third order differential equation:

$$\{w; x\} = 2kf_0(x) \qquad (k=\text{constant}) \tag{2.11}$$

is related to the Riccati equation:

$$r'(x) = f_0(x) + kr^2(x) \tag{2.12}$$

if we set:

$$r(x) = \frac{1}{2k}[\log w'(x)]' \tag{2.13}$$

Figure 1 summarizes these properties for the Riccati equation (2.12).

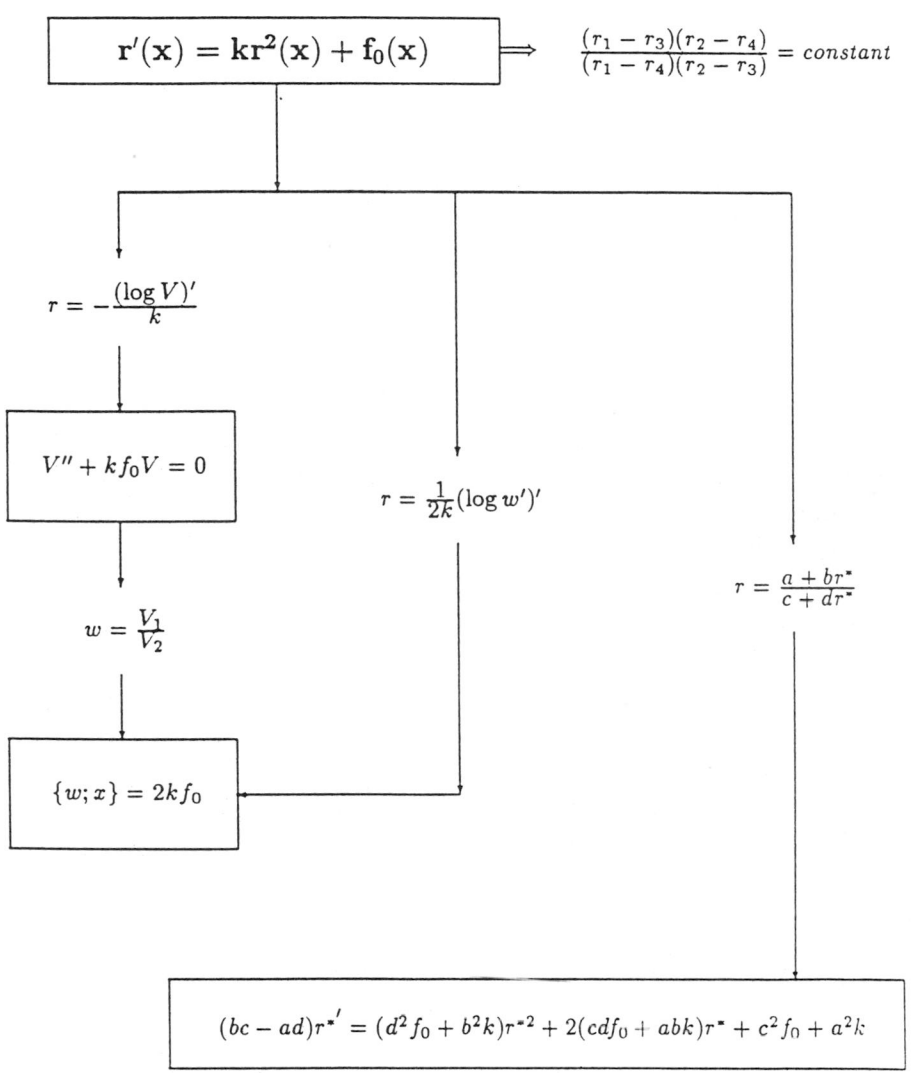

Figure 1

2.2   Second-order scattering

Let us consider a NLE equation in 1+1 dimensions:

$$q_t = H(q, q_x, q_{xx}, q_{xxx}, \ldots) \tag{2.14}$$

We look for a pseudopotential $u = u(x, t)$ such that:

$$u_x = ku^2 + F_1(q)u + F_0(q) \tag{2.15a}$$

$$u_t = [G_1(q, q_x, \ldots)u + G_0(q, q_x, \ldots)]_x \tag{2.15b}$$

with the requirement that $u_{xt} = u_{tx}$ whenever (2.14) is satisfied. If such a pseudopotential exists, we obtain the Lax equations from (2.15) by means of:

$$u = -\frac{1}{k}(\log \psi)_x \tag{2.16}$$

with ψ the spectral function. (2.16) comes from the linearizing transformation (2.3) of the Riccati equation. If the equation satisfied by u, obtained by eliminating q from (2.15), is invariant under the Möbius group, i.e.:

$$u = \frac{a + bu^*}{c + du^*} \tag{2.17}$$

then, combining:

$$u_x^* = ku^{*2} + F_1(q^*)u^* + F_0(q^*) \tag{2.18}$$

and (2.15a) through (2.17), we obtain the spatial part of an auto-Bäcklund transformation for (2.14), being q^* another solution of (2.14). The time part is obtained from (2.15b) by a similar approach. In [5] the recurrence of the cross ratio property both in Riccati equations and auto-Bäcklund transformations was noticed. If there exists a pseudopotential u such that (2.15a) is given by:

$$u_x = ku^2 + F_0(q) \tag{2.19}$$

then the singularity manifold equation with dependent variable $\phi = \phi(x, t)$ is obtained in either two ways. It can be derived from the Lax equations by means of:

$$\phi = \frac{\psi_1}{\psi_2} \tag{2.20}$$

with ψ_1 and ψ_2 two linear independent solutions of the equation satisfied by ψ. (2.20) comes from the transformation (2.8) between the linear second order equation (2.7) and the third order "Schwarzian" equation (2.9). Also, the singularity manifold equation can be obtained from (2.15) by means of:

$$u = \frac{1}{2k}(\log \phi_x)_x \qquad (2.21)$$

(2.21) comes from the transformation (2.13) between the Riccati equation (2.12) and the third order "Schwarzian" equation (2.11). It should be noticed that (2.16) and (2.21) imply:

$$\phi_x = \psi^{-2} \qquad (2.22)$$

A very instructive example of this technique is the **Korteweg-de Vries equation:**

$$q_t = q_{xxx} - 6qq_x \qquad (2.23)$$

In [28], a pseudopotential u was found such that:

$$u_x = -u^2 + q + \lambda \qquad (2.24a)$$

$$u_t = [-2(q - 2\lambda)u + q_x]_x \qquad (2.24b)$$

where λ is an arbitrary constant playing the rôle of the spectral parameter. The Lax equations [10]:

$$\psi_{xx} = (q + \lambda)\psi \qquad (2.25a)$$

$$\psi_t = -2(q - 2\lambda)\psi_x + q_x\psi \qquad (2.25b)$$

are easily obtained by applying:

$$u = (\log \psi)_x \qquad (2.26)$$

to (2.24). The equation satisfied by u is:

$$u_t = (u_{xx} - 2u^3 + 6\lambda u)_x \qquad (2.27)$$

which is invariant under the transformation:

$$u^* = -u \qquad \lambda^* = \lambda \tag{2.28}$$

Then, combining:

$$u_x^* = -u^{*2} + q^* + \lambda^* \tag{2.29}$$

and (2.24a) through (2.28), we obtain the spatial part of the well-known auto-Bäcklund transformation for (2.23), i.e.:

$$q + q^* = \frac{1}{2}\left[\int (q - q^*)\,dx\right]^2 - 2\lambda \tag{2.30}$$

The singularity manifold equation for (2.23) [38], i.e.:

$$\frac{\phi_t}{\phi_x} = \{\phi; x\} + 6\lambda \tag{2.31}$$

is derived by applying either (2.20) to the Lax equations or (2.21)[1] to (2.24). Figure 2 summarizes these results. Other examples shown in Figures 3 through 5 include the Harry Dym, sine-Gordon, and nonlinear Schrödinger equations, respectively.

More examples can be found in [25]-[28].

It should be noticed that the auto-Bäcklund transformation of the Harry Dym equation involves the independent variables [36]. See Section 4 for more details.

In [31], a different pseudopotential u was found for the sine-Gordon equation. It yields Riccati-type equations with a non-constant coefficient for u^2 and gives rise to the well-known auto-Bäcklund transformation of the sine-Gordon equation.

In Figure 5, the Riccati-type pseudopotential of the nonlinear Schrödinger equation suggests the following transformation[2]:

$$q_1 = \frac{q_x^c}{q^c} \qquad q_2 = qq^c \tag{2.32}$$

which leads to:

$$q_{1,t} = -i(q_{1,x} + 2q_2 + q_1^2)_x \tag{2.33a}$$

---

[1]$k = -1$.

[2]χ^c is the complex conjugate of χ.

$$q_{2,t} = i(q_{2,x} - 2q_1 q_2)_x \qquad\qquad (2.33\text{b})$$

a conservative form of the nonlinear Schrödinger equation. Note that the singularity manifold equation is obtained only by means of (2.20).

KORTEWEG–DE VRIES EQUATION

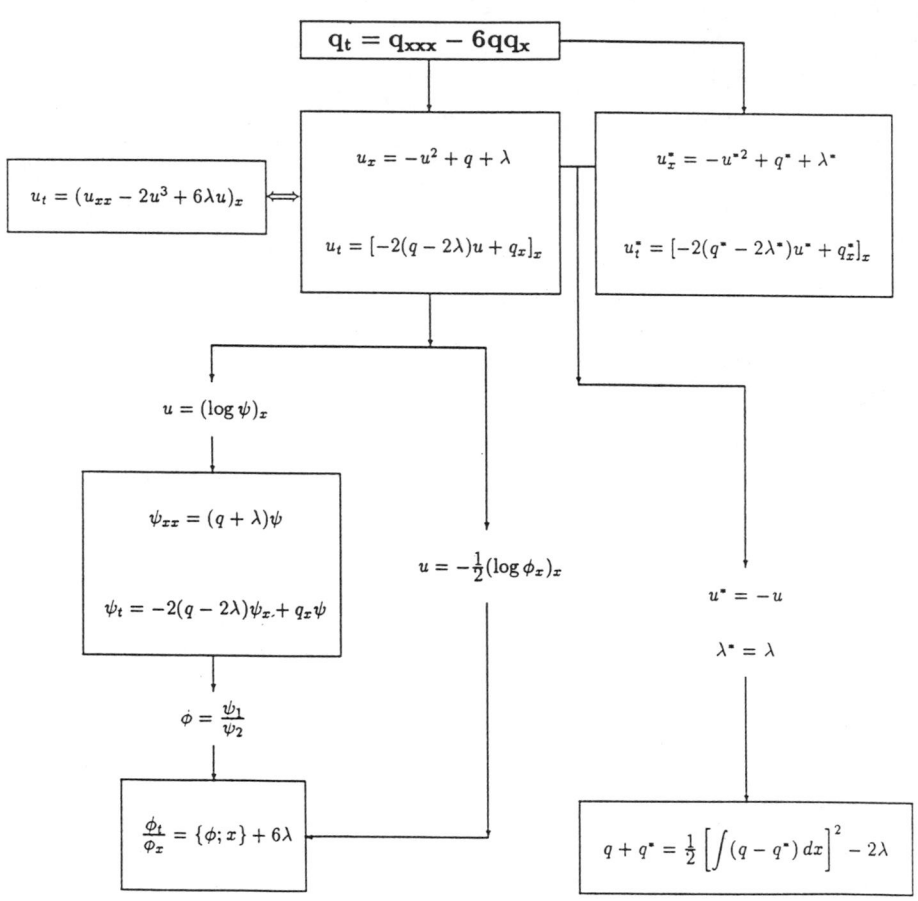

$$q_t = q_{xxx} - 6qq_x$$

$$u_t = (u_{xx} - 2u^3 + 6\lambda u)_x$$

$$u_x = -u^2 + q + \lambda$$

$$u_t = [-2(q - 2\lambda)u + q_x]_x$$

$$u_x^* = -u^{*2} + q^* + \lambda^*$$

$$u_t^* = [-2(q^* - 2\lambda^*)u^* + q_x^*]_x$$

$$u = (\log \psi)_x$$

$$\psi_{xx} = (q + \lambda)\psi$$

$$\psi_t = -2(q - 2\lambda)\psi_x + q_x\psi$$

$$u = -\tfrac{1}{2}(\log \phi_x)_x$$

$$u^* = -u$$

$$\lambda^* = \lambda$$

$$\dot{\phi} = \frac{\psi_1}{\psi_2}$$

$$\frac{\phi_t}{\phi_x} = \{\phi; x\} + 6\lambda$$

$$q + q^* = \tfrac{1}{2}\left[\int (q - q^*)\,dx\right]^2 - 2\lambda$$

Figure 2

HARRY DYM EQUATION

$$q_t = q^3 q_{xxx}$$

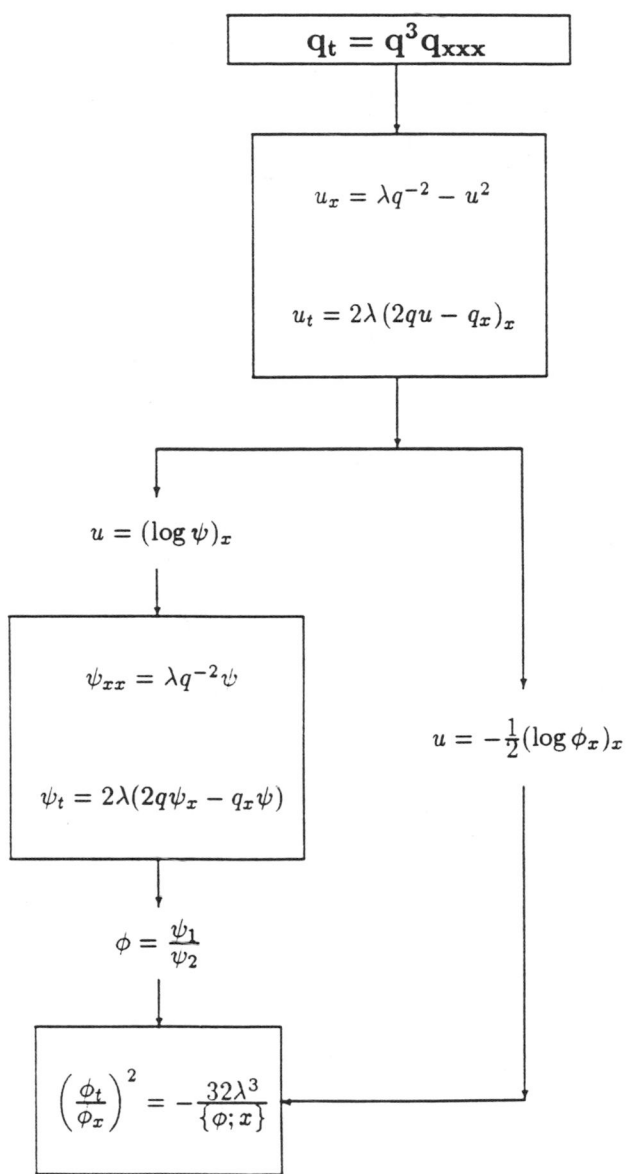

$$u_x = \lambda q^{-2} - u^2$$

$$u_t = 2\lambda \left(2qu - q_x\right)_x$$

$$u = (\log \psi)_x$$

$$\psi_{xx} = \lambda q^{-2}\psi$$

$$\psi_t = 2\lambda(2q\psi_x - q_x\psi)$$

$$u = -\tfrac{1}{2}(\log \phi_x)_x$$

$$\phi = \frac{\psi_1}{\psi_2}$$

$$\left(\frac{\phi_t}{\phi_x}\right)^2 = -\frac{32\lambda^3}{\{\phi; x\}}$$

Figure 3

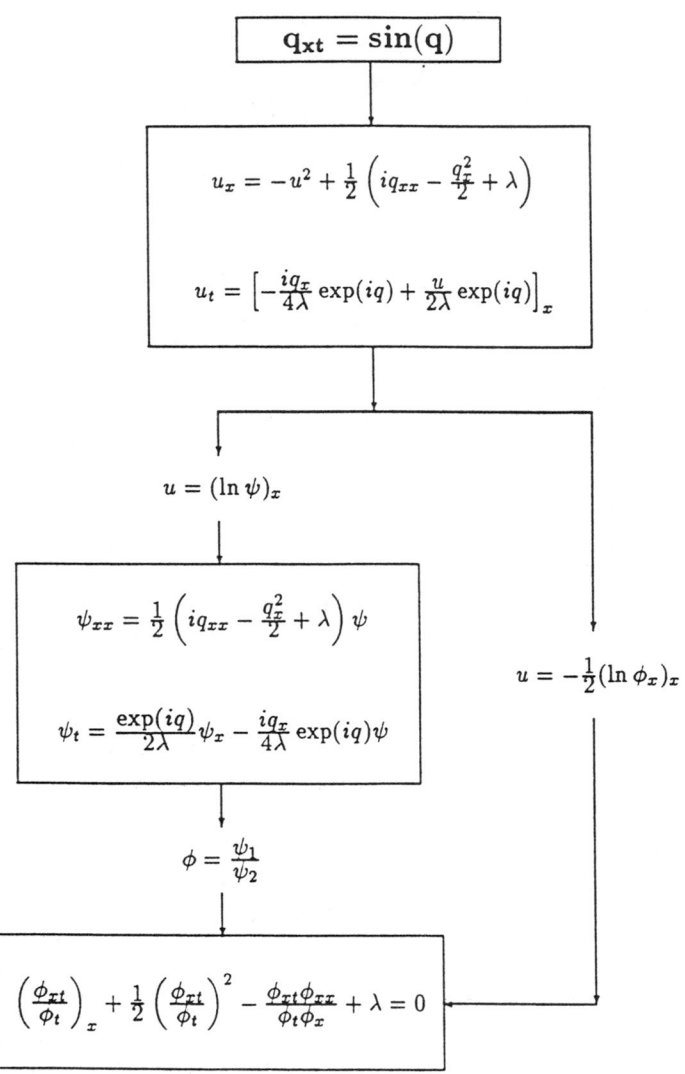

$$\boxed{\mathbf{q_{xt} = \sin(q)}}$$

$$u_x = -u^2 + \frac{1}{2}\left(iq_{xx} - \frac{q_x^2}{2} + \lambda\right)$$

$$u_t = \left[-\frac{iq_x}{4\lambda}\exp(iq) + \frac{u}{2\lambda}\exp(iq)\right]_x$$

$$u = (\ln\psi)_x$$

$$\psi_{xx} = \frac{1}{2}\left(iq_{xx} - \frac{q_x^2}{2} + \lambda\right)\psi$$

$$\psi_t = \frac{\exp(iq)}{2\lambda}\psi_x - \frac{iq_x}{4\lambda}\exp(iq)\psi$$

$$u = -\frac{1}{2}(\ln\phi_x)_x$$

$$\phi = \frac{\psi_1}{\psi_2}$$

$$\left(\frac{\phi_{xt}}{\phi_t}\right)_x + \frac{1}{2}\left(\frac{\phi_{xt}}{\phi_t}\right)^2 - \frac{\phi_{xt}\phi_{xx}}{\phi_t\phi_x} + \lambda = 0$$

Figure 4

NONLINEAR SCHRÖDINGER EQUATION

$$q_t = i(q_{xx} + 2q^2q^c)$$

$$u_x = u^2 + \left(\frac{q_x^c}{q^c} + \lambda\right)u + qq^c$$

$$u_t = i\left[-\left(\frac{q_x^c}{q^c} - \lambda\right)u + qq^c\right]_x$$

$$u = -(\ln\psi)_x$$

$$\psi_{xx} = \left(\frac{q_x^c}{q^c} + \lambda\right)\psi_x - qq^c\psi$$

$$\psi_t = -i\left(\frac{q_x^c}{q^c} - \lambda\right)\psi_x - iqq^c\psi$$

$$\phi = \frac{\psi_1}{\psi_2}$$

$$\left(\frac{\phi_t}{\phi_x}\right)_t + \left[\{\phi; x\} - \frac{3}{2}\left(\frac{\phi_t}{\phi_x}\right)^2 + 3\lambda^2 + 4i\lambda\frac{\phi_t}{\phi_x}\right]_x = 0$$

Figure 5

2.3   Riccati-chain

In [3] the Riccati-chain was given. It was noticed that any of its elements can be linearized. In Table 1 some of them are displayed up to the fourth-order.

$$\frac{V'}{V} = r$$

$$\frac{V''}{V} = r' + r^2$$

$$\frac{V'''}{V} = r'' + 3rr' + r^3$$

$$\frac{V^{(iv)}}{V} = r''' + 4rr'' + 6r^2r' + 3r'^2 + r^4$$

$$\frac{V^{(v)}}{V} = r^{(iv)} + 5rr''' + 10r'r'' + 10r^2r'' + 15r'^2r + 10r^3r' + r^5$$

Table 1: Linear transformation and Riccati-chain.

If $r = r(x)$ is a solution of any element of the Riccati-chain, the corresponding linear equation has solution $V = V(x)$ (see Table 1) given by:

$$r = (\log V)' \tag{2.34}$$

This is the transformation between the Riccati equation:

$$r' + r^2 = P(x) \tag{2.35}$$

and the linear second order equation:

$$V'' = P(x)V \tag{2.36}$$

Note that the elements of the Riccati-chain are formally invariant under the simple transformation $u^* = -u$. The transformation:

$$r = -\frac{1}{2}(\log w')' \tag{2.37}$$

leads to equations involving the Schwarzian derivative, but these novel equations (order ≥ 4) are not invariant under the Möbius group. In Table 2 the transformation (2.37) is applied to some elements of the Riccati-chain up to the fourth-order.

$$-\frac{w''}{2w'} = r$$

$$-\frac{1}{2}\{w; x\} = r' + r^2$$

$$-\frac{1}{2}\left[\{w; x\}' - \frac{1}{2}\frac{w''}{w'}\{w; x\}\right] = r'' + 3rr' + r^3$$

$$-\frac{1}{2}\left[\{w; x\}'' - \frac{1}{2}\{w; x\}^2 - \frac{w''}{w'}\{w; x\}'\right] = r''' + 4rr'' + 6r^2r' + 3r'^2 + r^4$$

$$-\frac{1}{2}\left[\{w; x\}''' - 2\{w; x\}\{w; x\}' - \frac{3}{2}\frac{w''}{w'}\{w; x\}'' + \frac{1}{4}\frac{w''}{w'}\{w; x\}^2\right] =$$
$$= r^{(iv)} + 5rr''' + 10r'r'' + 10r^2r'' + 15r'^2r + 10r^3r' + r^5$$

Table 2: Schwarzian transformation and Riccati-chain.

2.4 Higher-order (≥ 3) scattering

Now let us consider a NLE evolution equation:

$$q_t = H(q, q_x, q_{xx}, \ldots) \tag{2.38}$$

We assume there exists a pseudopotential $u = u(x, t)$ such that:

$$\begin{aligned}
\mathbf{R}_n[u; x] &= F_{n-2}(q, q_x, \ldots)\mathbf{R}_{n-2}[u; x] + \cdots + \\
&\quad + F_0(q, q_x, \ldots)\mathbf{R}_0[u; x] + F(q, q_x, \ldots)
\end{aligned} \tag{2.39a}$$

$$\begin{aligned}
u_t &= (G_{n-1}(q, q_x, \ldots)\mathbf{R}_{n-1}[u; x] + \cdots + \\
&\quad + G_0(q, q_x, \ldots)\mathbf{R}_0[u; x] + G(q, q_x, \ldots))_x
\end{aligned} \tag{2.39b}$$

with the integrability condition:

$$u_{\underbrace{x \cdots x}_{n} t} = u_{t \underbrace{x \cdots x}_{n}} \qquad (n \geq 2) \qquad (2.40)$$

whenever (2.38) is satisfied. Here $\mathbf{R}_i[u;x]$ $(i = 0,1,\ldots,n)$ is any element of the Riccati-chain of order n with respect to u as a function of x, i.e.:

$$
\begin{aligned}
\mathbf{R}_0[u;x] &= u \\
\mathbf{R}_1[u;x] &= u_x + u^2 \\
\mathbf{R}_2[u;x] &= u_{xx} + 3uu_x + u^3 \\
\mathbf{R}_3[u;x] &= u_{xxx} + 4uu_{xx} + 6u^2 u_x + 3u_x^2 + u^4 \\
&\vdots \qquad \vdots
\end{aligned}
\qquad (2.41)
$$

and $F_{n-2},\ldots,F_0,F,G_{n-1},\ldots,G_0,G$ are functions of $q, q_x, \ldots$ to be determined by (2.38) and (2.40). If such a pseudopotential u can be found, then the Lax equations ($n + 1$-order scattering problem) are easily obtained from (2.39) by means of the transformation (2.34), i.e.:

$$u = (\log \psi)_x \qquad (2.42)$$

where $\psi = \psi(x,t)$ is the spectral function. If we eliminate $q, q_x, \ldots$ from (2.39), then the equation satisfied by u is obtained. A simple invariance of this equation, e.g.:

$$u^* = -u \qquad (2.43)$$

can generate an auto-Bäcklund transformation of (2.38) by eliminating u and u^* in (2.39) and:

$$
\begin{aligned}
\mathbf{R}_n[u^*;x] = \ &F_{n-2}(q^*, q_x^*, \ldots)\mathbf{R}_{n-2}[u^*;x] + \cdots + \\
&+ F_0(q^*, q_x^*, \ldots)\mathbf{R}_0[u^*;x] + \\
&+ F(q^*, q_x^*, \ldots)
\end{aligned}
\qquad (2.44a)
$$

$$
u_t^* = \ (G_{n-1}(q^*, q_x^*, \ldots)\mathbf{R}_{n-1}[u^*;x] + \cdots + \\
+ G_0(q^*, q_x^*, \ldots)\mathbf{R}_0[u^*;x] + G(q^*, q_x^*, \ldots))_x \qquad (2.44b)
$$

where $q^* = q^*(x, t)$ is another solution of (2.38).

In the case of a second-order scattering problem, the transformation:

$$u = -\frac{1}{2}(\log \phi_x)_x \qquad (2.45)$$

applied to the Riccati-type defining equations of the pseudopotential u, gives rise to the singularity manifold equation of (2.38), the dependent variable of which is $\phi = \phi(x, t)$. This is not true, however, for the higher-order members of the Riccati-chain. In some cases, the singularity manifold equation is derived by the transformation (2.20) applied to the Lax equations [30].

Applications of this technique can be found in [30]. A very interesting example is given by the **Bullough-Dodd equation**:

$$q_{xt} = e^{2q} - e^{-q} \qquad (2.46)$$

For this equation no pseudopotential u exists so that the defining equations are of the Riccati-type. However, a pseudopotential u is found such that:

$$u_{xx} = -3uu_x - u^3 + (-q_{xx} + q_x^2)u + \lambda \qquad (2.47a)$$

$$u_t = \left[\frac{e^{-q}}{\lambda}(u_x + u^2) + \frac{e^{-q}q_x}{\lambda}u\right]_x \qquad (2.47b)$$

which gives:

$$e^{-q} = u_t + u \int u_t dx \qquad (2.48)$$

$$u_t = \frac{1}{\lambda}\left(u_t u_x + u^3 \int u_t\, dx - uu_{xt}\right)_x \qquad (2.49)$$

where λ is a constant. The Lax equations [14]:

$$\psi_{xxx} = (-q_{xx} + q_x^2)\psi_x + \lambda\psi \qquad (2.50a)$$

$$\psi_t = \frac{e^{-q}}{\lambda}\psi_{xx} + \frac{e^{-q}q_x}{\lambda}\psi_x \qquad (2.50b)$$

are easily obtained by applying (2.42) to (2.47). If we combine (2.49) and (2.42), then the equation satisfied by the spectral function ψ is given by:

$$\lambda\psi\psi_t + \psi_{xxx}\psi_x - \psi_{xt}\left(\psi_{xx} + \frac{\psi_x^2}{\psi}\right) = 0 \qquad (2.51)$$

It should be remarked that in [15] a "generalized pseudopotential" was introduced for (2.46) which corresponds to the spectral function ψ in our notation.

The equation satisfied by u (2.49) is invariant under the transformation:

$$u^* = -u, \qquad \lambda^* = -\lambda \qquad (2.52)$$

Another solution of (2.46) will correspond to:

$$u_{xx}^* = -3u^*u_x^* - u^{*3} + (-q_{xx}^* + q_x^{*2})u^* + \lambda^* \qquad (2.53a)$$

$$u_t^* = \left[\frac{e^{-q^*}}{\lambda^*}(u_x^* + u^{*2}) + \frac{e^{-q^*}q_x^*}{\lambda^*}u^*\right]_x \qquad (2.53b)$$

If we combine (2.47a), (2.53a), (2.52) and eliminate u, then the following spatial part of an auto-Bäcklund transformation for (2.46) is found [16]:

$$
\begin{aligned}
Q_{xx}^* - Q_{xx} + 2QQ_x - 2Q^*Q_x^* + \\
+ \tfrac{1}{36}\left[Q^* - Q + \int(Q^2 - Q^{*2})\,dx\right]^3 + \\
+ \tfrac{1}{2}(Q_x^* + Q_x - Q^2 + Q^{*2})\left[Q^* - Q + \int(Q^2 - Q^{*2})\,dx\right] \\
- 6\lambda = 0
\end{aligned}
\qquad (2.54)
$$

with $Q = q_x$, $Q^* = q_x^*$. The time part is obtained from (2.47b) and (2.53b) by a similar approach. Figure 6 summarizes the properties of (2.46). Note that neither the equation satisfied by ψ nor the equation satisfied by ϕ, which is obtained by substituting (2.45) into (2.49), i.e.:

$$8\lambda\frac{\phi_{xt}}{\phi_x} - 5\frac{\phi_{xx}^3\phi_{xt}}{\phi_x^4} - 4\frac{\phi_{xx}\phi_{xxxt}}{\phi_x^2} + 4\frac{\phi_{xxx}\phi_{xxt}}{\phi_x^2} + 4\frac{\phi_{xxt}\phi_{xx}^2}{\phi_x^3} = 0 \quad (2.55)$$

are invariant under the Möbius group:

$$\phi^* = \frac{a\phi + b}{c\phi + d} \tag{2.56}$$

This is in agreement with the results in [43], where it was shown that the overdetermined system of equations for the "singularity function" has only trivial solutions.

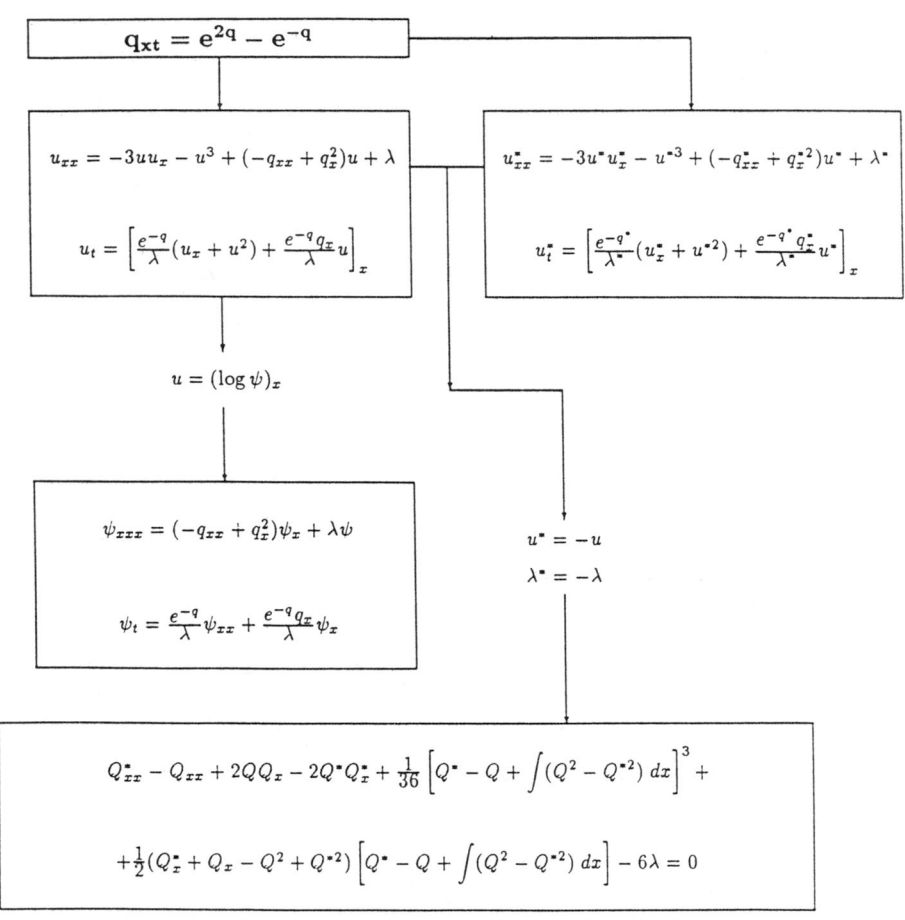

BULLOUGH–DODD EQUATION

$$q_{xt} = e^{2q} - e^{-q}$$

$$u_{xx} = -3uu_x - u^3 + (-q_{xx} + q_x^2)u + \lambda$$

$$u_t = \left[\frac{e^{-q}}{\lambda}(u_x + u^2) + \frac{e^{-q}q_x}{\lambda}u\right]_x$$

$$u_{xx}^* = -3u^*u_x^* - u^{*3} + (-q_{xx}^* + q_x^{*2})u^* + \lambda^*$$

$$u_t^* = \left[\frac{e^{-q^*}}{\lambda^*}(u_x^* + u^{*2}) + \frac{e^{-q^*}q_x^*}{\lambda^*}u^*\right]_x$$

$$u = (\log \psi)_x$$

$$\psi_{xxx} = (-q_{xx} + q_x^2)\psi_x + \lambda\psi$$

$$\psi_t = \frac{e^{-q}}{\lambda}\psi_{xx} + \frac{e^{-q}q_x}{\lambda}\psi_x$$

$$u^* = -u$$
$$\lambda^* = -\lambda$$

$$Q_{xx}^* - Q_{xx} + 2QQ_x - 2Q^*Q_x^* + \frac{1}{36}\left[Q^* - Q + \int(Q^2 - Q^{*2})\,dx\right]^3 +$$

$$+\frac{1}{2}(Q_x^* + Q_x - Q^2 + Q^{*2})\left[Q^* - Q + \int(Q^2 - Q^{*2})\,dx\right] - 6\lambda = 0$$

Figure 6

3   Riccati-type pseudopotentials for NLE equations in 2+1 dimensions

The 2+1 dimensional case is rather more complicated. Let us consider nonlinear evolution equations of the form:

$$q_t = H\left(q, q_x, q_y, q_{xx}, q_{xy}, \int q_y\, dx, \dots\right) \tag{3.1}$$

We assume that there exists a pseudopotential $u = u(t, x, y)$ such that:

$$
\begin{aligned}
u_x &= ku^2 + F_1\left(q, \int u_y\, dx, \int u\, dx\right) u \\
&\quad + F_0\left(q, \int u_y\, dx, \int u\, dx\right)
\end{aligned}
\tag{3.2a}
$$

$$
\begin{aligned}
u_t &= \left[G_1\left(q, q_x, \int u_y\, dx, \dots\right) u \right.\\
&\quad \left. + G_0\left(q, q_x, \int u_y\, dx, \dots\right)\right]_x
\end{aligned}
\tag{3.2b}
$$

with the requirement that $u_{xt} = u_{tx}$ whenever (3.1) is satisfied. Then (2.16) could be applied to (3.2), which will give the Lax equations for (3.1). If we can find another pseudopotential u^* which has a simple transformation to u, e.g.:

$$u^* = -u \tag{3.3}$$

then we can easily obtain an auto-Bäcklund transformation for (3.1). Finally, we get the singularity manifold equation for (3.1) from the Lax equations through (2.20), as in 1+1 dimensions. It should be remarked that (2.22) does not hold true in 2+1 dimensions, because the Wronskian of ψ_1 and ψ_2 with respect to x is never equal to one.

We now apply this technique to the **Kadomtsev-Petviashvili equation**:

$$q_t = q_{xxx} - 6qq_x + 3\int q_{yy}\, dx \tag{3.4}$$

We look for a pseudopotential $u = u(t, x, y)$ such that:

$$u_x = -u^2 + F_0\left(q, \int u_y\, dx\right) \qquad (3.5a)$$

$$u_t = \left[G_1\left(q, q_x, u_y, \int u_y\, dx, \int q_y\, dx\right) u + \right.$$
$$\left. + G_0\left(q, q_x, u_y, \int u_y\, dx, \int q_y\, dx\right)\right]_x \qquad (3.5b)$$

with the requirement that $u_{xt} = u_{tx}$ whenever (3.4) is satisfied. We obtain two cases. The first case is given by:

$$u_x = q - u^2 - \int u_y\, dx + \lambda \qquad (3.6a)$$

$$u_t = \left(q_x - 2uq + 4\lambda u - 4u_y - 4u\int u_y\, dx + 3\int q_y\, dx\right)_x \qquad (3.6b)$$

which means that u satisfies:

$$u_t = \left[u_{xx} - 2u^3 + 6\lambda u - 6u\int u_y\, dx + 6\int uu_y\, dx + \right.$$
$$\left. + 3\int\left(\int u_{yy}\, dx\right) dx\right]_x \qquad (3.7)$$

The second case is given by:

$$u_x^* = q^* - u^{*2} + \int u_y^*\, dx + \lambda^* \qquad (3.8a)$$

$$u_t^* = \left(q_x^* - 2u^*q^* + 4\lambda^* u^* + 4u_y^* + \right.$$
$$\left. + 4u^*\int u_y^*\, dx - 3\int q_y^*\, dx\right)_x \qquad (3.8b)$$

which means that u^* satisfies:

$$u_t^* = \left[u_{xx}^* - 2u^{*3} + 6\lambda u^* + 6u^*\int u_y^*\, dx - \right.$$
$$\left. - 6\int u^* u_y^*\, dx + 3\int\left(\int u_{yy}^*\, dx\right) dx\right]_x \qquad (3.9)$$

It easy to show that:

$$u^* = -u \qquad \lambda^* = \lambda \qquad (3.10)$$

Then the following auto-Bäcklund transformation for (3.4) is obtained [9]:

$$(Q^* - Q)^2 - 2(Q_x^* + Q_x) - 2\int (Q_y^* - Q_y)\,dx - 4\lambda = 0 \qquad (3.11)$$

where $Q_x = q$ and $Q_x^* = q^*$. If we apply:

$$u = (\ln \psi)_x \qquad (3.12)$$

to (3.6), we obtain the Lax equations [20]:

$$\psi_{xx} = (q + \lambda)\psi - \psi_y \qquad (3.13a)$$

$$\psi_t = -2(q - 2\lambda)\psi_x + \left(q_x + 3\int q_y\,dx\right)\psi - 4\psi_{xy} \qquad (3.13b)$$

In [24], the differential form approach was used to generate the Lax equations for (3.4). If we apply (2.20) to the Lax equations, we obtain the singularity manifold equation of (3.4) [42]:

$$\left[\frac{\phi_t}{\phi_x} - \{\phi; x\} - \frac{3}{2}\left(\frac{\phi_y}{\phi_x}\right)^2\right]_x - 3\left(\frac{\phi_y}{\phi_x}\right)_y = 0 \qquad (3.14)$$

Figure 7 summarizes these results. Figure 8 gives the Lax equations, auto-Bäcklund transformation and singularity manifold equation of the Boussinesq equation. This equation is not a (2+1)-dimensional NLE equation, but it can be derived from the Kadomtsev-Petviashvili equation if we apply the following transformation on the independent variables:

$$t = 0 \qquad y = \tau \qquad (3.15)$$

It should be noticed that the Boussinesq equation gives rise to a third-order scattering problem. In fact, the defining equations of its pseudopotential correspond to R_2 in the Riccati-chain. Furthermore, the singularity manifold equation can be obtained by (2.20). In [23], the differential form approach was used to generate the Lax equations of the Boussinesq equation.

Other examples in 2+1 dimensions can be found in [27], [28].

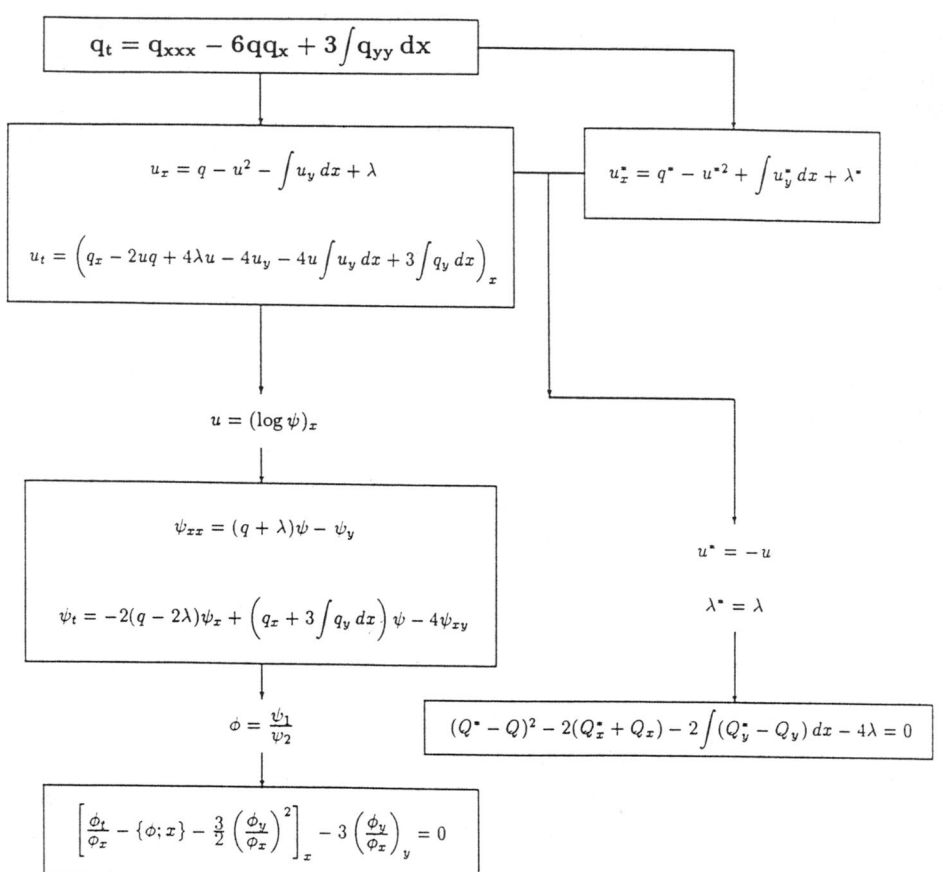

Figure 7

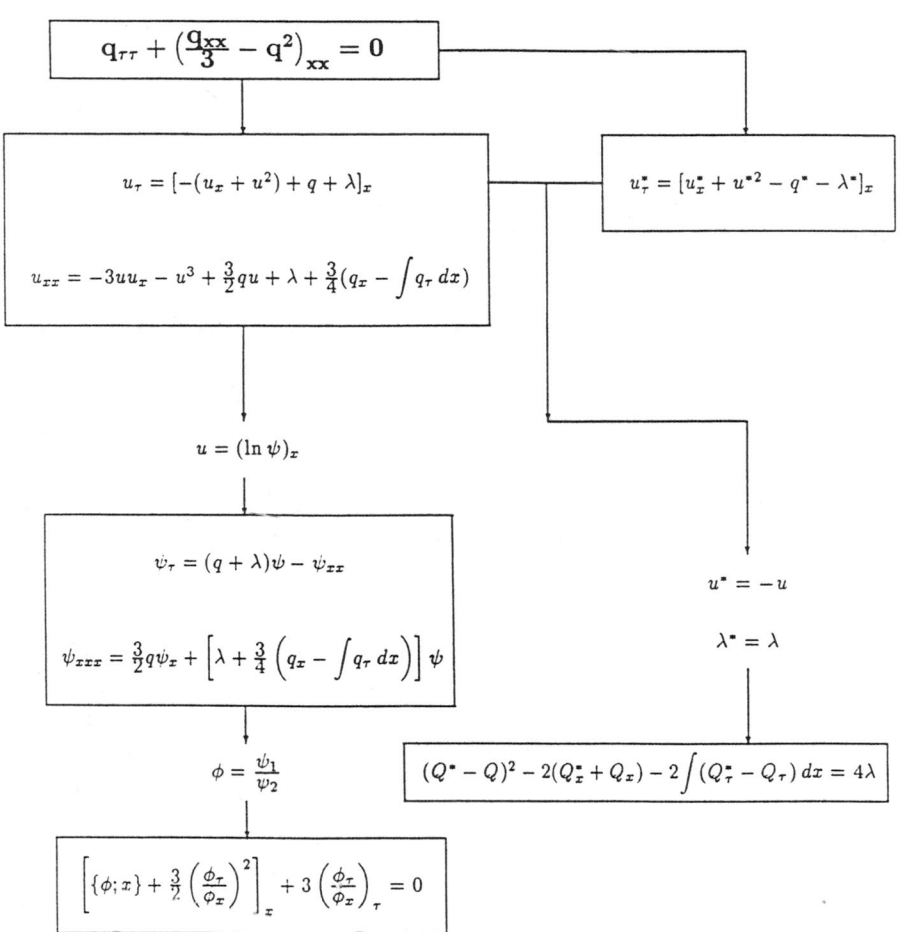

Figure 8

4   Singularity manifold equations and reciprocal auto-Bäcklund transformations

In 1983 the concept of the singularity manifold equation was introduced [44]. While it plays an important rôle in establishing if a partial differential equation possesses the Painlevé property, the singularity manifold equation did not seem to have other applications. In an appendix to one of his papers [38], Weiss noticed that the singularity manifold equation of the Korteweg-de Vries equation could be transformed into the Harry Dym equation via an inverse transformation plus a change of dependent variable. In [36], Rogers and Wong derived a reciprocal auto-Bäcklund transformation [19] for the Harry Dym equation. All these results were found to be true for the Korteweg-de Vries hierarchy [34]: the Harry Dym hierarchy was obtained by applying a reciprocal Bäcklund transformation [19] to the equations satisfied by the potentials of the singularity manifold functions. In 2+1 dimensions, substituting the Kadomtsev-Petviashvili equation for the Korteweg-de Vries equation, Rogers [32] obtained the (2+1)-dimensional Harry Dym equation and its auto-Bäcklund transformation using the same approach. He noticed that the reciprocal auto-Bäcklund transformation was induced by the invariance of the singularity manifold equation under the Möbius group. In [33], analogous results were obtained for both the Caudrey-Dodd-Gibbon and Kaup-Kuperschmidt hierarchies and two different hierarchies based on Kawamoto-type equations were derived. In [29], it was shown that these results are not exclusive to few cases but they are generalizable so that any singularity manifold equation generates a novel equation and its auto-Bäcklund transformation. Let us see how this works. In 1+1 dimensions, a singularity manifold equation in $\phi(x,t)$ is made of combinations of terms such as:

$$\frac{\phi_t}{\phi_x}, \qquad \{\phi;x\} = \left(\frac{\phi_{xx}}{\phi_x}\right)_x - \frac{1}{2}\left(\frac{\phi_{xx}}{\phi_x}\right)^2 \qquad (4.1)$$

and their derivatives, which are invariant under the Möbius group:

$$\phi^* = \frac{a\phi + b}{c\phi + d} \qquad (4.2)$$

$(ad - bc \neq 0; a, b, c, d = constants)$.
As in [38], if we apply an inverse transformation [4], i.e.:

$$x \to \phi, \qquad t \to t, \qquad \phi \to x \qquad (4.3)$$

the following equalities are true:

$$\frac{\phi_t}{\phi_x} = -x_t, \qquad \{\phi; x\} = -\{x; \phi\}/x^2,$$
$$\partial_x = \frac{1}{x_\phi}\partial_\phi, \qquad \partial_t = -\frac{x_t}{x_\phi}\partial_\phi + \partial_t \qquad (4.4)$$

Now we have an equation in $x = x(\phi, t)$. If we make the substitution:

$$x = \int v^{-1}\, d\phi \qquad (4.5)$$

which implies:

$$x_{\phi t} = -v^{-2}v_t, \qquad \{x; \phi\}/x^2 = -vv_{\phi\phi} + v_\phi^2/2 \qquad (4.6)$$

we obtain a novel equation (N.E.) in v. Finally, the invariance of the singularity manifold equation under the Möbius group (4.2) induces the following auto-Bäcklund transformation:

$$v^* = \frac{(ad - bc)v}{(c\int v^{-1}\, d\phi + d)^2} \qquad (4.7)$$

coupled with (4.2).

In 2+1 dimensions, we use the same procedure, via (4.3) coupled with:

$$y \to y \qquad (4.8)$$

which yields:

$$\frac{\phi_y}{\phi_x} = -x_y, \qquad \partial_y = -\frac{x_y}{x_\phi}\partial_\phi + \partial_y \qquad (4.9)$$

We will now consider the singularity manifold equation (S.M.E.) of the Burgers equation. In [26], the **Burgers equation:**

$$q_t = q_{xx} + qq_x \qquad (4.10)$$

was shown to possess a pseudopotential $u = u(x,t)$ such that[3]:

$$u_x = \frac{q_x}{2} + \frac{u^2}{2} - \frac{q^2}{8} \tag{4.11a}$$

$$u_t = \left(\frac{q_x + uq}{2} \right)_x \tag{4.11b}$$

In [28], the transformation:

$$u = (\ln \phi_x)_x \tag{4.12}$$

was used to obtain the singularity manifold equation for (4.10):

$$\left(\frac{\phi_t}{\phi_x} \right)_t - \frac{1}{2} \left(\frac{\phi_t^2}{\phi_x^2} \right)_x - 2 \left(\frac{\phi_t}{\phi_x} \right)_{xx} + \{\phi; x\}_x = 0 \tag{4.13}$$

If we apply the inverse transformation (4.3) to (4.13), and use (4.4), we obtain:

$$x_{tt} - 2\frac{x_{t\phi\phi}}{x_\phi^2} + 2\frac{x_{t\phi}x_{\phi\phi}}{x_\phi^3} + \frac{1}{x_\phi} \left(\frac{\{x;\phi\}}{x^2} \right)_\phi = 0 \tag{4.14}$$

Substituting (4.5) and deriving by ϕ will give a N.E. in $v(\phi,t)$:

$$v_{tt} = 2\frac{v_t^2}{v} + v^2 \left(\frac{2vv_{\phi t} - 2v_t v_\phi - v^3 v_{\phi\phi\phi}}{v} \right)_\phi \tag{4.15}$$

An auto-Bäcklund transformation for (4.15) is given by (4.2) and (4.7), where $v^*(\phi^*, t)$ is another solution of (4.15). Figure 9 summarizes these results for the Burgers equation. Other examples given in Figures 10 through 12 are the sine-Gordon, nonlinear Schrödinger, and (2+1)-dimensional Sawada-Kotera equations[4], respectively.

More examples can be found in [29].

---

[3]We assume $\lambda = 0$ to simplify the calculations.
[4]Note that the diagram in Figure 12 is not complete, because of the small space available.

BURGERS EQUATION

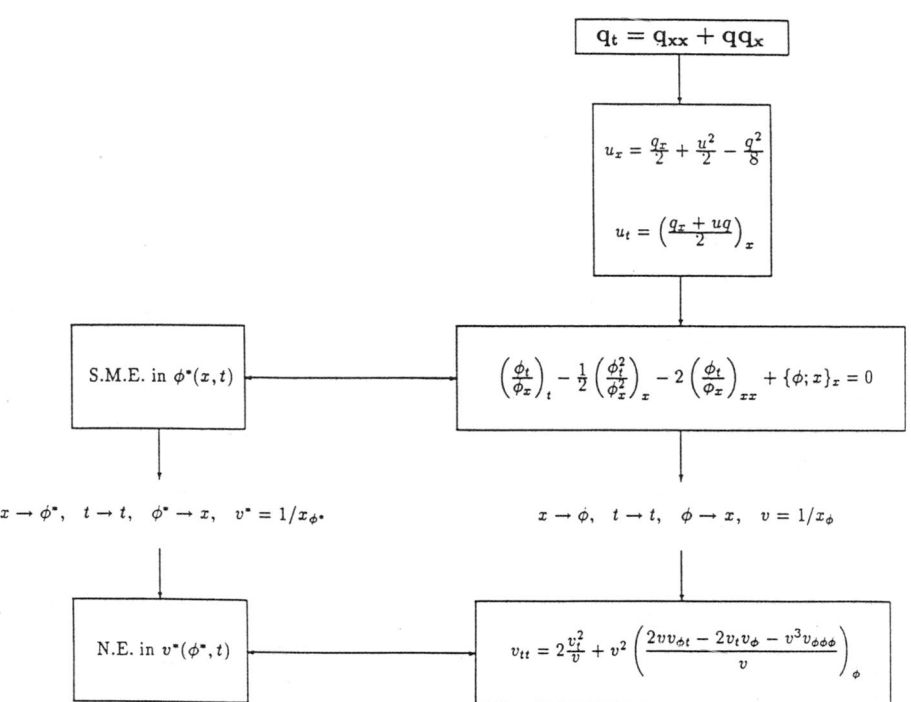

$$q_t = q_{xx} + qq_x$$

$$u_x = \frac{q_x}{2} + \frac{u^2}{2} - \frac{q^2}{8}$$

$$u_t = \left(\frac{q_x + uq}{2}\right)_x$$

S.M.E. in $\phi^*(x,t)$

$$\left(\frac{\phi_t}{\phi_x}\right)_t - \frac{1}{2}\left(\frac{\phi_t^2}{\phi_x^2}\right)_x - 2\left(\frac{\phi_t}{\phi_x}\right)_{xx} + \{\phi;x\}_x = 0$$

$x \to \phi^*, \quad t \to t, \quad \phi^* \to x, \quad v^* = 1/x_{\phi^*}.$ $\qquad\qquad$ $x \to \phi, \quad t \to t, \quad \phi \to x, \quad v = 1/x_\phi$

N.E. in $v^*(\phi^*,t)$

$$v_{tt} = 2\frac{v_t^2}{v} + v^2\left(\frac{2vv_{\phi t} - 2v_t v_\phi - v^3 v_{\phi\phi\phi}}{v}\right)_\phi$$

Figure 9

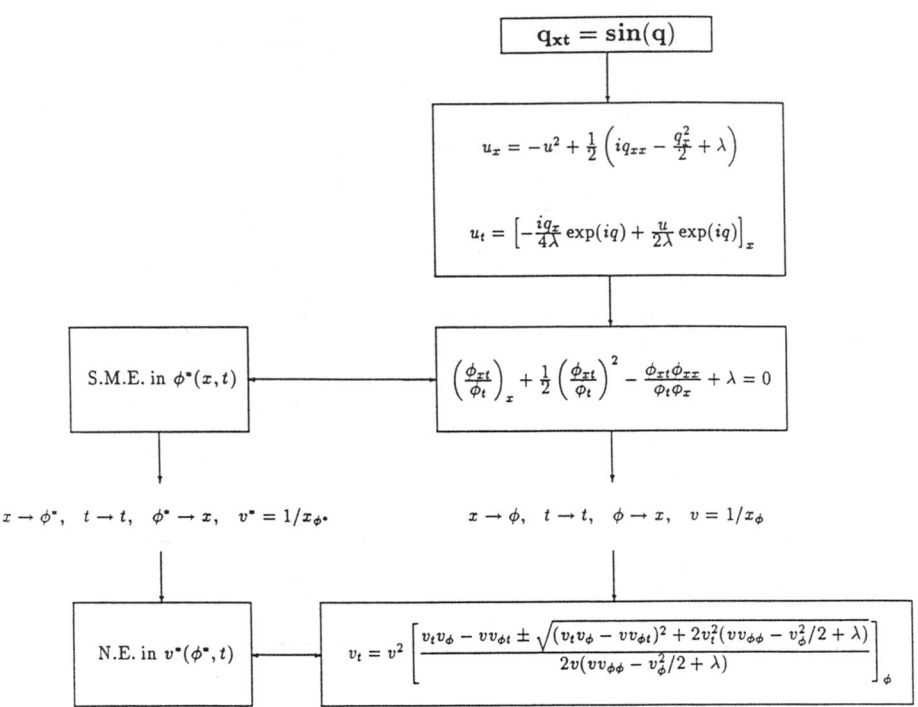

Figure 10

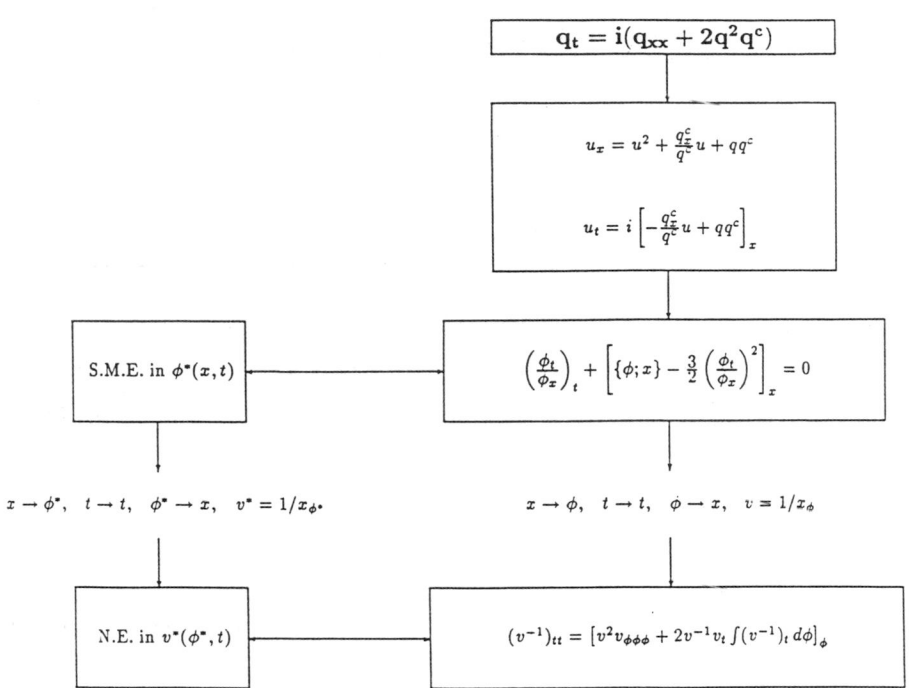

Figure 11

$$\boxed{\text{(2+1 DIM.)–SAWADA–KOTERA EQUATION}}$$

$$q_t = \left(q_{xxxx} + 5qq_{xx} + \frac{5}{3}q^3 + 5q_{xy}\right)_x - 5\int q_{yy}\,dx + 5qq_y + 5q_x\int q_y\,dx$$

$$u_x = q + u^2 - e^{\int u\,dx}\int\left(\int u_y\,dx\right)e^{-\int u\,dx}\,dx$$

$$u_t = \left(q_{xxx} + 2uq_{xx} + uq^2 + qq_x + 4q_y + 5u\int q_y\,dx - 3q\int u_y\,dx + 3q_x e^{\int u\,dx}\int\left(\int u_y\,dx\right)e^{-\int u\,dx}\,dx\right.$$

$$\left. - 9e^{\int u\,dx}\int\left[\int u_{yy}\,dx - \left(\int u_y\,dx\right)^2\right]e^{-\int u\,dx}\,dx\right)_x$$

$$\left[\frac{\phi_t}{\phi_x} - \{\phi;x\}_{xx} - 4\{\phi;x\}^2 - 5\left(\frac{\phi_y}{\phi_x}\right)_{xx}\right]_x - 10\{\phi;x\}\left(\frac{\phi_y}{\phi_x}\right)_x + 5\left(\frac{\phi_y}{\phi_x}\right)_y = 0$$

$$x \to \phi, \quad y \to y, \quad t \to t, \quad \phi \to x, \quad v = 1/x_\phi$$

$$v_t = v^5 v_{\phi\phi\phi\phi\phi} + 5v^4 v_\phi v_{\phi\phi\phi\phi} + 10v^4 v_{\phi\phi} v_{\phi\phi\phi} + 5v^2 v_{\phi\phi y} - 5vv_\phi v_{\phi y} + 5vv_y v_{\phi\phi} - 5v_y\int\frac{v_y}{v^2}\,d\phi - 5v\int\left(\frac{v_y}{v^2}\right)_y\,d\phi$$

Figure 12

5 Riccati-type pseudopotentials and non-integrability

Riccati-type pseudopotentials always exist for any NLE equation in 1+1 dimensions:

$$q_t = H(q, q_x, q_{xx}, q_{xxx}, \ldots) \tag{5.1}$$

which has local conservation laws of the form:

$$f_t = g_x \tag{5.2}$$

where f and g are functions of q and its derivatives by x. If (5.1) is not S-integrable, then the only Riccati-type pseudopotentials involve expressions of the local conservation laws, i.e.[5]:

$$u_x = -u^2 + \left(\frac{f_x}{f} + \mu f - 2\eta\right) u + \lambda f^2 + \eta\left(\frac{f_x}{f} + \mu f - \eta\right) \tag{5.3a}$$

$$u_t = \left(\frac{g+\rho}{f} u + \eta\frac{g+\rho}{f} + \nu\right)_x \tag{5.3b}$$

where $\mu, \eta, \lambda, \rho, \nu$ are arbitrary constants. If we apply:

$$u = (\log \psi)_x \tag{5.4}$$

then we obtain the Lax equations given in [8]. Note that for $\eta = 0$, these Lax equations reduce to:

$$\psi_{xx} = \left(\frac{f_x}{f} + \mu f\right) \psi_x + \lambda f^2 \psi \tag{5.5a}$$

$$\psi_t = \frac{g+\rho}{f} \psi_x + \nu\psi \tag{5.5b}$$

which were shown to be trivial in [8]. As an example we consider the Kuramoto-Sivashinsky equation [12]:

$$q_t = (q_{xxx} + q_x + q^2)_x \tag{5.6}$$

A pseudopotential u given by (2.15), i.e.:

$$u_x = -u^2 + F_1(q)u + F_0(q) \tag{5.7a}$$

---

[5] We assume $k = -1$.

$$u_t = [G_1(q, q_x, q_{xx})u + G_0(q, q_x, q_{xx})]_x \qquad (5.7b)$$

does not exist. If we look for a pseudopotential u such that:

$$u_x = -u^2 + F_1(q, q_x)u + F_0(q, q_x) \qquad (5.8a)$$

$$u_t = [G_1(q, q_x, q_{xx}, q_{xxx})u + G_0(q, q_x, q_{xx}, q_{xxx})]_x \qquad (5.8b)$$

we find only (5.3) with $f = q + c_1$ and $g = q_{xxx} + q_x + q^2 + c_2$ (c_1 and c_2 are arbitrary constants). Other examples can be found in [8].

Bibliography

[1] M. J. Ablowitz, D. J. Kaup, A. C. Newell and H. Segur, "Nonlinear evolution equations of physical significance", Phys. Rev. Lett. **31** (1973) 125–127.

[2] M. J. Ablowitz and H. Segur, *Solitons and Inverse Scattering Transform*, SIAM, Philadelphia (1981).

[3] W.F. Ames, *Nonlinear Ordinary Differential Equations in Transport Processes*, Academic Press, New York (1968).

[4] W.F. Ames, *Nonlinear Partial Differential Equations in Engineering, Vol. 2*, Academic Press, New York (1972).

[5] R.L. Anderson and N.H. Ibraghimov, *Lie-Bäcklund Transformation in Applications*, SIAM, Philadelphia (1979).

[6] F. Calogero, "Why are certain nonlinear PDEs both widely applicable and integrable?", in *What is integrability (for nonlinear PDEs)?* (ed. V.E. Zakharov) Springer-Verlag, Berlin (1988) to be published.

[7] F. Calogero and A. Degasperis, *Spectral transform and solitons: tools to solve and investigate nonlinear evolution equations, Vol. 1*, North Holland, Amsterdam (1982).

[8] F. Calogero and M.C. Nucci, "Lax pairs galore", J. Math. Phys. (1990) to be published.

[9] H.-H. Chen, "A Bäcklund transformation in two dimensions", J. Math. Phys. **16** (1975) 2382–2383.

[10] H.-H. Chen, "Relation between Bäcklund transformations and inverse scattering problems", in *Bäcklund Transformations, the Inverse Scattering Method, Solitons, and Their Applications* (ed. R.M. Miura) (*Lecture Notes in Mathematics* **515**) Springer-Verlag, Berlin (1976) 241–252.

[11] P.A. Clarkson, "The Painlevé conjecture, the Painlevé property for partial differential equations and complete integrability", Physica D **18** (1986) 209–210.

[12] R. Conte and M. Musette, "Painlevé analysis and Bäcklund transformation in the Kuramoto-Sivashinsky equation", J. Phys. A: Math. Gen. **22** (1989) 169–177.

[13] J.P. Corones and F.J. Testa, "Pseudopotentials and their Applications", in *Bäcklund Transformations, the Inverse Scattering Method, Solitons, and Their Applications* (ed. R.M. Miura) (*Lecture Notes in Mathematics* **515**) Springer-Verlag, Berlin (1976) 184–198.

[14] P. Fordy and J. Gibbons, "Integrable nonlinear Klein-Gordon equations and Toda lattices", Commun. Math. Phys. **77** (1980) 21–30.

[15] B. Gaffet, "A class of 1-D gas flows soluble by the inverse scattering transform", Physica D **26** (1987) 123–139.

[16] B. Gaffet, "Common structure of several completely integrable non-linear equations", J. Phys. A: Math. Gen. **21** (1988) 2491–2531.

[17] A.C. Hearn, *REDUCE User's Manual. Version 3.3*, The Rand Corporation, Santa Monica (1987).

[18] E. Hille, *Ordinary Differential Equations in the Complex Domain*, Wiley & Sons, New York (1976).

[19] J.G. Kingston and C. Rogers, "Reciprocal Bäcklund transformations of conservation laws", Phys. Lett. A **92** (1982) 261–264.

[20] B.G. Konopelchenko and V.G. Dubrovsky, "Some new integrable nonlinear evolution equations in $2+1$ dimensions", Phys. Lett. A **102** (1984) 15–17.

[21] P. D. Lax, "Integrals of nonlinear equations of evolution and solitary waves", Comm. Pure Appl. Math. **21** (1968) 467–490.

[22] R. M. Miura, "The Korteweg-de Vries equation: a survey of results", SIAM Rev. **18** (1976) 412–459.

[23] H.C. Morris, "Prolongation structures and a generalized inverse scattering problem", J. Math. Phys. **17** (1976) 1867–1869.

[24] H.C. Morris, "Prolongation structures and nonlinear evolution equations in two spatial dimensions", J. Math. Phys. **17** (1976) 1870–1872.

[25] M.C. Nucci, "Pseudopotentials, Lax equations and Bäcklund transformations for non-linear evolution equations", J. Phys. A: Math. Gen. **21** (1988) 73–79.

[26] M.C. Nucci, "Pseudopotentials and integrability properties of the Burgers' equation", Atti Sem. Mat. Fis. Univ. Modena **38** (1990) 313–317.

[27] M.C. Nucci, "Pseudopotentials for non-linear evolution equations in $2 + 1$ dimensions", Int. J. Non-Linear Mech. **23** (1988) 361–367.

[28] M.C. Nucci, "Painlevé property and pseudopotentials for non-linear evolution equations ", J. Phys. A: Math. Gen. **22** (1989) 2897–2913.

[29] M.C. Nucci, "Reciprocal auto-Bäcklund transformations via the Möbius group", Atti Sem. Mat. Fis. Univ. Modena (1090) to be published.

[30] M.C. Nucci, "Riccati-chain type pseudopotentials, higher-order scattering and Bäcklund transformations for non-linear evolution equations", Preprint GT Math:052190-040 (1990).

[31] M. Omote, "Prolongation structures of nonlinear equations and infinite-dimensional algebras", J. Math. Phys. **27** (1986) 2853–2860.

[32] C. Rogers, "The Harry-Dym equation in 2+1 dimensions: a reciprocal link with the Kadomtsev-Petviashvili equation", Phys. Lett. A **120** (1987) 15–18.

[33] C. Rogers and S. Carillo, "On reciprocal properties of the Caudrey-Dodd-Gibbon and Kaup-Kupershmidt hierarchies", Physica Scripta **36** (1987) 865–869.

[34] C. Rogers and M.C. Nucci, "On reciprocal auto-Bäcklund transformations and the Korteweg-de Vries hierarchy", Physica Scripta **33** (1986) 289–292.

[35] C. Rogers and W.F. Shadwick, *Bäcklund Transformations and Their Applications*, Academic Press, New York (1982).

[36] C. Rogers and P. Wong, "On reciprocal Bäcklund transformations of inverse scattering schemes", Physica Scripta **30** (1984) 10–13.

[37] H.D. Wahlquist and F.B. Estabrook, "Prolongation structures of nonlinear evolution equations", J. Math. Phys. **16** (1975) 1–7.

[38] J. Weiss, "The Painlevé property for partial differential equations. II: Bäcklund transformations, Lax pairs, and the Schwarzian derivative", J. Math. Phys. **24** (1983) 1405–1413.

[39] J. Weiss, "On classes of integrable systems and the Painlevé property", J. Math. Phys. **25** (1984) 13–24.

[40] J. Weiss, "The sine-Gordon equations: Complete and partial integrability", J. Math. Phys. **25** (1984) 2226–2235.

[41] J. Weiss, "The Painlevé property and Bäcklund transformations for the sequence of Boussinesq equations", J. Math. Phys. **26** (1985) 258–269.

[42] J. Weiss, "Modified equations, rational solutions, and the Painlevé property for the Kadomtsev-Petviashvili and Hirota-Satsuma equations", J. Math. Phys. **26** (1985) 2174–2180.

[43] J. Weiss, "Bäcklund transformation and the Painlevé property", J. Math. Phys. **27** (1986) 1293–1305.

[44] J. Weiss, M. Tabor and G. Carnevale, "The Painlevé property for partial differential equations" J. Math. Phys. **24** (1983) 522–526.

Nonlinear Elasticity: Incremental Equations and Bifurcation Phenomena

R. W. Ogden
University of Glasgow

Abstract

This paper begins with a summary of the basic equations governing nonlinear quasi-static deformations of an elastic solid. Attention is then focussed on the linearized equations associated with incremental deformations superimposed on an underlying finite deformation. The question of stability of the underlying deformation is then examined and the connection between stability and uniqueness of solution of incremental boundary-value problems discussed. Non-uniqueness of solution is illustrated in two simple examples—bifurcation from a homogeneously deformed configuration under dead loading, and surface deformations on a pre-stressed half-space.

1   Introduction

The mathematical theory of elasticity has a long history and has its roots in engineering mechanics; in particular, the *linear* theory has been developed over a period of more than 300 years. As well as providing many challenges for mathematicians, the linear theory has, within certain limitations, satisfied the needs of applications in, for example, structural mechanics, fracture mechanics and seismology. The linear theory remains a very active area of research and still offers a wide range of opportunities for mathematical development.

It is only relatively recently that the *nonlinear* theory of elasticity has received comparable attention; the theory has developed rapidly since the 1940's, stimulated by the need to describe and explain the mechanical behaviour of materials, such as natural rubber, which are capable of undergoing large elastic deformations. The fabrication of synthetic rubber-like materials has provided further stimulus, and, more recently, application of the theory to the study of the mechanics of biological tissues has been much in evidence. The range of applications is growing, and nonlinear elasticity can now be regarded as very much an interdisciplinary subject area, impinging on engineering, materials science, chemistry, physics and biology. The inherent nonlinearity of the mathematical theory presents many difficulties but it also offers a wealth of exciting and challenging opportunities for researchers in a diversity of fields, but notably in theoretical mechanics, applied mathematics, pure mathematics, numerical analysis and computational mechanics.

In this article the intention is not to provide a comprehensive account of nonlinear elasticity but to concentrate on the main ingredients of the theory and to illustrate how the nonlinearity influences the nature of the solution of some very simple boundary-value problems. Different flavours of the theory, with varying degrees of mathematical sophistication, are encountered in the literature, and the reader is referred to the texts by Truesdell and Noll [20], Marsden and Hughes [11], Ogden [14] and Ciarlet [6], in particular, for more details of the theory. Only a limited number of selected papers are cited here and in general, rather than give full documentation, we point to the above-mentioned monographs for references to original sources; see also the recent review article by Beatty [4].

2   Basic Equations

2.1   Kinematics

We consider a continuous body which occupies a connected open subset of a three-dimensional Euclidean point space, and we refer to such a subset as a *configuration* of the body. We identify an arbitrary configuration as a *reference configuration* and denote this by $\mathcal{B}_r$. Let points in $\mathcal{B}_r$ be labelled by their position vectors $\mathbf{X}$ relative to an arbitrarily chosen origin and let ∂B_r denote the boundary of B_r. Now suppose that the body is deformed quasi-statically from B, so that it occupies a new configuration, B say, with boundary ∂B — we refer to B as the *current configuration* of the body. The deformation can be represented by the mapping $\chi\colon B_r \to B$ which takes points $\mathbf{X} \in B_r$ to points $\mathbf{x} \in B$ according to

$$\mathbf{x} = \chi(\mathbf{X}) \quad \mathbf{X} \in B_r, \tag{1}$$

where $\mathbf{x}$ is the position vector of the point $\mathbf{X}$ in B. The mapping χ is called the *deformation* from B_r to B. We require χ to be one-to-one and, for purposes of this article, twice continuously differentiable.

For simplicity, we consider only Cartesian coordinate systems and let $\mathbf{X}$ and $\mathbf{x}$ respectively have coordinates X_α and x_i, where $\alpha, i \in \{1, 2, 3\}$, so that $x_i = \chi_i(X_\alpha)$. Greek and Roman indices refer respectively to B_r and B, and the usual summation convention will be used.

The *deformation gradient tensor* A is given by

$$\mathsf{A} = \mathrm{Grad}\,\mathbf{x} \tag{2}$$

and has Cartesian components $A_{i\alpha} = \partial x_i / \partial X_\alpha$, Grad being the gradient operator in B_r. Local invertibility of χ requires that A is nonsingular, and we adopt the usual convention $\det \mathsf{A} > 0$ and the notation

$$J = \det \mathsf{A}. \tag{3}$$

Locally, A describes the deformation in the neighbourhood of the point $\mathbf{X}$.

The *polar decompositions*

$$\mathsf{A} = \mathsf{R}\mathsf{U} = \mathsf{V}\mathsf{R}, \tag{4}$$

where R is proper orthogonal and U, V are positive definite and symmetric, are unique. Respectively, U and V are called the *right* and *left stretch tensors*.

We also have the *spectral decomposition*

$$\mathsf{U} = \sum_{i=1}^{3} \lambda_i \mathbf{u}^{(i)} \otimes \mathbf{u}^{(i)}, \tag{5}$$

where $\lambda_i > 0, i \in \{1,2,3\}$, are the *principal stretches* and $\mathbf{u}^{(i)}$, the (unit) eigenvectors of U, are called the *Lagrangean principal axes*. Similarly, V has the spectral decomposition

$$\mathsf{V} = \sum_{i=1}^{3} \lambda_i \mathbf{v}^{(i)} \otimes \mathbf{v}^{(i)}, \tag{6}$$

where $\mathbf{v}^{(i)} = \mathsf{R}\mathbf{u}^{(i)}, i \in \{1,2,3\}$. It follows from (3)–(5) that

$$J = \lambda_1 \lambda_2 \lambda_3. \tag{7}$$

Let $\mathbf{dA} \equiv \mathbf{N}dA$ denote a vector surface area element on $\partial \mathcal{B}_r$, where $\mathbf{N}$ is the unit outward normal to the surface, and $\mathbf{da} \equiv \mathbf{n}da$ the corresponding area element on $\partial \mathcal{B}$. Then, the area elements are connected according to *Nanson's formula*

$$\mathbf{n}da = J\mathsf{A}^{-\mathsf{T}}\mathbf{N}dA, \tag{8}$$

where $\mathsf{A}^{-\mathsf{T}} = (\mathsf{A}^{-1})^{\mathsf{T}}$ and ^{T} denotes the transpose. If dV and dv denote volume elements in $\mathcal{B}_r$ and $\mathcal{B}$ respectively then we also have

$$dv = JdV. \tag{9}$$

For an *isochoric* (volume-preserving) deformation we have

$$\det \mathsf{A} \equiv \lambda_1 \lambda_2 \lambda_3 = 1. \tag{10}$$

At this point we also note that the *mass conservation equation* may be written in the form

$$\rho_r = \rho J, \tag{11}$$

where ρ_r and ρ are the *mass densities* in $\mathcal{B}_r$ and $\mathcal{B}$ respectively.

2.2   Stress and Balance Laws

We use the notation $\mathbf{t}$ for the (surface) force per unit area on the vector area element $\mathbf{da}$. Then $\mathbf{t}$ depends on $\mathbf{n}$ according to the formula

$$\mathbf{t} = \mathsf{T}^{\mathsf{T}}\mathbf{n}, \tag{12}$$

where T, a second-order tensor independent of $\mathbf{n}$, is called the *Cauchy stress tensor*, while $\mathbf{t}$ is also referred to as the *stress vector*.

Using (8) we may write the force on $\mathbf{da}$ as

$$\mathbf{t}da = \mathsf{S}^\mathsf{T}\mathbf{N}dA, \tag{13}$$

where the *nominal stress tensor* S is related to T by $\mathsf{S} = J\mathsf{A}^{-1}\mathsf{T}$.

Let $\mathbf{b}$ denote the body force per unit mass. Then, if the body $\mathcal{B}$ is held in equilibrium by the combination of body forces in $\mathcal{B}$ and surface forces on $\partial\mathcal{B}$, we obtain

$$\int_{\partial\mathcal{B}} \mathsf{T}^\mathsf{T}\mathbf{n}da + \int_{\mathcal{B}} \rho\mathbf{b}dv = \mathbf{0}, \tag{14}$$

or, equivalently,

$$\int_{\partial\mathcal{B}_r} \mathsf{S}^\mathsf{T}\mathbf{N}dA + \int_{\mathcal{B}_r} \rho_r\mathbf{b}dV = \mathbf{0}. \tag{15}$$

We require (14) and (15) to hold for every subbody. Hence, we deduce from the divergence theorem that

$$\operatorname{div}\mathsf{T} + \rho\mathbf{b} = \mathbf{0}, \tag{16}$$

or, equivalently,

$$\operatorname{Div}\mathsf{S} + \rho_r\mathbf{b} = \mathbf{0}, \tag{17}$$

assuming that the left-hand sides of (16) and (17) are continuous, where div and Div denote the divergence operators in $\mathcal{B}$ and $\mathcal{B}_r$ respectively. Equations (16) and (17) are alternative forms of the *equilibrium equations* for $\mathcal{B}$. In components (17) has the form

$$\frac{\partial S_{\alpha i}}{\partial X_\alpha} + \rho_r b_i = 0, \tag{18}$$

and similarly for (16), where $S_{\alpha i}$ are the components of S.

Balance of the moments of the forces acting on the body yields simply

$$\mathsf{T}^\mathsf{T} = \mathsf{T}, \tag{19}$$

which may also be expressed as $\mathsf{S}^\mathsf{T}\mathsf{A}^\mathsf{T} = \mathsf{A}\mathsf{S}$.

Henceforth, we confine attention to the Lagrangean formulation based on the use of S with $\mathbf{X}$ as the independent variable.

2.3   Elasticity

We motivate the definition of elasticity by considering the work done by the surface and body forces in a virtual displacement $\delta\mathbf{x}$ from the current configuration $\mathcal{B}$. By use of the divergence theorem and equation (17) we obtain the *virtual work* equation

$$\int_{\mathcal{B}_r} \rho_r \mathbf{b} \cdot \delta\mathbf{x}\,dV + \int_{\partial\mathcal{B}_r} (\mathsf{S}^{\mathsf{T}}\mathbf{N}) \cdot \delta\mathbf{x}\,dA = \int_{\mathcal{B}_r} \mathrm{tr}(\mathsf{S}\delta\mathsf{A})\,dV, \qquad (20)$$

where the left-hand side of (20) represents the virtual work and tr denotes the trace of a second-order tensor. This work is converted into stored energy if there exists a scalar function, W say, defined on the set of deformation gradients, such that $\delta W \approx \mathrm{tr}(\mathsf{S}\delta\mathsf{A})$. It follows that

$$\mathsf{S} = \frac{\partial W}{\partial\mathsf{A}}, \qquad (21)$$

or, in components, $S_{\alpha i} = \partial W/\partial A_{i\alpha}$. This is the *stress-deformation relation* or *constitutive equation* for an elastic material which possesses a *strain-energy function* W, W being defined per unit volume in $\mathcal{B}_r$.

Thus, for purposes of this article, an elastic material is characterized by the existence of a strain-energy function such that (21) holds. At this point there is no restriction on the form of the function W, but there are a number of factors limiting the class of functions that are acceptable for the description of the elastic behaviour of materials, as we see shortly. First, however, we examine how (21) is modified if the material is subject to internal constraints. We illustrate this by considering the *incompressibility constraint*, so that (10) holds at each point $\mathbf{X} \in \mathcal{B}_r$. Using the fact that $\partial J/\partial\mathsf{A} = \mathsf{A}^{-1}$ we see that (21) is replaced by

$$\mathsf{S} = \frac{\partial W}{\partial\mathsf{A}} - p\mathsf{A}^{-1} \quad \det\mathsf{A} = 1, \qquad (22)$$

where p is a Lagrange multiplier associated with the constraint; p is often referred to as an arbitrary hydrostatic pressure. Thus, (22) is the stress-deformation relation for an *incompressible elastic* material.

The corresponding Cauchy stress tensor is given by

$$\mathsf{T} = \mathsf{A}\frac{\partial W}{\partial \mathsf{A}} - p\mathsf{I} \quad \det \mathsf{A} = 1,$$

where I is the identity tensor.

Now consider a rigid deformation superimposed on the deformation (1); this is defined by the transformation $\mathbf{x} \mapsto \mathbf{x}' = \mathsf{Q}\mathbf{x} + \mathbf{c}$, where Q is a rotation tensor and $\mathbf{c}$ is a translation (both independent of $\mathbf{x}$). The resulting deformation gradient (relative to $\mathcal{B}_r$) is $\mathsf{A}' = \mathsf{QA}$. We require the stored elastic energy to be independent of such superimposed deformations and it therefore follows that

$$W(\mathsf{QA}) = W(\mathsf{A}) \tag{23}$$

for *all* rotations Q. A strain-energy function satisfying this requirement is said to be *objective*.

Use of the polar decomposition (4) and the choice $\mathsf{Q} = \mathsf{R}^{\mathsf{T}}$ in (23) shows that

$$W(\mathsf{A}) = W(\mathsf{U}). \tag{24}$$

Thus, W depends on A only through the stretch tensor U and is therefore defined on the class of positive definite symmetric tensors. This motivates the introduction of the symmetric stress tensor T^1 defined by

$$\mathsf{T}^1 = \frac{\partial W}{\partial \mathsf{U}} \tag{25}$$

for an unconstrained material and by

$$\mathsf{T}^1 = \frac{\partial W}{\partial \mathsf{U}} - p\mathsf{U}^{-1} \quad \det \mathsf{U} = 1 \tag{26}$$

for an incompressible material. This is called the *Biot stress tensor*.

Material Symmetry

Further restrictions on the form of W are obtained if the material possesses symmetries in the reference configuration. A material possesses a symmetry if its response is unchanged after a non-trivial

change of reference configuration. Suppose that the reference config-
uration $\mathcal{B}_r$ is changed to $\mathcal{B}'_r$ with $\mathbf{X}$ changing to $\mathbf{X}'$; then, we denote
by $\mathbf{P}$ the deformation gradient $\text{Grad}\mathbf{X}'$ of $\mathcal{B}'_r$ relative to $\mathcal{B}_r$ and by
$\mathbf{A}'$ the deformation gradient of $\mathcal{B}$ relative to $\mathcal{B}'_r$ so that $\mathbf{A} = \mathbf{A}'\mathbf{P}$. In
general, the strain-energy function of the material relative to $\mathcal{B}'_r$ is
different from that relative to $\mathcal{B}_r$. If, however, the change of reference
configuration just described does not affect the material response
then we write W also for the strain-energy function relative to $\mathcal{B}'_r$,
and hence

$$W(\mathbf{A}'\mathbf{P}) = W(\mathbf{A}) = W(\mathbf{A}') \tag{27}$$

for all deformation gradients $\mathbf{A}'$.

The set of $\mathbf{P}$ for which (27) holds forms a group; we refer to this
group as the *symmetry group of the material relative to* $\mathcal{B}_r$.

Isotropy

To be specific we confine attention to *isotropic elastic materials*, for
which the symmetry group is the *proper orthogonal group*. Then, we
have

$$W(\mathbf{A}\mathbf{Q}) = W(\mathbf{A}) \tag{28}$$

for *all* rotations $\mathbf{Q}$. The combination of (23) and (28) then yields

$$W(\mathbf{Q}\mathbf{U}\mathbf{Q}^\mathsf{T}) = W(\mathbf{U}) \tag{29}$$

for all rotations $\mathbf{Q}$; note that the rotations in (23) and (28) are independent. Equation (29) states that W is an *isotropic function* of $\mathbf{U}$.
It follows from the spectral decomposition (5) that W depends on $\mathbf{U}$
only through the principal stretches $\lambda_1, \lambda_2, \lambda_3$. To avoid introducing
additional notation we express this dependence as $W(\lambda_1, \lambda_2, \lambda_3)$; by
selecting appropriate values for $\mathbf{Q}$ in (29) we deduce that W depends
symmetrically on $\lambda_1, \lambda_2, \lambda_3$. Thus,

$$W(\lambda_1, \lambda_2, \lambda_3) = W(\lambda_1, \lambda_3, \lambda_2) = W(\lambda_2, \lambda_1, \lambda_3). \tag{30}$$

A consequence of isotropy is that $\mathbf{T}^1$ is *coaxial* with $\mathbf{U}$ and hence,
paralleling (5), we have

$$\mathbf{T}^1 = \sum_{i=1}^{3} t_i^1 \mathbf{u}^{(i)} \otimes \mathbf{u}^{(i)}, \tag{31}$$

where, for an unconstrained material,

$$t_i^1 = \frac{\partial W}{\partial \lambda_i}. \tag{32}$$

For an incompressible material the latter is replaced by

$$t_i^1 = \frac{\partial W}{\partial \lambda_i} - p\lambda_i^{-1} \qquad \lambda_1\lambda_2\lambda_3 = 1. \tag{33}$$

It is worth noting in this case that

$$\mathbf{S} = \mathbf{T}^1\mathbf{R}^\mathsf{T} = \sum_{i=1}^{3} t_i^1 \mathbf{u}^{(i)} \otimes \mathbf{v}^{(i)}, \tag{34}$$

the first of which is analogous to $(4)_1$.

2.4   Boundary-value Problems

If the body force in (20) is conservative then we may write

$$\mathbf{b} = -\mathrm{grad}\phi, \tag{35}$$

where ϕ is a scalar field defined on points $\mathbf{x} \in \mathcal{B}$. Given that the material is elastic with strain-energy function W, it follows that (20) can be expressed in the form

$$\delta \int_{\mathcal{B}_r} (W + \rho_r\phi)dV - \int_{\partial\mathcal{B}_r} (\mathbf{S}^\mathsf{T}\mathbf{N}) \cdot \delta\mathbf{x}dA = 0. \tag{36}$$

At this point we consider the equilibrium equation (17) together with the stress-deformation relation (21) for an unconstrained material, and the deformation gradient (2) coupled with (1):

$$\mathrm{Div}\left(\frac{\partial W}{\partial \mathbf{A}}\right) + \rho_r\mathbf{b} = \mathbf{0}, \quad \mathbf{A} = \mathrm{Grad}\mathbf{x} \quad \mathbf{X} \in \mathcal{B}_r. \tag{37}$$

We supplement (37) with boundary conditions, typical of those arising in problems of nonlinear elasticity, in which $\mathbf{x}$ is specified on part of the boundary, $\partial \mathcal{B}_r^x \subset \partial \mathcal{B}_r$ say, and the stress vector on the remainder, $\partial \mathcal{B}_r^\sigma$, so that $\partial \mathcal{B}_r^x \cup \partial \mathcal{B}_r^\sigma = \partial \mathcal{B}_r$ and $\partial \mathcal{B}_r^x \cap \partial \mathcal{B}_r^\sigma = \emptyset$. We write

$$\mathbf{x} = \boldsymbol{\xi}(\mathbf{X}) \quad \text{on } \partial \mathcal{B}_r^x, \tag{38}$$

$$\mathbf{S}^\mathsf{T} \mathbf{N} = \boldsymbol{\sigma}(\mathbf{X}) \quad \text{on } \partial \mathcal{B}_r^\sigma, \tag{39}$$

where $\boldsymbol{\xi}$ and $\boldsymbol{\sigma}$ are specified functions. More generally, $\boldsymbol{\sigma}$ can be allowed to depend on the deformation but, for simplicity, we restrict attention here to the dependence expressed in (39). The surface stress defined by (39) is referred to as a *dead-load traction*. The basic boundary-value problem of nonlinear elasticity is characterized by (37)–(39). In order to analyse this problem additional information about the nature of the function W is needed; this important aspect of the theory is not addressed to any great extent in this article. Reference can be made to, for example, [6] and [14] for discussion of this matter.

In view of the boundary condition (39) we now have $\delta \mathbf{x} = \mathbf{0}$ on $\partial \mathcal{B}_r^x$, and (36) becomes

$$\delta \int_{\mathcal{B}_r} (W + \rho_r \phi) dV - \int_{\partial \mathcal{B}_r^\sigma} \boldsymbol{\sigma} \cdot \delta \mathbf{x} dA = 0. \tag{40}$$

Since $\boldsymbol{\sigma}$ is independent of the deformation, we write $\boldsymbol{\sigma} = \mathrm{grad}(\boldsymbol{\sigma} \cdot \mathbf{x})$, and (40) then becomes

$$\delta \left\{ \int_{\mathcal{B}_r} (W + \rho_r \phi) dV - \int_{\partial \mathcal{B}_r^\sigma} \boldsymbol{\sigma} \cdot \mathbf{x} dA \right\} = 0. \tag{41}$$

If we interpret $\delta \chi$ as a variation of the function χ, then (41) provides a variational formulation of the boundary-value problem (37)–(39) and can be written $\delta E = 0$, where E is the functional defined by

$$E\{\chi\} = \int_{\mathcal{B}_r} \{W(\mathrm{Grad}\chi) + \rho_r \phi(\chi)\} dV - \int_{\partial \mathcal{B}_r^\sigma} \boldsymbol{\sigma} \cdot \chi dA. \tag{42}$$

In (42) χ is taken to be in some appropriate class of mappings, which need not be specified here, and similarly for admissible variations $\delta \chi$

subject to $\delta\chi = 0$ on $\partial\mathcal{B}_r^x$. A similar variational statement can be formulated if the boundary traction in (39) is replaced by a hydro-static pressure, $\mathsf{T}\mathbf{n} = -P\mathbf{n}$ say, in which case σ depends on the deformation in the form

$$\sigma = -JPA^{-\mathsf{T}}\mathbf{N} \quad \text{on} \quad \partial\mathcal{B}_r^\sigma. \tag{43}$$

As we shall see in Section 3, the energy functional plays on important role in the analysis of stability and bifurcation. Detailed discussion of variational principles in the context of nonlinear elasticity can be found in [14]; for the more technical mathematical aspects the reader is referred to [6,11] and the papers by Ball [1,2].

For an incompressible material, the functional E is modified by the addition of a constraint term $-p\{\det(\text{Grad}\chi) - 1\}$.

In components, equation (37) can be written as

$$\frac{\partial^2 W}{\partial A_{i\alpha}\partial A_{j\beta}} \frac{\partial^2 x_j}{\partial X_\alpha\partial X_\beta} + \rho_r b_i = 0 \tag{44}$$

for $i \in \{1,2,3\}$. When coupled with suitable boundary conditions, this forms a system of quasi-linear partial differential equations for x_i. The coefficients $\partial^2 W/\partial A_{i\alpha}\partial A_{j\beta}$ are, in general, nonlinear functions of the components of the deformation gradient. Only a very limited number of boundary-value problems have been solved for un-constrained materials, and only then for very special choices of the form of W; some of these are discussed in [14], for example. For incompressible materials the corresponding equations, obtained by substituting (22) into (17), have yielded more success, and again we refer to [14] for pointers to the literature. It is not our intention in this article to discuss such solutions, rather to examine certain aspects of the structure of the equations.

Equations (44) are *strongly elliptic* if

$$\mathcal{A}^1_{\alpha i\beta j}m_i m_j N_\alpha N_\beta > 0 \tag{45}$$

for all non-zero vectors $\mathbf{m}$ and $\mathbf{N}$, where we have introduced the notation

$$\mathcal{A}^1_{\alpha i\beta j} = \frac{\partial^2 W}{\partial A_{i\alpha}\partial A_{i\beta}}. \tag{46}$$

3   Incremental Deformations

Suppose that a solution χ to the boundary-value problem (37)–(39) is known and consider the problem of finding solutions near to χ when the boundary conditions are perturbed. Let χ' be a solution for the perturbed problem and write $\mathbf{x}' = \chi'(\mathbf{X})$. Also, we write

$$\dot{\mathbf{x}} = \mathbf{x}' - \mathbf{x} = \chi'(\mathbf{X}) - \chi(\mathbf{X}) \equiv \dot{\chi}(\mathbf{X}). \tag{47}$$

Then

$$\mathrm{Grad}\,\dot{\chi} = \mathrm{Grad}\,\chi' - \mathrm{Grad}\,\chi \equiv \dot{\mathbf{A}}.$$

In the above and henceforth a superposed dot represents the difference

$$(\dot{\ }) = (\ )' - (\ ),$$

while $\delta(\cdot)$ represents a variation of $(\cdot)$. Note that $\dot{\mathbf{A}}$ is linear in $\dot{\chi}$.

The nominal stress difference is

$$\dot{\mathbf{S}} = \mathbf{S}' - \mathbf{S} = \frac{\partial W}{\partial \mathbf{A}}(\mathbf{A}') - \frac{\partial W}{\partial \mathbf{A}}(\mathbf{A}). \tag{48}$$

When $\dot{\mathbf{A}}$ is small (in some appropriate sense) this can be approximated as

$$\dot{\mathbf{S}} = \mathcal{A}^1 \dot{\mathbf{A}} + \frac{1}{2}\mathcal{A}^2[\dot{\mathbf{A}}, \dot{\mathbf{A}}] + \dots, \tag{49}$$

where

$$\mathcal{A}^1 = \frac{\partial^2 W}{\partial \mathbf{A}^2}, \quad \mathcal{A}^2 = \frac{\partial^3 W}{\partial \mathbf{A}^3}. \tag{50}$$

In components, the first of the expressions (50) is given by (46) while the second can be written

$$\mathcal{A}^2_{\alpha i \beta j \gamma k} = \frac{\partial^3 W}{\partial A_{i\alpha} \partial A_{j\beta} \partial A_{k\gamma}},$$

and similarly for higher-order terms if required (given that W is sufficiently regular). The component form of (49) is

$$\dot{S}_{\alpha i} = \mathcal{A}^1_{\alpha i \beta j} \dot{A}_{j\beta} + \frac{1}{2}\mathcal{A}^2_{\alpha i \beta j \gamma k} \dot{A}_{j\beta} \dot{A}_{k\gamma} + \dots, \tag{51}$$

and this serves to define the products appearing in (49).

For our purposes it suffices to consider the linear approximation to (48), namely

$$\dot{S} = \mathcal{A}^1 \dot{A}.\tag{52}$$

We note that for an incompressible material this is modified to

$$\dot{S} = \mathcal{A}^1 \dot{A} - \dot{p} A^{-1} + p A^{-1} \dot{A} A^{-1}.\tag{53}$$

From the equilibrium equation (17) and its counterpart for χ', namely

$$\mathrm{Div} S' + \rho_r \mathbf{b}' = \mathbf{0},$$

we obtain, by subtraction,

$$\mathrm{Div}\dot{S} + \rho_r \dot{\mathbf{b}} = \mathbf{0}.\tag{54}$$

This is *exact*, but in the linear approximation it becomes

$$\mathrm{Div}(\mathcal{A}^1 \dot{A}) + \rho_r \dot{\mathbf{b}} = \mathbf{0},\tag{55}$$

with $\dot{\mathbf{b}}$ linearized in $\dot{\chi}$.

Let ξ' and σ' be the prescribed data for χ' so that

$$\mathbf{x}' = \xi' \quad \text{on} \quad \partial \mathcal{B}_r^x,$$

$$S'^{\mathrm{T}} \mathbf{N} = \sigma' \quad \text{on} \quad \partial \mathcal{B}_r^\sigma,$$

and hence

$$\dot{\mathbf{x}} = \dot{\xi} \quad \text{on} \quad \partial \mathcal{B}_r^x,\tag{56}$$

$$\dot{S}^{\mathrm{T}} \mathbf{N} = \dot{\sigma} \quad \text{on} \quad \partial \mathcal{B}_r^\sigma.\tag{57}$$

We now consider the linearized problem (55)–(57) for $\dot{\chi}$, with $\dot{\sigma}$ linearized if it depends on the deformation. For definiteness and simplicity we now assume that there are no body forces and that σ is a dead-load traction, as in (39); thus

$$\mathrm{Div}\dot{S} = \mathbf{0},\tag{58}$$

$$\dot{S} = \mathcal{A}^1 \dot{A},\tag{59}$$

$$\dot{\mathbf{A}} = \mathrm{Grad}\dot{\chi}. \tag{60}$$

Taken together, equations (56)–(60) constitute the basic boundary-value problem of incremental elasticity when the underlying deformation χ is known. We next examine the question of uniqueness of solution of the incremental problem and the associated question of stability of the deformation χ.

3.1   Uniqueness and Stability

Let $\dot{\chi}$ and $\dot{\chi}^*$ be two possible solutions of the incremental problem and let

$$\Delta\dot{\chi} = \dot{\chi}^* - \dot{\chi} \tag{61}$$

denote their difference. More generally, we adopt the notation $\Delta(\cdot)$ to represent $(\cdot)^* - (\cdot)$. Then

$$\Delta\dot{\mathbf{A}} = \mathrm{Grad}\Delta\dot{\chi}, \quad \Delta\dot{\mathbf{S}} = \mathcal{A}^1\Delta\dot{\mathbf{A}}, \tag{62}$$

and hence

$$\mathrm{Div}\Delta\dot{\mathbf{S}} = \mathbf{0} \quad \text{in } \mathcal{B}_r, \tag{63}$$

$$\Delta\dot{\chi} = \mathbf{0} \quad \text{on } \partial\mathcal{B}_r^x, \tag{64}$$

$$\Delta\dot{\mathbf{S}}^{\mathrm{T}}\mathbf{N} = \mathbf{0} \quad \text{on } \partial\mathcal{B}_r^\sigma. \tag{65}$$

On use of the divergence theorem it follows that

$$\int_{\mathcal{B}_r} \mathrm{tr}(\Delta\dot{\mathbf{S}}\Delta\dot{\mathbf{A}})dV = 0. \tag{66}$$

Clearly, a *sufficient condition* for uniqueness of solution of the incremental problem is

$$\int_{\mathcal{B}_r} \mathrm{tr}(\Delta\dot{\mathbf{S}}\Delta\dot{\mathbf{A}})dV > 0, \tag{67}$$

for *all* $\Delta\dot{\chi} \neq \mathbf{0}$ in $\mathcal{B}_r$ and satisfying $\Delta\dot{\chi} = \mathbf{0}$ on $\partial\mathcal{B}_r^x$, or, equivalently,

$$\int_{\mathcal{B}_r} \mathrm{tr}\{(\mathcal{A}^1\Delta\dot{\mathbf{A}})\Delta\dot{\mathbf{A}}\}dV > 0. \tag{68}$$

Note that this sufficient condition applies when the boundary conditions are inhomogeneous (i.e. $\dot{\xi} \neq 0$, $\dot{\sigma} \neq 0$ in (56)–(57)) so in considering incremental uniqueness it suffices henceforth to examine the homogeneous boundary-value problem

$$\mathrm{Div}\dot{S} = 0 \quad \text{in } \mathcal{B}_r, \tag{69}$$

$$\dot{\chi} = 0 \quad \text{on } \partial\mathcal{B}_r^x, \tag{70}$$

$$\dot{S}^T N = 0 \quad \text{on } \partial\mathcal{B}_r^\sigma, \tag{71}$$

one solution of which is $\dot{\chi} = 0$. Thus, on setting $\dot{\chi}^* = 0$, the inequality (67) becomes

$$\int_{\mathcal{B}_r} \mathrm{tr}(\dot{S}\dot{A})dV > 0, \tag{72}$$

and (68) similarly becomes

$$\int_{\mathcal{B}_r} \mathrm{tr}\{(\mathcal{A}^1\dot{A})\dot{A}\}dV > 0 \tag{73}$$

for all $\dot{\chi} \not\equiv 0$ in $\mathcal{B}_r$ satisfying $\dot{\chi} = 0$ on $\partial\mathcal{B}_x^\sigma$.

If (73) holds in the configuration $\mathcal{B}$ then the only solution of the homogeneous incremental problem is the trivial solution $\dot{\chi} \equiv 0$; correspondingly, the solution of the inhomogeneous problem is unique (assuming it exists). Expressed otherwise, (73), or equivalently (67), excludes bifurcation from the solution χ of the underlying boundary-value problem. The inequality (73), or equivalently (72), is therefore referred to as the *exclusion condition* (for the dead-load traction boundary condition). The exclusion condition requires modification if σ is allowed to depend on the deformation.

To see why we take ' > 0 ' rather than ' < 0 ' in (73) we consider the change in the energy functional due to the change in deformation from χ to χ'. On use of the boundary condition (70) and the divergence theorem this is seen to be

$$E\{\chi'\} - E\{\chi\} = \int_{\mathcal{B}_r} \{W(A') - W(A) - \mathrm{tr}(\dot{S}\dot{A})\}dV. \tag{74}$$

By application of the Taylor expansion to $W(A')$ this then becomes

$$E\{\chi'\} - E\{\chi\} = \int_{\mathcal{B}_r} \{\frac{1}{2}\mathrm{tr}[(\mathcal{A}^1\dot{A})\dot{A}] + \ldots\}dV \qquad (75)$$

Thus, to the second order in $\dot{A}$, (73) implies that $E\{\chi'\} > E\{\chi\}$ for all admissible $\dot{\chi} \not\equiv 0$ satisfying (70). This inequality states that χ is *stable* with respect to perturbations $\dot{\chi}$ from χ, and that χ is a *local minimiser* of the energy functional. For full discussion of *infinitesimal* stability we refer to [20].

Suppose now that we restrict attention to the all-round dead-load problem so that $\partial\mathcal{B}_r^x = \emptyset$, $\partial\mathcal{B}_r = \partial\mathcal{B}_r^\sigma$. If the underlying deformation is homogeneous then A is independent of X and, therefore, so is $\mathcal{A}^1$. It then follows that (73) is equivalent to the local condition

$$\mathrm{tr}\{(\mathcal{A}^1\dot{A})\dot{A}\} > 0 \qquad (76)$$

for all $\dot{A} \neq 0$, i.e. $\mathcal{A}^1$ is positive definite at each point $X \in \mathcal{B}_r$.

In order to illustrate loss of uniqueness and stability we now concentrate attention on the dead-load traction boundary-value problem in which the deformation is homogeneous, so that (76) is the relevant exclusion condition.

4   Bifurcation from a Homogeneous Deformation under Dead-load

Let us imagine a path of dead-loading from the unstressed configuration in S-space resulting in a path of deformation in A-space along which $\mathcal{A}^1$ is positive definite, i.e. along which χ is locally (infinitesimally) stable so that (76) holds. (Note, however, that, as we see shortly, (76) does not hold at the origin $A = I$ in A-space from which the path starts.) The inequality first fails, if at all, when a point is reached where

$$\mathrm{tr}\{(\mathcal{A}^1\dot{A})\dot{A}\} \geq 0 \qquad (77)$$

for all $\dot{A}$, with equality holding for some $\dot{A} \neq 0$. By the extremal property of quadratic forms it follows that

$$\mathcal{A}^1\dot{A}_e = 0 \qquad (78)$$

for the $\dot{A}_e$ which minimises (77). Note, however, that in general there may be more than one $\dot{A}_e$ satisfying (78). Points in A-space where (78) holds are given by

$$\det(\mathcal{A}^1) = 0, \qquad (79)$$

where we are regarding $\mathcal{A}^1$ as a linear mapping on the vector space of second-order tensors $\dot{A}$. Equation (79) describes a hypersurface in A-space and can be regarded as the stability limit or bifurcation surface since, on this surface,

$$\dot{S}_e \equiv \mathcal{A}^1 \dot{A}_e = 0. \qquad (80)$$

Following the terminology of Hill [8,9] we refer to $\dot{A}_e$ as an *eigenmode* (an eigenvector of $\mathcal{A}^1$ associated with zero eigenvalue) and the surface as an *eigensurface*. Bifurcation occurs since, to the first order in $\dot{A}$, A and $A' \equiv A + \dot{A}$ correspond to the same state of stress S. In Figure 1 we depict the stable region of A-space bounded by surfaces on which (79) holds together with a path of deformation emanating from $A = I$ and terminating at a point on such a surface.

Clearly, in the dead-load traction problem, *local bifurcation* occurs where $\mathcal{A}^1$ is *singular*. Since, in (77), equality holds for $\dot{A} = \dot{A}_e$ the stability criterion (76) is inappropriate to determine the stability status of points on the eigensurface. Referring to (75) we see that stability is determined by the sign of the integrand for $\dot{A} = \dot{A}_e$, and that this is dependent on third- and higher-order terms in $\dot{A}_e$; we do not pursue the details here. *Globally*, different branches of the solution of the dead-load problem can be found by inverting the stress-deformation relation (21) for an unconstrained material or (22) for an incompressible material. Before examining this inversion we consider the singularities of $\mathcal{A}^1$ for an isotropic elastic material.

For an isotropic elastic material the components of $\mathcal{A}^1$ referred to the principal axes $\mathbf{u}^{(i)}$ and $\mathbf{v}^{(i)}$ are given by

$$\mathcal{A}^1_{iijj} = W_{ij}, \qquad (81)$$

$$\mathcal{A}^1_{ijij} - \mathcal{A}^1_{ijji} = \frac{W_i + W_j}{\lambda_i + \lambda_j} \quad i \neq j, \qquad (82)$$

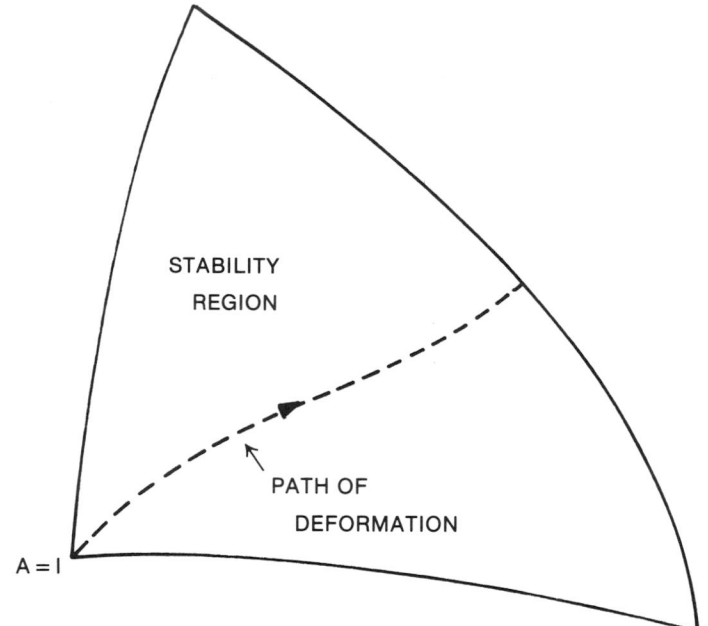

Figure 1: Depiction of the region of stability in A-space.

$$\mathcal{A}^1_{ijij} + \mathcal{A}^1_{ijji} = \frac{W_i - W_j}{\lambda_i - \lambda_j} \quad i \neq j, \lambda_i \neq \lambda_j, \tag{83}$$

$$\mathcal{A}^1_{ijij} + \mathcal{A}^1_{ijji} = W_{ii} - W_{ij} \quad i \neq j, \lambda_i = \lambda_j, \tag{84}$$

where $W_i = \partial W/\partial\lambda_i$, $W_{ij} = \partial^2 W/\partial\lambda_i\partial\lambda_j$ and $i,j \in \{1,2,3\}$, and no summation is implied by the repetition of indices. (We have now dropped our convention of using Greek letters for indices relating to Lagrangean components.) For details of the derivation of these components we refer to [14].

In the classical theory of elasticity, corresponding to the situation in which there is no underlying deformation, the components of $\mathcal{A}^1$ can be written compactly in the form

$$\mathcal{A}^1_{ijkl} = \lambda\delta_{ij}\delta_{kl} + \mu\delta_{ik}\delta_{jl} + \mu\delta_{il}\delta_{jk}, \tag{85}$$

where λ and μ are the classical *Lamé* moduli of elasticity and δ_{ij} is the Kronecker delta. The values of W_{ij} when $\lambda_i = 1$ for $i \in \{1,2,3\}$ should be consistent with (85). Also, we take $W_i = 0$ when $\lambda_j = 1$

for $i, j \in \{1, 2, 3\}$ so that the configuration $\mathcal{B}_r$ is stress free; such a configuration is called a *natural configuration.*

Necessary and sufficient conditions for (76) to hold are

$$W_{ij} \text{ is positive definite,} \tag{86}$$

$$W_i + W_j > 0 \quad i, j \in \{1, 2, 3\}, \, i \neq j, \tag{87}$$

$$\frac{W_i - W_j}{\lambda_i - \lambda_j} > 0 \quad i \neq j \tag{88}$$

jointly, while singularities of $\mathcal{A}^1$ are identified by one or more of the equations

$$\det(W_{ij}) = 0, \tag{89}$$

$$W_i + W_j = 0 \quad i \neq j, \tag{90}$$

$$\frac{W_i - W_j}{\lambda_i - \lambda_j} = 0 \quad i \neq j. \tag{91}$$

The form of W is critical in deciding whether (89) or (91) can hold, but (90) must hold on some surfaces in A-space irrespective of the form of W since the principal Biot stresses $t_i^1 = W_i$ can take both positive and negative values (i.e. an elastic material can be subjected to both tensile and compressive stresses). In particular, as indicated above, (90) holds in the natural configuration $A = I$. The purpose of the above discussion was to show that $\mathcal{A}^1$ can possess singularities, and we do not examine here the associated eigenmodes; the latter are studied in detail in [14, page 369].

For incompressible materials, bearing in mind the difference between (52) and (53), the components of $\mathcal{A}^1$ have the same form as in (81)–(84), but are subject to (10). In this case, however, when $\lambda_i = 1$ for $i \in \{1, 2, 3\}$ the components are given by

$$A^1_{iiii} = A^1_{ijij} = \mu \quad i \neq j, \tag{92}$$

$$A^1_{iijj} = A^1_{ijji} = 0 \quad i \neq j, \tag{93}$$

and $W_{ii} = W_i = \mu$, $W_{ij} = 0$, where μ is the *shear modulus* in $\mathcal{B}_r$.

The analogue of the stability inequality (76) for an incompressible material is

$$\mathrm{tr}\{(\mathcal{A}^1\dot{A})\dot{A}\} + p\,\mathrm{tr}(A^{-1}\dot{A}A^{-1}\dot{A}) > 0. \tag{94}$$

Necessary and sufficient conditions for this to hold for an isotropic elastic material are given in [14, page 450].

4.1 Global Aspects of Non-uniqueness

The singularities of $\mathcal{A}^1$ considered above are local manifestations of non-uniqueness in the relationship between S and A expressed through the constitutive equation

$$S = \frac{\partial W}{\partial A}(A) \qquad (95)$$

(for an unconstrained material). We now examine briefly the global inversion of (95).

First, we have from $(34)_1$

$$S = T^1 R^T, \qquad (96)$$

from which it follows that

$$(T^1)^2 = SS^T = \sum_{i=1}^{3}(t_i^1)^2 u^{(i)} \otimes u^{(i)} \qquad (97)$$

and

$$\det S = \det T = t_1^1 t_2^1 t_3^1. \qquad (98)$$

Unlike (4), however, the polar decomposition (96) is not in general unique since T^1 is not sign definite. For a given S there are in general four distinct polar decompositions of the form (96), and infinitely many when $(t_i^1)^2 = (t_j^1)^2$ $(i \neq j)$, but at most one of these satisfies the stability inequality (87). For each such polar decomposition R is determined (at least to within an arbitrary rotation about a principal axis) and U can in principle be found by inverting (25), an inversion which itself may not be unique. The resulting deformation gradients are then calculated from $A = RU$. For further discussion of these points we refer to [14, pages 358–361].

In order to illustrate the non-uniqueness more explicitly it suffices to take $R = I$ and to examine the inversion of (25) or (26). More particularly, we consider the pure strain $A = U$ in which $u^{(i)}$ are fixed in direction as the load increases. The problem then reduces to inverting the scalar equations (32) or (33). The mathematics can be associated with the physical problem in which a cuboid of elastic material is subjected to normal forces t_i^1 per unit reference area on its three pairs of faces, as in the following example.

4.2   A Two-dimensional Example

For purposes of this example we consider an incompressible material and rewrite (33) in the form

$$\lambda_i t_i^1 = \lambda_i \frac{\partial W}{\partial \lambda_i} - p \quad i \in \{1,2,3\}, \tag{99}$$

with

$$\lambda_1 \lambda_2 \lambda_3 = 1. \tag{100}$$

On elimination of p from (99) we obtain

$$\lambda_1 t_1^1 - \lambda_1 \frac{\partial W}{\partial \lambda_1} = \lambda_2 t_2^1 - \lambda_2 \frac{\partial W}{\partial \lambda_2} = \lambda_3 t_3^1 - \lambda_3 \frac{\partial W}{\partial \lambda_3}. \tag{101}$$

Equations (100)–(101) provide three equations for $\lambda_1, \lambda_2, \lambda_3$ when t_1^1, t_2^1, t_3^1 are prescribed. Alternatively, if λ_3 is given, along with t_1^1 and t_2^1, then λ_1 and λ_2 and hence t_3^1 can be found. For simplicity we consider this latter situation with $\lambda_3 = 1$, so that it remains to determine λ_1 and λ_2 from the two equations

$$\lambda_1 t_1^1 - \lambda_1 \frac{\partial W}{\partial \lambda_1} = \lambda_2 t_2^1 - \lambda_2 \frac{\partial W}{\partial \lambda_2}, \quad \lambda_1 \lambda_2 = 1. \tag{102}$$

On introducing the change of variables $\lambda_1 = \lambda, \lambda_2 = \lambda^{-1}$ and the notation

$$\hat{W}(\lambda) = W(\lambda, \lambda^{-1}, 1) \tag{103}$$

we can reduce equations (102) to

$$\lambda t_1^1 - \lambda^{-1} t_2^1 = \lambda \frac{d\hat{W}}{d\lambda}, \tag{104}$$

subject to

$$\hat{W}'(1) = 0, \tag{105}$$

where the prime indicates differentiation with respect to λ. The further specialization $t_1^1 = t_2^1 \, (= t^1$, say) reduces (104) to

$$(\lambda - \lambda^{-1})t^1 = \lambda\hat{W}'(\lambda). \tag{106}$$

The problem is now two-dimensional and we take the cuboid to have a square cross-section in the $(1,2)$-plane. We require to determine λ when the dead-load traction t^1 is given.

Noting (105), we see that, independently of t^1, $\lambda = 1$ is always one solution of (106); this corresponds to the situation in which the square cross-section remains unchanged as t^1 increases from zero. We wish, however, to consider the possible emergence of solutions of (106) for which $\lambda \neq 1$. To this end we rewrite (106) in the form

$$t^1 = \frac{\lambda^2\hat{W}'(\lambda)}{(\lambda^2 - 1)}. \tag{107}$$

In the limit $\lambda \to 1$ this becomes

$$t_c^1 = \frac{1}{2}\hat{W}''(1) = \mu, \tag{108}$$

which is the (positive) shear modulus of the material. The stress t_c^1 is a *critical value* of t^1 in the sense that it is the value at which bifurcation from the symmetrical solution $\lambda = 1$ can occur. This can be seen by expanding (107) as a Taylor series near $\lambda = 1$ to give

$$t^1 = t_c^1 + \eta(\lambda - 1)^2 + \ldots, \tag{109}$$

where

$$\eta = \frac{1}{12}\hat{W}''''(1) - \hat{W}''(1). \tag{110}$$

Note that it can be shown that $dt^1/d\lambda = 0$ at $\lambda = 1$.

Equation (109) shows that a symmetry-breaking bifurcation can emerge in which the square cross-section becomes rectangular. (We do not consider here possible complications associated with rotation of the principal axes and the inclusion of R in the equations.) The stability inequality (94) for this two-dimensional problem can be shown to reduce to

$$\lambda^2\hat{W}''(\lambda) > \frac{2\lambda\hat{W}'(\lambda)}{(\lambda^2 - 1)} > 0, \tag{111}$$

remembering that we are ignoring rotations. (Inclusion of rotations would impose the additional stability requirement

$$(t_1^1 - t_2^1)/(\lambda_1 - \lambda_2) > 0$$

which, of course, is violated on the asymmetrical path.)

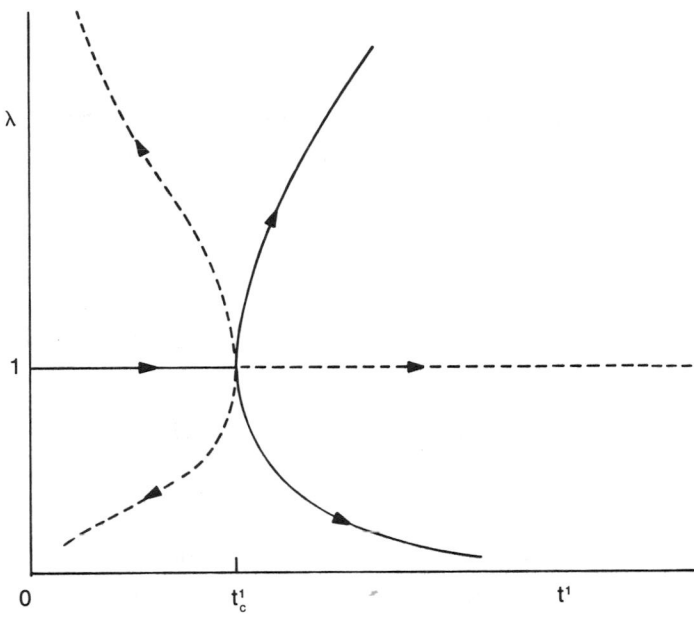

Figure 2: Bifurcation Diagram Showing λ as a Function of t^1: Stable Solutions – Continuous Curves; Unstable Solutions – Broken Curves.

Finally, we illustrate the above results by choosing a strain energy of the form

$$W(\lambda_1, \lambda_2, \lambda_3) = \mu(\lambda_1^m + \lambda_2^m + \lambda_3^m - 3)/m^2, \qquad (112)$$

so that

$$\hat{W}(\lambda) = \mu(\lambda^m + \lambda^{-m} - 2)/m^2. \qquad (113)$$

From (110) we obtain

$$\eta = \frac{1}{6}\mu(m^2 - 1), \qquad (114)$$

and the inequalities (111) show that the solution branches corresponding to $\lambda \neq 1$ are stable for $m^2 > 1$ and unstable for $m^2 < 1$. Moreover, the symmetrical solution $\lambda = 1$ is unstable for $t^1 > t_c^1$. We emphasize that bifurcation is predicted for *every* form of W consistent with the requirement $\hat{W}''(1) = 2\mu$ of the classical theory. We illustrate these results in Figure 2.

Energy Considerations

For the dead-load traction boundary-value problem considered here the energy functional (42) can be written, on use of the divergence theorem, as

$$E = \int_{\mathcal{B}_r} \{W - \text{tr}(\mathsf{SA})\}dV, \tag{115}$$

arguments being omitted. Since the deformation is homogeneous it suffices to consider the integrand of (115), which we write as

$$\bar{E} = W - \text{tr}(\mathsf{SA}) = W - \sum_{i=1}^{3} t_i^1 \lambda_i \tag{116}$$

for an isotropic material. This is the energy per unit volume of $\mathcal{B}_r$. The additional term associated with the incompressibility constraint is omitted since it plays no part in the following analysis.

The two-dimensional counterpart of (116), obtained by a minor change in the definition of $\bar{E}$ (λ_3 is replaced by $\lambda_3 - 1$), can be expressed as

$$\bar{E} = W(\lambda_1, \lambda_2, 1) - t_1^1 \lambda_1 - t_2^1 \lambda_2 \tag{117}$$

on taking $\lambda_3 = 1$. Now let E_{as} and E_s respectively denote the energies (per unit cross-sectional area and per unit length in the 3-direction) on the asymmetric and symmetric solution paths discussed above. Then

$$E_{as} = \hat{W}(\lambda) - t^1(\lambda + \lambda^{-1}), \quad E_s = -2t^1,$$

and hence

$$E_{as} - E_s = \hat{W}(\lambda) - t^1(\lambda + \lambda^{-1} - 2). \tag{118}$$

On the asymmetric path we can regard λ as a function of t^1 through (107); note that $E_{as} = E_s$ when $t^1 = t_c^1$. On use of (107) we then obtain

$$\frac{d}{dt^1}(E_{as} - E_s) = -(\lambda^{\frac{1}{2}} - \lambda^{-\frac{1}{2}})^2 \le 0. \tag{119}$$

This shows that on a stable (asymmetric) path of deformation with t^1 increasing from t_c^1 the energy is less than on the corresponding symmetric path. Thus, the asymmetric path provides the (global) energy-minimising solution. In particular, near $t^1 = t_c^1$, we have

$$E_{as} - E_s = -\frac{1}{\eta}(t^1 - t_c^1)^2 + \dots, \tag{120}$$

where η is given by (110). If $\eta > 0$ the asymmetric solution is stable and branches to the right, while if $\eta < 0$ it is unstable and branches to the left (Figure 2). In the latter case the symmetric path provides the energy minimiser.

To be more explicit, if we take $m = 2$ in (113) then $\eta > 0$,

$$t^1 = \frac{1}{2}\mu(\lambda + \lambda^{-1}), \tag{121}$$

and (118) reduces to

$$E_{as} - E_s = -\frac{1}{\mu}(\mu - t^1)^2. \tag{122}$$

For $m = 2$ the strain-energy function (112) is called the *neo-Hookean* strain-energy function and this provides a model for the behaviour of rubber-like solids at moderate deformations.

For further discussion of this and related problems we refer to, for example, [3,10,14,15,16,19].

Clearly, the degree of non-uniqueness in the relation between t_i^1 and λ_i, as reflected in that between S and A will have implications for the solution of more general boundary-value problems than the homogeneous dead-load problem discussed in this section. Reference to other problems in which non-uniqueness is evident can be found in, for example, [6,14]. In Section 5, we consider one further example of the onset of bifurcation from a homogeneously-deformed configuration.

5   Surface Deformations of a Half-space

We now consider a pure homogeneous deformation of the form

$$x_1 = \lambda_1 X_1, \quad x_2 = \lambda_2 X_2, \quad x_3 = \lambda_3 X_3,$$

and we suppose that $\mathcal{B}$ corresponds to the region $x_2 < 0$ with boundary $x_2 = 0$. Let t_1, t_2, t_3 denote the principal components of the Cauchy stress tensor T. An incremental deformation of the form (47) is superimposed on the above deformation, and we denote the displacement vector $\dot{\mathbf{x}}$ by $\mathbf{u}$. At this point we update the reference configuration so that $\mathcal{B}_r$ coincides with $\mathcal{B}$ and we regard $\mathbf{u}$ as a function of $\mathbf{x}$ through (47) and the inverse of (1).

Let $\boldsymbol{\Gamma} = \text{grad}\,\mathbf{u}$ denote the *displacement gradient*, where grad is the gradient operator in $\mathcal{B}$. The components $\Gamma_{ij} = \partial u_i/\partial x_j$ of $\boldsymbol{\Gamma}$ are denoted by $u_{i,j}$ for compactness, where $,i$ represents $\partial/\partial x_i$. We also take the material to be incompressible so that $\lambda_1 \lambda_2 \lambda_3 = 1$, while, for the incremental deformation,

$$\text{tr}(\boldsymbol{\Gamma}) \equiv \text{div}\,\mathbf{u} = 0. \tag{123}$$

The updating of the reference configuration modifies the incremental stress-deformation relation (53), which now becomes

$$\dot{\mathsf{S}}_0 = \mathcal{A}_0^1 \boldsymbol{\Gamma} + p\boldsymbol{\Gamma} - \dot{p}\mathsf{I}, \tag{124}$$

where the zero subscript indicates evaluation in $\mathcal{B}$. In the absence of body forces, the equilibrium equations (54) become

$$\text{div}\,\dot{\mathsf{S}}_0 = \mathbf{0}. \tag{125}$$

In components, equations (124) and (123) together can be written

$$\dot{S}_{0ji} = \mathcal{A}_{0jilk}^1 u_{k,l} + p u_{j,i} - \dot{p}\delta_{ij}, \quad u_{i,i} = 0, \tag{126}$$

while the incremental traction $\dot{\mathsf{S}}_0^{\mathsf{T}} \mathbf{n}$ per unit area of $\partial\mathcal{B}$ has components

$$\dot{S}_{0ji} n_j \equiv (\mathcal{A}_{0jilk}^1 + p\delta_{jk}\delta_{il}) u_{k,l} n_j - \dot{p} n_i. \tag{127}$$

For an incompressible material, the components of $\mathcal{A}_0^1$ are given by

$$\mathcal{A}_{0ijkl}^1 = \lambda_i \lambda_k \mathcal{A}_{ijkl}^1 \tag{128}$$

(no summation), where the components of $\mathcal{A}^1$ are obtained from (81)–(84) subject to $\lambda_1 \lambda_2 \lambda_3 = 1$. We also note that the strong ellipticity condition (45) becomes

$$\mathcal{A}_{0jilk}^1 n_j n_l m_i m_k > 0 \tag{129}$$

for all non-zero vectors $\mathbf{m}$ and $\mathbf{n}$ satisfying the condition $\mathbf{m} \cdot \mathbf{n} = 0$ imposed by the incompressibility constraint.

We now specialize to two-dimensional incremental deformations with $u_3 = 0$ and u_1, u_2 independent of x_3 so that (123) reduces to

$$u_{1,1} + u_{2,2} = 0, \tag{130}$$

from which we deduce the existence of a function ψ such that

$$u_1 = \psi_{,2}, \quad u_2 = -\psi_{,1}. \tag{131}$$

Substitution of (131) into the appropriate specialization of (125) yields (after some rearrangement [7]) an equation for ψ, namely

$$\alpha \psi_{,1111} + 2\beta \psi_{,1122} + \gamma \psi_{,2222} = 0 \quad x_2 < 0. \tag{132}$$

If we take the incremental boundary traction to vanish on ∂B then use of (127) leads to the following boundary conditions on ψ:

$$\gamma(\psi_{,22} - \psi_{,11}) + t_2 \psi_{,11} = 0 \quad \text{on } x_2 = 0, \tag{133}$$

$$(2\beta + \gamma - t_2)\psi_{,112} + \gamma \psi_{,222} = 0 \quad \text{on } x_2 = 0. \tag{134}$$

See [7] for details in a slightly more general setting. In (132)–(134) we have introduced the notation

$$\alpha = \mathcal{A}_{01212}^1, \quad \gamma = \mathcal{A}_{02121}^1, \tag{135}$$

$$2\beta = \mathcal{A}_{01111}^1 + \mathcal{A}_{02222}^1 - 2\mathcal{A}_{01122}^1 - 2\mathcal{A}_{01221}^1. \tag{136}$$

The strong ellipticity condition (129), appropriately specialized to the two-dimensional situation considered here, reduces simply to

$$\alpha > 0, \quad \beta > -(\alpha\gamma)^{\frac{1}{2}}. \tag{137}$$

We seek incremental deformations which decay as $x_2 \to -\infty$; such a deformation is regarded as a *surface deformation*. For simplicity, we take ψ to have the form

$$\psi = A\exp(skx_2 - ikx_1), \tag{138}$$

where A, s and k are constants. Substitution into (132) yields a quadratic equation for s^2, namely

$$\gamma s^4 - 2\beta s^2 + \alpha = 0. \tag{139}$$

Denoting the roots of (139) by s_1^2 and s_2^2, we deduce that

$$s_1^2 s_2^2 = \alpha/\gamma = \lambda_1^2/\lambda_2^2. \tag{140}$$

Taking s_1 and s_2 to be the solutions of (139) with positive real part, we write the general solution of (132) with the required properties in the form

$$\psi = (Ae^{s_1 kx_2} + Be^{s_2 kx_2})e^{-ikx_1}, \tag{141}$$

where B is another constant.

When substituted into the boundary conditions (133) and (134) the expression (141) leads to

$$(\gamma s_1^2 + \gamma - t_2)A + (\gamma s_2^2 + \gamma - t_2)B = 0, \tag{142}$$

$$(2\beta + \gamma - t_2 - \gamma s_1^2)s_1 A + (2\beta + \gamma - t_2 - \gamma s_2^2)s_2 B = 0. \tag{143}$$

For a non-trivial solution A, B of these equations the determinant of coefficients must vanish. After elimination of a factor $s_1 - s_2$, whose vanishing yields a trivial solution, we obtain

$$\gamma\alpha + 2(\beta + \gamma - t_2)(\gamma\alpha)^{\frac{1}{2}} - (\gamma - t_2)^2 = 0. \tag{144}$$

Thus, surface deformations with ψ given by (141) are possible for values of $\lambda_1, \lambda_2, \lambda_3$ and t_2 satisfying (144). In particular, when the normal surface stress t_2 vanishes equation (144) becomes

$$\gamma\{\alpha - \gamma + 2(\alpha/\gamma)^{\frac{1}{2}}(\beta + \gamma)\} = 0. \tag{145}$$

In the undeformed stress-free configuration $\gamma = \mu$ and the factor in braces in (145) is 4μ. Since μ is positive then, by continuity, we must have

$$\gamma > 0, \quad \alpha - \gamma + 2(\alpha/\gamma)^{\frac{1}{2}}(\beta + \gamma) > 0 \tag{146}$$

for surface deformations to be excluded on a path of homogeneous deformation from this stress-free configuration. We note that the inequalities (146) imply the strong-ellipticity inequalities (137). When $t_2 \neq 0$ the corresponding exclusion condition is obtained by replacing '=' in (144) by '>', and this yields a range of values for t_2 which simplifies to $-2\mu < t < 2\mu$ when the stress is hydrostatic ($t_i = t$, $i \in \{1,2,3\}$).

In terms of the strain-energy function W, expressed as a function of two independent stretch variables through

$$\tilde{W}(\lambda_1, \lambda_3) = W(\lambda_1, \lambda_1^{-1}\lambda_3^{-1}, \lambda_3), \tag{147}$$

the second inequality in (146) becomes simply

$$\lambda_1^2 \tilde{W}_{11} + \lambda_2 \tilde{W}_1 > 0. \tag{148}$$

A further specialization, using the notation (103) with $\lambda_3 = 1$, enables (148) to be written

$$\lambda^3 \hat{W}''(\lambda) + \hat{W}'(\lambda) > 0, \tag{149}$$

while $\gamma > 0$ requires that

$$\frac{\hat{W}'(\lambda)}{(\lambda - 1)} > 0. \tag{150}$$

Particular cases of the criterion (144) for the onset of surface deformations (or instabilities) have been obtained by a number of

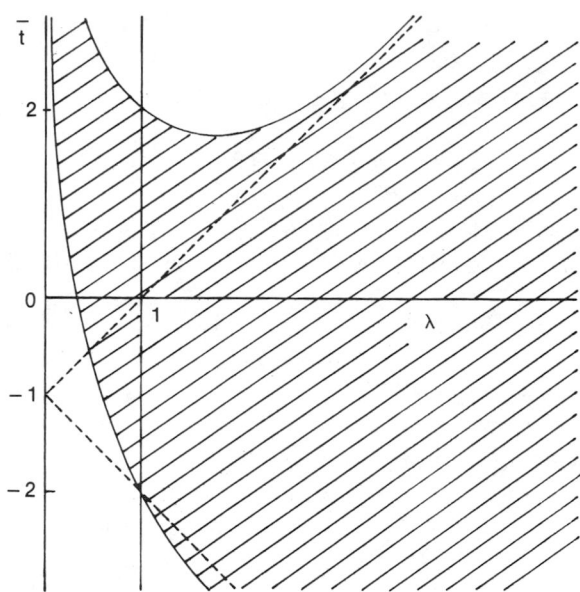

Figure 3: Region of Stability in $(\lambda, \bar{t})$-space (shaded) for the strain-energy function (113) with $m = 2$.

authors; see, for example, [5,12,13,17,18,21]. Detailed discussion is contained in [7], which is set in the framework of surface wave propagation. For illustration, in Figure 3 we show a plot of the region of $(\lambda, \bar{t})$-space in which the exclusion condition holds in respect of the neo-Hookean strain-energy function for $\lambda_3 = 1$, where $\bar{t} = t_2/\mu$. The boundary of this region corresponds to values of λ and $\bar{t}$ for which surface deformations can appear.

Bibliography

[1] J. M. Ball, *Convexity conditions and existence theorems in non-linear elasticity,* Arch. Rat. Mech. Anal. **63**, 1977, 337–403.

[2] J. M. Ball, *Discontinuous equilibrium solutions and cavitation in non-linear elasticity,* Phil. Trans. R. Soc. Lond. A**306**, 1982, 557–611.

[3] J. M. Ball and D. G. Schaeffer, *Bifurcation and stability of homogeneous equilibrium configurations of an elastic body under dead-load tractions,* Math. Proc. Cambridge Philos. Soc. **94**, 1983, 315–339.

[4] M. F. Beatty, *Topics in finite elasticity: hyperelasticity of rubbers, elastomers and biological tissues—with examples,* Appl. Mech. Rev. **40**, 1987, 1699–1734.

[5] M. A. Biot, *Mechanics of Incremental Deformations.* Wiley, New York, 1965.

[6] P. G. Ciarlet, *Mathematical Elasticity Vol.1: Three-dimensional Elasticity.* North Holland, Amsterdam, 1988.

[7] M. A. Dowaikh and R. W. Ogden, *On surface waves and deformations in a pre-stressed incompressible elastic solid,* I. M. A. J. Appl. Math. **44**, 1990, 261–284.

[8] R. Hill, *On uniqueness and stability in the theory of finite elastic strain,* J. Mech. Phys. Solids **5**, 1957, 229–241.

[9] R. Hill, *Aspects of invariance in solid mechanics,* Adv. Appl. Mech. **18**, 1978, 1–75.

[10] G. P. MacSithigh, *Energy-minimal finite deformations of a symmetrically loaded elastic sheet,* Quart. J. Mech. Appl. Math. **39**, 1986, 111–123.

[11] J. E. Marsden and T. J. R. Hughes, *Mathematical Foundations of Elasticity.* Prentice Hall, Englewood Cliffs, 1983.

[12] J. L. Nowinski, *On the surface instability of an isotropic highly elastic half-space,* Indian J. Math. Mech. **18**, 1969, 1–10.

[13] J. L. Nowinski, *Surface instability of a half-space under high two-dimensional compression,* J. Franklin Inst. **288**, 1969, 367–376.

[14] R. W. Ogden, *Non-linear Elastic Deformations.* Ellis Horwood, Chichester, 1984.

[15] R. W. Ogden, *On non-uniqueness in the traction boundary-value problem for a compressible elastic solid,* Quart. Appl. Math. **42**, 1984, 337–344.

[16] R. W. Ogden, *Local and global bifurcation phenomena in plane strain finite elasticity,* Int. J. Solids Structures **21**, 1985, 121–132.

[17] B. D. Reddy, *Surface instabilities on an equibiaxially stretched half-space,* Math. Proc. Cambridge Philos. Soc. **91**, 1982, 491–501.

[18] B. D. Reddy, *The occurrence of surface instabilities and shear bands in plane-strain deformation of an elastic half-space,* Quart. J. Mech. Appl. Math. **36**, 1983, 337–350.

[19] R. S. Rivlin, *Stability of pure homogeneous deformations of an elastic cube under dead loading,* Quart. Appl. Math. **32**, 1974, 265–271.

[20] C. A. Truesdell and W. Noll, *The Nonlinear Field Theories of Mechanics.* Handbuch der Physik Vol. III/3 (Ed. S. Flügge). Springer, Berlin, 1965.

[21] S. A. Usmani and M. F. Beatty, *On the surface instability of a highly elastic half-space,* J. Elasticity 4, 1974, 249–263.

Index

469

Mathematics in Science and Engineering

Edited by William F. Ames, *Georgia Institute of Technology*